John
Alexander

THE CERTIFIED RELIABILITY ENGINEER HANDBOOK

Also available from ASQ Quality Press:

Reliability Data Analysis with Excel and Minitab
Kenneth S. Stephens

Achieving a Safe and Reliable Product: A Guide to Liability Prevention
E. F. "Bud" Gookins

Product Safety Excellence: The Seven Elements Essential for Product Liability Prevention
Timothy A. Pine

The Weibull Analysis Handbook, Second Edition
Bryan Dodson

The Certified Quality Technician Handbook, Second Edition
H. Fred Walker, Donald W. Benbow, and Ahmad K. Elshennawy

The Certified Quality Engineer Handbook, Third Edition
Connie M. Borror, editor

The Certified Quality Inspector Handbook, Second Edition
H. Fred Walker, Ahmad K. Elshennawy, Bhisham C. Gupta, and Mary McShane Vaughn

The Metrology Handbook, Second Edition
Jay L. Bucher, editor

Quality Audits for Improved Performance, Third Edition
Dennis R. Arter

HALT, HASS, and HASA Explained: Accelerated Reliability Techniques, Revised Edition
Harry W. McLean

The Quality Toolbox, Second Edition
Nancy R. Tague

Root Cause Analysis: Simplified Tools and Techniques, Second Edition
Bjørn Andersen and Tom Fagerhaug

The Certified Manager of Quality/Organizational Excellence Handbook, Third Edition
Russell T. Westcott, editor

To request a complimentary catalog of ASQ Quality Press publications, call 800-248-1946, or visit our website at http://www.asq.org/quality-press.

The Certified Reliability Engineer Handbook

Second Edition

Donald W. Benbow and Hugh W. Broome

ASQ Quality Press
Milwaukee, Wisconsin

American Society for Quality, Quality Press, Milwaukee 53203

Published 2013
Printed in the United States of America
19 18 17 16 15 14 13   5 4 3 2 1

Library of Congress Cataloging-in-Publication Data

Benbow, Donald W., 1936–
The certified reliability engineer handbook / Donald W. Benbow and Hugh W. Broome
—Second edition.
p. cm.
Includes bibliographical references and index.
ISBN 978-0-87389-837-9 (hard cover : alk. paper)
1. Reliability (Engineering) 2. Quality control. 3. Industrial engineers—Certification.
I. Broome, Hugh W., 1936– II. Title.

TS173.B42 2012
620'.00452—dc23   2012042201

ISBN: 978-0-87389-837-9

Publisher: William A. Tony
Acquisitions Editor: Matt T. Meinholz
Project Editor: Paul Daniel O'Mara
Production Administrator: Randall Benson

ASQ Mission: The American Society for Quality advances individual, organizational, and community excellence worldwide through learning, quality improvement, and knowledge exchange.

Attention Bookstores, Wholesalers, Schools, and Corporations: ASQ Quality Press books, video, audio, and software are available at quantity discounts with bulk purchases for business, educational, or instructional use. For information, please contact ASQ Quality Press at 800-248-1946, or write to ASQ Quality Press, P.O. Box 3005, Milwaukee, WI 53201-3005.

To place orders or to request ASQ membership information, call 800-248-1946. Visit our website at http://www.asq.org/quality-press.

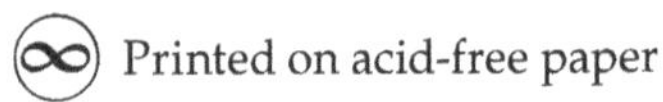

Quality Press
600 N. Plankinton Ave.
Milwaukee, WI 53203-2914
E-mail: authors@asq.org
The Global Voice of Quality™

Table of Contents

Part III Reliability in Design and Development

Part IV Reliability Modeling and Predictions

Part V Reliability Testing

Part VIII Appendices

CD-ROM Contents

Study questions have been provided on the accompanying CD-ROM, listed by Parts (I–VII) of the Body of Knowledge.

A simulated exam approximately half the size of the actual exam, with questions distributed (Parts I–VII) approximately proportional to the information contained in the Body of Knowledge, has also been provided.

List of Figures and Tables

Preface

We decided to number chapters and sections by the same method used in the Body of Knowledge (BoK) specified for the Certified Reliability Engineer examination. This made for some awkward placement and, in some cases, redundancy. We thought the ease of access for readers, who might be struggling with some particular point in the BoK, would more than balance these disadvantages.

The enclosed CD-ROM contains supplementary problems covering each chapter, and a simulated exam containing problems distributed among chapters according to the information in the BoK. It is suggested that the reader study a particular section, repeating any calculations independently, and then do the supplementary problems for that section. After attaining success with all chapters, the reader may complete the simulated exam to confirm mastery of the entire Body of Knowledge.

Acknowledgments

We'd like to thank Paul O'Mara, Randall Benson, Matt Meinholz, and the entire ASQ Quality Press staff for their help and patience. We also appreciate the fine copyediting and typesetting by New Paradigm Prepress and Graphics, and the reviews done by Anton Cherepache, James McLinn, and Daniel Zrymiak.

Part I

Reliability Management

Chapter 1

A. Strategic Management

The structure of this book is based on that of the Body of Knowledge specified by ASQ for the Certified Reliability Engineer. Before the formal Body of Knowledge is approached, a definition of reliability is needed.

Reliability is defined as the probability that an item will perform a required function without failure under stated conditions for a specified period of time.

A statement of reliability has four key components:

- *Probability.* For example, a timing chain might have a reliability goal of .9995. This would mean that at least 99.95 percent are functioning at the end of the stated time.
- *Required function.* This should be defined for every part, subassembly, and product. The statement of the required function should explicitly state or imply a failure definition. For example, a pump's required function might be moving at least 20 gallons per minute. The implied failure definition would be moving fewer than 20 gallons per minute.
- *Stated conditions.* These include environmental conditions, maintenance conditions, usage conditions, storage and moving conditions, and possibly others.
- *Specified period of time.* For example, a pump might be designed to function for 10,000 hours. Sometimes it is more appropriate to use some other measure of stress than time. A tire's reliability might be stated in terms of miles, and that of a laundry appliance in terms of cycles.

1. BENEFITS OF RELIABILITY ENGINEERING

> Demonstrate how reliability engineering techniques and methods improve programs, processes, products, systems, and services. (Understand)
>
> Body of Knowledge I.A.1

The following are among the influences that have increased the importance of the study of reliability engineering:

- Customers expect products to not only meet the specified parameters upon delivery, but to function throughout what they perceive as a reasonable lifetime.
- As products become more complex, the reliability requirements of individual components increase. Suppose, for instance, that a system has 1000 independent components that must function in order for the system to function. Further suppose that each component has a reliability of 99.9 percent. The system would have a reliability of $0.999^{1000} = .37$, an obviously unacceptable value.
- An unreliable product often has safety and health hazards.
- Reliability values are used in marketing and warranty material.
- Competitive pressures require increased emphasis on reliability.
- An increasing number of contracts specify reliability requirements.

The study of reliability engineering responds to each of these influences by helping designers determine and increase the useful lifetime of products, processes, and services.

2. INTERRELATIONSHIP OF QUALITY AND RELIABILITY

> Define and describe the relationships among safety, quality, and reliability. (Understand)
>
> **Body of Knowledge I.A.2**

In most organizations the quality assurance function is designed to continually improve the ability to produce products and services that meet or exceed customer requirements. Narrowly construed, this means, in the manufacturing industries, producing parts with dimensions that are within tolerance. Quality engineering must expand this narrow construction to include reliability considerations, and all quality engineers should have a working knowledge of reliability engineering. What, then, is the distinction between these two fields?

- Once an item has been successfully manufactured, the traditional quality assurance function has done its job (although the search for ways to improve is continuous). The reliability function's principal focus is on what happens next. Answers are sought to questions such as:
  - Are components failing prematurely?
  - Was burn-in time sufficient?

  - Is the constant failure rate acceptable?
  - What changes in design, manufacturing, installation, operation, or maintenance would improve reliability?

- Another way to delineate the difference between quality and reliability is to note how data are collected. In the case of manufacturing, data for quality engineering are generally collected during the manufacturing process. Inputs such as voltages, pressures, temperatures, and raw material parameters are measured. Outputs such as dimensions, acidity, weight, and contamination levels are measured. The data for reliability engineering generally are collected after a component or product is manufactured. For example, a switch might be toggled repeatedly until it fails, and the number of successful cycles noted. A pump might be run until its output in gallons per minute falls below a defined value, and the number of hours recorded.
- Quality and reliability engineers provide different inputs into the design process. Quality engineers suggest changes that permit the item to be produced within tolerance at a reasonable cost. Reliability engineers make recommendations that permit the item to function correctly for a longer period of time.

The preceding paragraphs show that although the roles of quality and reliability are different, they do interrelate. For example, in the product design phase both quality and reliability functions have the goal of proposing cost-effective ways to satisfy and exceed customer expectations. This often mandates that the two functions work together to produce a design that both works correctly and performs for an acceptable period of time. When processes are designed and operated, the quality and reliability engineers work together to determine the process parameters that impact the performance and longevity of the product so that those parameters can be appropriately controlled. A similar interrelationship holds as specifications are developed for packaging, shipment, installation, operation, and maintenance.

Reliability will be impacted by product design and by the processes used in the product's manufacture. Therefore, the designers of products and processes must understand and use reliability data as design decisions are made. Generally, the earlier reliability data are considered in the design process, the more efficient and effective their impact will be.

Safety considerations pervade all aspects of both the quality engineering and reliability engineering fields. When a process/product change is proposed, the proposal should be accompanied by a thorough study of the impact the improvement will have on safety. Questions to investigate include:

- *Could this change make the production process less safe? How will this be mitigated?*

  Example: Workers accustomed to doing things the old way may be more at risk with the proposed changes.

- *Could this change make the use of the product less safe? How will this be mitigated?*

  Example: The new dishwasher latch that, if not engaged properly, allows steam to escape into the electronic timer, causing a fire hazard.

- With the failure of another component now more likely, are there new safety risks?

  Example: Proposals for increasing the useful life of a component should be accompanied by a study on the effect the increased life will have on other components.

- *As the product reaches its wear-out phase, could this change introduce safety risks? How will this be mitigated?*

  Example: The new lighting system contains chemical compounds that are toxic when improperly disposed of.

When conducting an FMEA study, all failure modes should be investigated for possible safety risks. And once a reliable product is designed, quality engineering techniques are used to make sure that the processes produce that product.

3. ROLE OF THE RELIABILITY FUNCTION IN THE ORGANIZATION

> Describe how reliability techniques can be applied in other functional areas of the organization, such as marketing, engineering, customer/product support, safety and product liability, etc. (Apply)
>
> **Body of Knowledge I.A.3**

The study of reliability engineering is usually undertaken primarily to determine and improve the useful lifetime of products. Data are collected on the failure rates of components and products, including those produced by suppliers. Competitors' products may also be subjected to reliability testing and analysis.

Reliability techniques can also help other facets of an organization:

- Reliability analysis can be used to improve product design. Reliability predictions, as discussed in Chapter 9, provide guidance as components are selected. Derating techniques, covered in Chapter 7, aid in increasing a product's useful lifetime. Reliability improvements can be effected through component redundancy.
- Marketing and advertising can be enhanced as warranty and other documents that inform customer expectations are prepared.

Warranties that are not supported by reliability data can cause extra costs and inflame customer ire.

- It is increasingly important to detect and prevent or mitigate product liability issues. Warnings and alarms should be incorporated into the design when hazards can't be eliminated. Products whose failure can introduce safety and health hazards need to be analyzed for reliability so that procedures can be put in place to reduce the probability that they will be used beyond their useful lifetime. As discussed in Chapter 2, failure rates typically escalate in the final phase of a product's life. Components whose useful lifetime is shorter than the product's should be replaced on a schedule that can be determined through reliability engineering techniques.
- Manufacturing processes can use reliability tools in the following ways:
  - The impact of process parameters on product failure rates can be studied.
  - Alternative processes can be compared for their effect on reliability.
  - Reliability data for process equipment can be used to determine preventive maintenance schedules and spare parts inventories.
  - The use of parallel process streams to improve process reliability can be evaluated.
  - Safety can be enhanced through the understanding of equipment failure rates.
  - Vendors can be evaluated more effectively.
- Every facet of an organization, including purchasing, quality assurance, packaging, field service, logistics, and so on, can benefit from a knowledge of reliability engineering. An understanding of the life cycles of the products and equipment they use and handle can improve the effectiveness and efficiency of their function.

4. RELIABILITY IN PRODUCT AND PROCESS DEVELOPMENT

Integrate reliability engineering techniques with other development activities, concurrent engineering, corporate improvement initiatives such as lean and Six Sigma methodologies, and emerging technologies. (Apply)

Body of Knowledge I.A.4

Some implementations of reliability engineering have consisted of testing products at the end of the manufacturing phase to determine their life cycle parameters. At this point it is, of course, too late to have much impact on those parameters.

Reliability engineering tools help the design engineer work more efficiently and effectively in various ways:

- Mean time between failure (MTBF) values for existing products can be determined and reasonable goals established.
- MTBF values for components and purchased parts can be determined.
- Failure types and times of occurrence can be anticipated.
- Optimal break-in/burn-in times can be determined.
- Recommendations for warranty times can be established.
- The impact of age and operating conditions on the life of the product can be studied.
- The effects of parallel or redundant design features can be determined.
- Accelerated life testing can be used to provide failure data.
- Field failure data can be analyzed to help evaluate product performance.
- Concurrent engineering can improve the efficiency and effectiveness of product development by scheduling design tasks in parallel rather than sequentially.
- Reliability engineering can provide information to individual teams about failure rates of their proposed components.
- Cost accounting estimates can be improved through the use of life cycle cost analysis using reliability data.
- When management employs FMEA/FMECA techniques, reliability engineering provides essential input, as described in Chapter 6.

Lean manufacturing, with its emphasis on workplace organization and standardized work procedures, can reduce product variation. This in turn makes a more predictable life cycle, and helps the reliability engineer provide more useful data to the design function. The term "lean thinking" refers to the use of ideas originally employed in lean manufacturing to improve functions in all departments of an enterprise. The National Institute of Standards and Technology (NIST), through its Manufacturing Extension Partnership, defines *lean* as:

> *A systematic approach to identifying and eliminating waste (non-value-added activities) through continuous improvement by flowing the product at the pull of the customer in pursuit of perfection.*

ASQ defines the phrase "non-value-added" as:

> *A term that describes a process step or function that is not required for the direct achievement of process output. This step or function is identified and examined for potential elimination.*

This represents a shift in focus for manufacturing engineering, which has traditionally studied ways to improve value-added functions and activities (for example, how can this process run faster and more precisely?). Lean thinking doesn't ignore the value-added activities, but it does shine the spotlight on waste.

Lean manufacturing seeks to eliminate or reduce waste by use of the following:

- *Teamwork* with well-informed, cross-trained employees who participate in the decisions that impact their function
- *Clean*, organized, and well-marked work spaces
- *Flow systems* instead of batch and queue (that is, reduce batch size toward its ultimate ideal—one)
- *Pull systems* instead of push systems (that is, replenish what the customer has consumed)
- *Reduced lead times* through more efficient processing, setups, and scheduling

The history of lean thinking may be traced to Eli Whitney, who is credited with spreading the concept of part interchangeability. Henry Ford, who went to great lengths to reduce cycle times, furthered the ideas of lean thinking. And later, the Toyota Production System (TPS) packaged most of the tools and concepts now known as *lean manufacturing.*

Six Sigma strategies support extensive data collection and analysis. This can aid the reliability engineering function by increasing emphasis on designed experiments and life cycle testing programs. A wide range of companies have found that when the Six Sigma philosophy is fully embraced, the enterprise thrives. What is this Six Sigma philosophy? Several definitions have been proposed, with the following common threads:

- Use of teams that are assigned well-defined projects that have direct impact on the organization's bottom line.
- Training in statistical thinking at all levels and providing key people with extensive training in advanced statistics and project management. These key people are designated "Black Belts."
- Emphasis on the DMAIC approach to problem solving, which executes these steps: define, measure, analyze, improve, and control.
- A management environment that supports these initiatives as a business strategy.

Opinions on the definition of Six Sigma differ:

- *Philosophy*—The philosophical perspective views all work as processes that can be defined, measured, analyzed, improved, and controlled (DMAIC). Processes require inputs and produce outputs. If inputs are controlled, the outputs will be controlled. This is generally expressed as the $y = f(x)$ concept.
- *Set of tools*—Six Sigma as a set of tools includes all the qualitative and quantitative techniques used by the Six Sigma expert to drive process

improvement. A few such tools include statistical process control (SPC), control charts, failure mode and effects analysis (FMEA), and process mapping. Six Sigma professionals do not completely agree as to exactly which tools constitute the set.

- *Methodology*—The methodological view of Six Sigma recognizes the underlying and rigorous approach known as DMAIC. DMAIC defines the steps a Six Sigma practitioner is expected to follow, starting with identifying the problem, and ending with implementing long-lasting solutions. While DMAIC is not the only Six Sigma methodology in use, it is certainly the most widely adopted and recognized.
- *Metrics*—In simple terms, six sigma quality performance means 3.4 defects per million opportunities (accounting for a 1.5-sigma shift in the mean).

Lean and Six Sigma have the same general purpose of providing the customer with the best possible quality, cost, delivery, and a newer attribute—nimbleness. The two initiatives approach their common purpose from slightly different angles:

- Lean focuses on waste reduction, whereas Six Sigma emphasizes variation reduction.
- Lean achieves its goals by using less technical tools such as kaizen, workplace organization, and visual controls, whereas Six Sigma tends to use statistical data analysis, design of experiments, and hypothesis tests.

5. FAILURE CONSEQUENCE AND LIABILITY MANAGEMENT

> Describe the importance of these concepts in determining reliability acceptance criteria. (Understand)
>
> **Body of Knowledge I.A.5**

Reliability analysis provides estimates of the probability of failure. The reliability engineer must go beyond these calculations and examine the consequences of failure. These consequences typically represent costs to the customer. The customer finds ways of sharing these costs with the producer through the warranty system, loss of business, decrease in reputation, or the civil litigation system. Therefore, an important reliability function is the anticipation of possible failures and the establishment of reliability acceptance goals that will limit their occurrence and consequent costs.

The acceptance criteria associated with these goals generally fall into three categories:

- *Functional requirements* (such as 20 gallons per minute for a pump)
- *Environmental requirements* (such as temperature, radiation, pH)
- *Time requirements* (such as failure rate during useful life, time elapsed before wear-out phase begins)

Once component, product, and system reliability goals have been set, a testing protocol should be implemented to provide validation that these goals will impact the failure rates and the associated consequences as planned. These reliability goals typically impose specifications on the product. In anticipation of the start of production, reliability engineers provide further testing procedures to provide *verification* that these specifications are being met.

6. WARRANTY MANAGEMENT

> Define and describe warranty terms and conditions, including warranty period, conditions of use, failure criteria, etc., and identify the uses and limitations of warranty data. (Understand)
>
> **Body of Knowledge I.A.6**

A good warranty program reaches into many different organizational functions:

- *Product pricing* must include allowances for warranty costs.
- *Advertising* can use warranties as inducements to purchase.
- *Supply chain agreements* should spell out warranty liabilities and procedures.
- *Accounting procedures* should include accruals or sinking funds to cover future warranty costs.
- *Operation instruction manuals* should include preventive maintenance recommendations that will mitigate failures.
- *Engineering and reliability functions* have responsibilities for determining causes and determining product/process corrective actions as required.

The breadth of warranty considerations implies that improvements in warranty management have the potential for substantial financial impact. Finding and executing these improvements begins with an understanding of terms.

Warranties should spell out:

- Effective initial date and duration of the warranty
- Definitions of parts and defects to be covered

- Warranty procedures, including return routes, replacement/reimbursement policies, and policies that will be followed when no defect is found
- Conditions of use such as environmental specifications (ranges for temperature, vibration, salinity, electronic parameters) and operation procedures
- Conditions of storage, transportation, and installation

If suppliers are involved, effective warranty management requires agreements as to all these aspects with each supplier, regardless of tier.

Collection and Analysis of Warranty Data

The data collection process should include an effective method for identifying the failure conditions, including the following information:

- Traceability of each component so that its supplier, date, and conditions of manufacture and assembly are available
- Conditions of use such as environment, handling, maintenance, and operation
- Failure parameters, including type of failure, symptoms, condition of other components, and action taken (replacement, repair, refund, and so on)

Effort should be made to get beyond the "125 failures during Q3" or "$7.5M in warranty costs" type of warranty data. Obtaining useful warranty information may require expenditure of considerable resources to determine the validity of the warranty claims and the causes of the failures. These expenditures can be justified if the follow-up action results in product improvements that avoid future costs. In addition, the data may yield cost avoidance if current practices result in paying for warranty claims that are not valid or that should properly be assessed to suppliers.

Investigations of causes may require extensive study using FMEA and other tools. Consideration should be given to analyzing data for correlations with geography, place of manufacture, installer, type of application, environmental conditions, and other groupings as appropriate. Trend analyses are also useful. In some situations, sampling techniques may be employed rather than studying the entire population.

Warranty information can be used in various ways to improve customer satisfaction, including:

- Product design changes
- Process/manufacturing changes
- Changes in installation/operation documentation
- Product recalls
- Improvement in supplier performance

- Improvement of FMEA processes

In addition, warranty data can be used to refine reliability prediction for future product designs.

7. CUSTOMER NEEDS ASSESSMENT

> Use various feedback methods (e.g., quality function deployment (QFD), prototyping, beta testing) to determine customer needs in relation to reliability requirements for products and services. (Apply)
>
> **Body of Knowledge I.A.7**

The current emphasis on listening to the *voice of the customer* (VOC) applies to internal as well as external customers. There is no substitute for close, face-to-face communication with those to whom products and services are provided. A number of tools can be used for measuring customer needs and desires.

The most elementary tool is the *customer satisfaction survey*. It has the advantage of being the simplest to use. The data obtained from such surveys are often of questionable validity due to the nonrandom nature of responses. Another disadvantage of such surveys is that they tend to be reactive rather than proactive. Some of the most innovative products and services were developed in anticipation of perceived needs rather than in response to them. Automotive pioneer Henry Ford once said, "If I'd asked people what they wanted, they would have said 'a faster horse.'" The following techniques represent attempts to anticipate customer reaction as part of the product and service design process.

Prototyping

This is the process of building a preliminary model of the product or service for the purpose of determining design features, reliability, usability, and user reactions. Examples:

- A supplier provides a model of a proposed oil filter to an automotive company so ease of filter changes can be determined.
- A hardware manufacturer provides a sample of proposed door hinges for laboratory reliability testing.

Production of prototypes provides the design team with a three-dimensional object they can examine and, in some cases, run through reliability tests. The main disadvantage of prototyping is the cost.

The term *rapid prototyping* is sometimes used to refer to a prototype that can be produced in a much shorter time than the standard production process, which may include die and fixture work. Researchers in the field point out, however, that the actual machining process in many cases is not very rapid. Some current work focuses on generating computer codes for a milling or turning machine from the data produced by a *computer aided design* (CAD) system. To date, the resultant program produces a process that tends to be very slow in execution.

Quality Function Deployment

Quality function deployment (QFD) provides a process for planning new or redesigned products and services. The input to the process is the voice of the customer. The QFD process requires that a team discover the needs and desires of their customer and study the organization's response to these needs and desires. The QFD matrix aids in illustrating the linkage between the VOC and the resulting technical requirements. A quality function deployment matrix consists of several parts. There is no standard format matrix or key for the symbols, but the example shown in Figure 1.1 is typical. A map of the various parts of Figure 1.1 is shown in Figure 1.2. The matrix is formed by first filling in the customer requirements ①, which are developed from analysis of the VOC. This section often includes a scale reflecting the importance of the individual entries. The technical requirements are established in response to the customer requirements and placed in area ②. The symbols on the top line in this section indicate whether lower (↓) or higher (↑) is better. A circle indicates that target is better. The relationship area ③ displays the connection between the technical requirements and the customer requirements. Various symbols can be used here. The most common are shown in Figure 1.1. Area ④ is not shown on all QFD matrices. It plots comparison with competition for the customer requirements. Area ⑤ provides an index to documentation concerning improvement activities. Area ⑥ is not shown on all QFD matrices. It plots comparison with competition for the technical requirements. Area ⑦ lists the target values for the technical requirements. Area ⑧ shows the co-relationships between the technical requirements. A positive co-relationship indicates that both technical requirements can be improved at the same time. A negative co-relationship indicates that improving one of the technical requirements will make the other one worse. The "column weights" shown at the bottom of the figure are optional. They indicate the importance of the technical requirements in meeting customer requirements. The values in the column weights row are obtained by multiplying the value in the "importance" column in the customer requirements section by values assigned to the symbols in the relationship matrix. These assigned values are arbitrary, and in the example a strong relationship was assigned a 9, moderate 3, and weak 1.

The completed matrix can provide a database for product development, serve as a basis for planning product or process improvements, and suggest opportunities for new or revised product or process introductions.

The customer requirements section is sometimes called the "what" information, while the technical requirements section is referred to as the "how" area. The basic QFD product planning matrix can be followed with similar matrices for planning the parts that make up the product and for planning the processes that will produce the parts. See Figure 1.3.

If a matrix has more than 25 customer voice lines, it tends to become unmanageable.

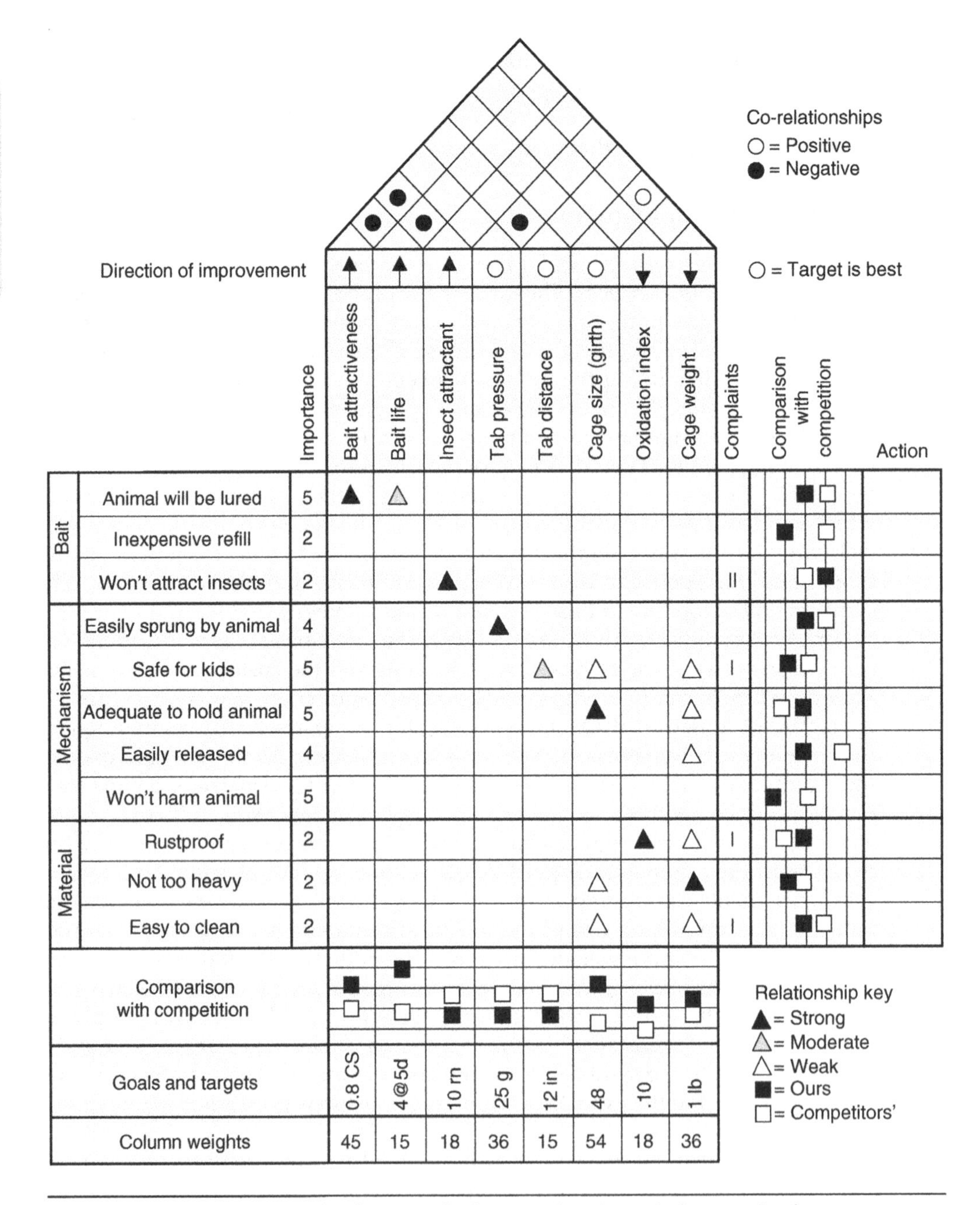

Figure 1.1 Example of a quality function deployment (QFD) matrix for an animal trap.

The release of a preliminary version of a product to a restricted set of users has come to be known as *beta testing*. A principal advantage of this technique is the exposure of the product to a larger audience with varied needs and levels of expertise who might detect flaws that in-house (alpha) testing missed. The customers

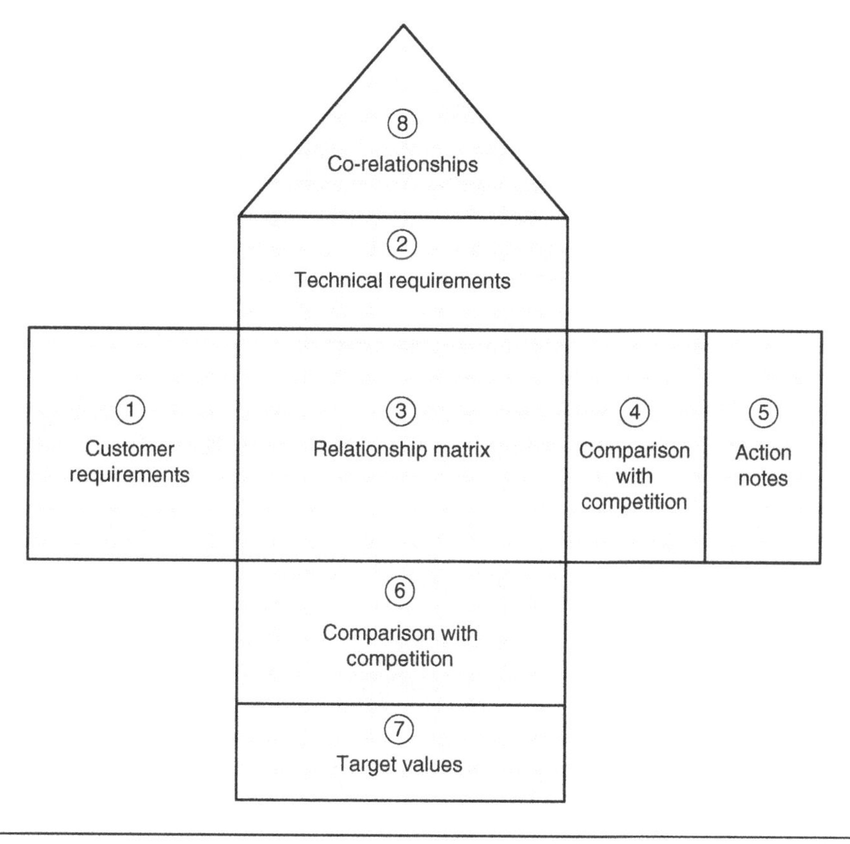

Figure 1.2 Map to the entries for the QFD matrix illustrated in Figure 1.1.

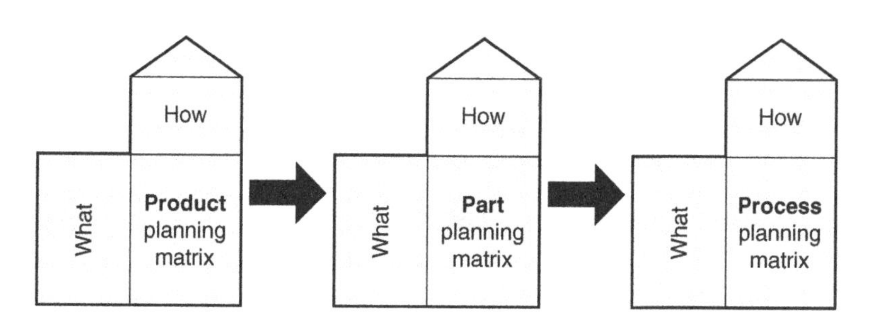

Figure 1.3 Sequence of QFD matrices for product, part, and process planning.

entrusted with the early designs are expected to report good and bad features and recommendations to the development team. This frequently results in the identification of potential corrections and improvements that can be factored into the final version. Beta testing tends to be more important with complex products for which unusual combinations of usage circumstances may not be envisioned by designers.

8. SUPPLIER RELIABILITY

> Define and describe supplier reliability assessments that can be monitored in support of the overall reliability program. (Understand)
>
> **Body of Knowledge I.A.8**

In an ideal world, every supplier would have an excellent reliability engineering program with regular, dependable reports to customers. While awaiting this state of affairs it is essential that the customer choose between three scenarios:

1. The customer assumes all responsibility for the reliability engineering function and requires supplier compliance with all specifications. With this arrangement the supplier must report any proposed changes in the process or product so that the potential impact on reliability can be studied. This option is more common in situations where the customer has full design responsibilities and outsources relatively minor components.
2. The supplier assumes responsibility for reliability engineering, and reports its analysis and decision-making process to the customer for agreement. The customer is, of course, ultimately responsible to its customers, but the supplier may share financial responsibility for warranty claims, and so on. This is more common when the supplier has design control of the supplied component.
3. Some sort of shared responsibility for the reliability engineering analysis and interpretation, perhaps involving a third party. Third-party involvement is more common when the supply chain is long geographically.

With any of these options the arrangement must be clearly spelled out in the contractual agreement between the parties. The customer will want to conduct assessments customized to that agreement.

Examples:

- For suppliers with a long-term relationship based on mutual trust and understanding, the reliability functions conducted by the supplier can be verified at the time of a quality audit. At least one auditing team member should be familiar with reliability engineering functions. The supplier's collection and analysis of life cycle cost data should be studied, and the mechanism for feedback of this information to the product/design functions should be confirmed.
- For suppliers without a strong favorable history with the components involved, the customer should consider performing actual testing and evaluation of the products. This could vary from a full-fledged

reliability program to less elaborate programs, depending on the situation.

The full reliability program would begin with establishing goals and translating goals into product/process design requirements, and continue to validation of production output. Less elaborate programs could consist of monitoring of the reliability engineering function at the supplier's location, training supplier personnel, and/or testing of random samples from production to assure that reliability requirements are being met.

Chapter 2

B. Reliability Program Management

1. TERMINOLOGY

> Explain basic reliability terms (e.g., MTTF, MTBF, MTTR, availability, failure rate, reliability, maintainability). (Understand)
>
> **Body of Knowledge I.B.1**

The *mean life* of a product is the average time to failure of identical products operating under identical conditions. Mean life is also referred to as the *expected time to failure*. Mean life is denoted by *mean time to failure* (MTTF) for non-repairable products and *mean time between failures* (MTBF) for repairable products. The reliability engineer should exercise care in the use of the terms MTBF and MTTF. These terms are usually used when the underlying failure distribution is the exponential and the failure rate is constant. The relationships given in the remainder of Chapter 2 are based on this assumption. MTTF and MTBF are often denoted with the letter m or the Greek theta (θ). "Time" as used here refers to some measure of life units for the product. In the case of automotive products, the life units may be miles. In other equipment, life units may be cycles, rounds fired, and so forth. Some documents, for instance, replace MTBF with MCBF (*mean cycles between failures*).

For a particular set of failure times, the mean life is obtained by averaging the failure times. This value serves as an estimate for θ and is sometimes denoted $\hat{\theta}$, and called "θ hat."

EXAMPLE 2.1

Ten randomly selected non-repairable products are tested to failure, and their failure times in hours are:

132 140 148 150 157 158 159 163 163 168

$$\text{MTTF} = \frac{132+140+148+150+157+158+159+163+163+168}{10} = 153.8 \text{ hours}$$

EXAMPLE 2.2

Suppose 100 repairable items are tested for 1000 hours each, and failed items are promptly repaired and returned to the test. Suppose 25 failures occurred during the test. Then,

$$\text{MTBF} = \theta \approx \hat{\theta} = \frac{100{,}000}{25} = 4000 \text{ hours}$$

If n items are tested to failure, the general formula is

$$\text{MTTF} = \hat{\theta} = \frac{\sum t_i}{n}$$

where t_i's are failure times.

Example 2.2 shows the approach to use if repairable items are repaired and placed back on test.

The general formula for the situation where a number of repairable items are tested for a given amount of time, with failed items being promptly replaced, is

$$\text{MTBF} = \frac{nm}{r}$$

where

n = Number of items

m = Number of hours in the test

r = Number of failures

Censored Data

There are four types of failure data:

1. *Exact failure times,* in which the exact failure time is known. Example 2.1 illustrates this type of data.
2. *Right-censored* data, in which it is known only that the failure happened or would have happened after a particular time. This occurs if an item is still functioning when the test is concluded.
3. *Left-censored* data, in which it is known only that the failure happened before a particular time. This occurs if the items are not checked prior to being tested but are periodically examined, and a failure is observed at the first examination.
4. *Interval-censored* data, in which it is known only that the failure happened between two times, for example, if the items are checked every five hours, and an item was functioning at hour 145 but had failed sometime before hour 150.

Note to Minitab users: If the data have exact failure times and right-censored data, use Minitab's right-censoring functions. If the data have exact failure times and a varied censoring scheme including right censoring, left censoring, and interval censoring, use Minitab's arbitrary censoring functions.

The *mean time to repair* (MTTR) is the average time it takes to return the product to operational status.

Failure rate is the reciprocal of the mean life. Failure rate is usually denoted by the letter *f* or the Greek letter lambda (λ). So,

$$\lambda = \frac{1}{\text{MTBF}}$$

or

$$\lambda = \frac{1}{\text{MTTF}}$$

and, of course,

$$\text{MTBF} = \frac{1}{\lambda}$$

and

$$\lambda = \frac{1}{\text{MTTF}}$$

Availability can be defined as the probability that a product is operable and in a committable state when needed. In other words, it is the probability that an item has not failed or is not undergoing repair. This measure takes into account an item's reliability and its maintainability. Another way to express this is the proportion of time a system is in a functioning condition. This can be written as the fraction

$$A = \frac{\text{Total time a functional unit is capable of being used during a given interval}}{\text{The length of the interval}}$$

If the product is repairable and needs no preventive maintenance, and if repair can begin immediately when failure occurs, availability can be defined as

$$A = \frac{\text{MTBF}}{\text{MTBF} + \text{MTTR}}$$

A more general formula for availability can be written as the ratio of the average value of the uptime of a system to the sum of the average values of uptime and downtime:

$$\text{Availability} = \frac{\text{Average uptime}}{\text{Average uptime} + \text{Average downtime}}$$

Dependability is a very similar concept. It is defined as the probability that a product will function at a particular point in time during a mission.

Maintainability is the probability that a failed product will be repaired within a given amount of time once it has failed. Thus, maintainability is a function of time. If there is a 95 percent probability that a product will be operable within three hours, then $M(3) = .95$. In defining maintainability it is necessary to describe exactly what is included in the maintenance action. The following items are typical: diagnosis time, part procurement time, teardown time, rebuild time, and verification time.

Preventive maintenance, that is, the replacement, at scheduled intervals, of parts or components that have not failed rather than waiting for a failure, is frequently more cost-effective. Preventive maintenance reduces the diagnosis and part procurement times and thus may improve maintainability.

2. ELEMENTS OF A RELIABILITY PROGRAM

> Explain how planning, testing, tracking, and using customer needs and requirements are used to develop a reliability program, and identify various drivers of reliability requirements, including market expectations and standards, as well as safety, liability, and regulatory concerns. (Understand)
>
> **Body of Knowledge I.B.2**

A reliability program should impact many functions in an enterprise, including research, purchasing, manufacturing, quality assurance, testing, shipping, and field service, among others. In order to accomplish this, a reliability program should have the following elements:

1. *Established reliability goals and requirements.* The general goal of reliability efforts is to delight customers by increasing the reliability of products. The reliability program accomplishes this goal by establishing reliability goals and meeting them. Customer input and market analysis typically determine minimum reliability requirements. In general, consumers have rising expectations for reliability. The minimum reliability requirements are time dependent because reliability changes throughout the life of the product.

2. *Product design.* The reliability program must have a mechanism for translating the minimum reliability requirements into design requirements. Reliability requirements should be documented for each stage of a product's design and for all subsystems and components.

3. *Process design.* As the product design firms up, attention can shift toward the design of the processes that will produce it. Reliability requirements must be finalized for components, whether produced in-house or from suppliers. These requirements must be linked to manufacturing process parameters by determining what processes and what settings will produce components with the required reliability.

4. *Validation and verification.* As either prototypes or the first production pieces become available, the reliability program must facilitate tests that are conducted to validate that the reliability requirements do indeed produce the desired product reliability. When these requirements have been validated it is necessary to verify that the production processes can produce products that meet these requirements.

5. *Post-production evaluation.* The reliability program must make provisions for collecting and analyzing data from products during their useful life:

   a. Random samples from regular production should be collected and tested for reliability.

   b. Customer feedback should be actively solicited and analyzed.

   c. Field service and warranty records should be studied.

6. *Training and education.* Although listed last, this is certainly not the least important element of a reliability program. No reliability program can succeed without a basic understanding of its elementary concepts by people at all levels. Support from key managers is essential because their cooperation is needed for the testing and analysis process. Top-level management must see the importance of the program to the success of the enterprise. So, this element of the reliability program must sometimes be given first priority if the rest of the program is to succeed.

Design requirements originate from customer needs. These requirements, as determined by the marketplace, customer input, organizational objectives, and other sources, must often be prioritized. A classic example is the NASA motto of the 1990s, "faster—better—cheaper," to which the engineers famously replied, "We can do any two."

An effective tool for managing requirements is the QFD matrix discussed in Chapter 1. Figure 1.2 shows that the customer requirements are listed in the area labeled ①. The design team then lists the technical requirements in area ②, which it plans to use to satisfy these customer requirements. Meeting a customer requirement is often a matter of degree, and in some cases may conflict with other requirements. For instance, in the example shown in Figure 1.1, it may be that the bait that best meets the first requirement, "Animal will be lured," is not the best bait for the third requirement, "Won't attract insects." These conflicts are shown as negative co-relationships in the triangular matrix at the top of the diagram.

The design team uses the QFD matrix to help manage the requirement conflicts in a number of ways:

- The team may use the "action" column to specify design activity. For example, "Find a bait with an animal attractiveness greater than 1.1 cs and an insect attractant number greater than 14 rn."
- The "importance" column shown in Figure 1.1 provides prioritization guidelines.
- The "comparison with competition" charts give the team guidance. If the product is already far ahead of the competition in meeting one requirement and far behind in meeting a conflicting requirement, it may be prudent to produce a design that works toward meeting the latter. This strategy must be used carefully because the competition is seldom a fixed target. It would be appropriate to add regulatory, liability, and safety requirements in the customer requirements section since these issues represent part of the set of customers to be satisfied.

Meeting reliability design requirements within time and resource constraints requires an efficient testing and documentation program. The tasks associated with the program must be accomplished in synchronization with other design, development, and manufacturing functions. These tasks must be given adequate priority at each stage of design if unpleasant late-term surprises are to be avoided. The reliability engineer should provide the project manager with time/resource needs during the project planning phase. The project manager is typically responsible for assuring that these needs are fulfilled at the appropriate stages.

3. TYPES OF RISK

> Describe the relationship between reliability and various types of risk, including technical, scheduling, safety, financial, etc. (Understand)
>
> **Body of Knowledge I.B.3**

Reliability engineering is primarily focused on testing and evaluation of technical properties of products. However, the reliability program must be cognizant of a broader view to include the various systems that bring the product to the customer. The best life cycle analysis may be made irrelevant if these systems fail. Examples of these issues follow:

- *Scheduling.* Are advance planning and provision for contingencies adequate? Will conflicts with schedules of other products impact delivery?
- *Safety.* Will various phases of the life cycle cause health and safety hazards to the user or others? Will proposed changes in processes,

materials, and so on, increase health and safety hazards to operators, installers, or others?

- *Financial.* How will changes in exchange rates for international transactions impact the organization? Is the information regarding creditworthiness of customers adequate?
- *Regulatory.* Will domestic/international political trends impact the organization? What expenses or delays will occur to meet proposed regulations?
- *Suppliers.* How would change of status (for example, buyout) of a supplier impact the delivery schedule? Will changes in the transportation infrastructure impact supplies?
- *Materials.* How will changes in cost/availability of key materials impact the product? What effect will supply/demand and international relations have on the product?

ISO Guide 73:2009 defines *risk* as the effect of uncertainty on objectives. Risk assessment/risk management uses qualitative and quantitative data to provide estimates of the impact of various adverse events. The purpose is to identify and prioritize vulnerabilities so that corrective action can be initiated. One approach is to form a matrix listing product and/or services in a vertical column, with an estimate of the importance of each in another vertical column. Across the top of the matrix a list of risk types can be listed. See the example in Figure 2.1.

A matrix such as that shown in Figure 2.1 typically provides for multiplying an estimated importance score of an adverse event by an estimate of the probability that the event will occur.

	Importance	Risk types																						
		Technical			Scheduling				Safety		Financial			Regulatory				Supplier			Materials			Etc.
		A	B	C	A	B	C	D	A	B	A	B	C	A	B	C	D	A	B	C	A	B	C	
Product 1																								
Product 2																								
Product 3																								
Service 1																								
Service 2																								

Figure 2.1 Example of risk assessment matrix.

Reliability engineering data can improve the risk assessment process by providing more-accurate values for probability of product failure at given points in the product's life cycle. Changes in reliability obviously impact the resulting risk assessment scores.

The risk assessment helps to prioritize vulnerabilities. The next step is to implement corrective actions, which may well involve improving product reliability.

4. PRODUCT LIFECYCLE ENGINEERING

> Describe the impact various lifecycle states (concept/design, introduction, growth, maturity, decline) have on reliability, and the cost issues (product maintenance, life expectation, software defect phase containment, etc.) associated with those stages. (Understand)
>
> **Body of Knowledge I.B.4**

Reliability engineering techniques help quantify the "pay me now or pay me later" concept. The goal is to determine the reliability level that will minimize the total life cycle cost of the product. The *life cycle cost* of a product includes the cost to purchase, operate, and maintain the product during its useful lifetime. In some cases, such as automotive products, where the customer seldom keeps the product for its entire useful lifetime, costs associated with depreciation may be factored into life cycle costs.

The real cost of failures is frequently underestimated. If a 90-cent natural gas valve component fails to function, the cost may far exceed the 90-cent replacement cost.

Reliability engineers take the long-term view and develop cost-effective ways to reduce life cycle costs. These may range from design techniques such as redundancy and derating to specification of manufacturing parameters such as burn-in time.

Increased reliability sometimes means increased manufacturing cost and selling price. Properly implemented, however, the result will be a decrease in life cycle cost. Consider, for instance, a national truck line that discovered that its most frequent cause of vehicle downtime was loss of a headlight bulb. This entailed stopping the truck at the side of the road and summoning a repair vehicle from the nearest company depot. The resultant delay caused late deliveries and dissatisfied customers. The trucking company determined that a much more reliable bulb reduced life cycle costs even though the new bulb had a considerably higher initial purchase price and required retrofitting a step-up transformer to obtain the required voltage. The company now specifies the more reliable bulb for new truck purchases. In this case the truck manufacturer can be faulted for producing a product that didn't have the lowest life cycle cost.

As component and product design decisions are made, the reliability engineer can aid in calculating the cost–benefit relationships by providing life expectancies for various design options.

Product Life Cycle

Reliability engineers identify three stages in the life cycle of a product:

1. The first stage is referred to variously as the *early failure* stage, the *infant mortality* stage, or the *decreasing failure rate* stage. The failures that occur during the early failure stage are usually associated with manufacturing rather than design. Examples of causes of failure include inadequate test or burn-in time, poor quality control, poor handling, weak materials or components, and human error in fabrication or assembly. Ideally, all these failures should occur in-house and be corrected before the customer takes possession.

2. The second stage is called the *constant failure rate* stage, the *random causes* stage, or the *useful life* stage. During the useful life stage the failure rate is approximately constant. Note that the failure rate is not necessarily zero. During this stage the failures have random causes and can't usually be assigned to production problems. Reducing the failure rate during this stage usually requires changes in product design.

3. The third stage is called the *wear-out* stage, *fatigue* stage, or the *increasing failure rate* stage. The wear-out stage is characterized by an increasing failure rate over time. These failures are caused by product or component fatigue.

These stages are depicted in block diagram format in Figure 2.2. A graph of the failure rate is shown as the *bathtub curve* in Figure 2.3.

Note: Although the *useful life* stage is sometimes referred to as the *random causes* stage, random causes are generally present during all three stages.

Knowing the locations of the three stages of the bathtub curve can help answer important questions such as:

- Is the quality control system doing a good job?
- What is the optimum break-in/burn-in time?

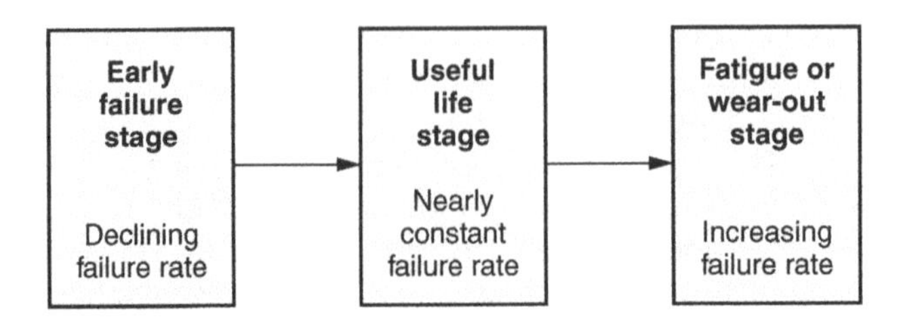

Figure 2.2 Block diagram of life cycle phases.

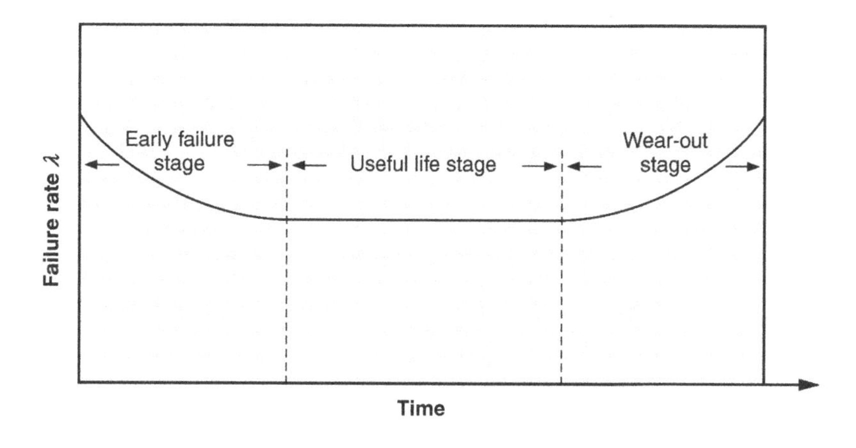

Figure 2.3 The reliability bathtub curve.

- What is the optimum warranty period?
- What are the optimum replacement times for various components?
- What are the spare parts requirements?
- Is the preventive maintenance schedule optimized?
- Is the failure rate during the useful life stage low enough to meet customer expectations?

Each of these items relates to costs in some way, so insight gained from the bathtub curve can have a direct impact on financial performance.

Data are used to determine the shape of the bathtub curve, including locating the boundaries between the stages. Reliability engineers try to change the curve by improving the product. They typically do one or more of the following:

1. Improve the early failure stage by shortening its length and giving it a flatter slope. This is usually accomplished by studying the processes used by the company and/or its suppliers to see where tighter control of process parameters is needed. Shortening the length of the early failure phase reduces costs by decreasing burn-in time and reducing the risk of early failure after installation.

2. Improve the useful life stage by decreasing the constant failure rate. Failure data are studied to determine the most frequent failure types. Reliability engineers work with product/process engineers to find changes that will decrease the failure rate. These activities reduce cost per unit of useful time by increasing life expectancy.

3. Improve the wear-out stage by delaying its onset and flattening the curve. The timing and steepness of the failure rate curve is generally a function of design. However, the wear-out phase can often be postponed somewhat and its slope reduced by more aggressive preventive maintenance and component replacement schedules. Delaying wear-out increases useful life and reduces cost per unit of useful time.

EXAMPLE 2.3

A batch of 364 light bulbs has a constant failure rate of .000058 failures per hour. Find the reliability after 1500 hours of service. About how many bulbs have burned out at the end of the 1500-hour period?

Solution:

$$R(1500) = e^{-(.000058)(1500)} \approx .917$$

This indicates that about 91.7 percent of the bulbs are still functioning, and 8.3 percent have burned out. 364(.083) ≈ 30 of the bulbs have burned out.

During the useful life (constant failure rate stage) of the product, the formula for product reliability is

$$R(t) = e^{-\lambda t}$$

where t is the time elapsed and λ is the constant failure rate.

Reliability Engineering for Software Products

The reliability effort for software products consists of three general phases:

- Error prevention
- Fault detection and removal
- Fault containment through redundancy

The software package is written, then tested against predetermined criteria. Typically, a relatively large number of faults are identified in the early stages of testing, and as this phase continues, the number of faults decreases. At some point the product is released to customers, at which point faults continue to crop up, sometimes at a higher rate than before release. The software package may continue to be used until other considerations force its obsolescence. The typical reliability curve for a software product is shown in Figure 2.4.

Software reliability engineering attempts to improve the reliability curve in Figure 2.4 by:

- Reducing the time and effort involved in the test phase (defect containment phase)
- Lowering the failure rate during the useful life of the product
- Employing redundant techniques to automatically detect and contain faults

Techniques that have been used include:

- Use of alternate formulas to calculate a value, and comparing results

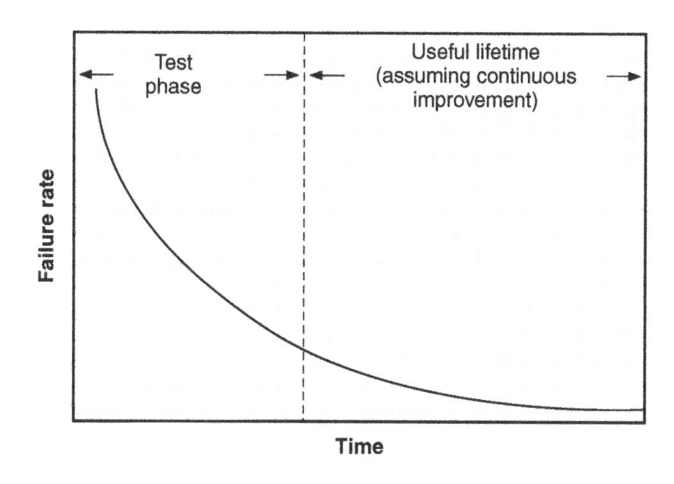

Figure 2.4 Reliability curve for a typical software product.

- Use of dual processors, and comparing results
- Establishing known limits on a result and flagging values outside the limits

Obviously, it is better to prevent errors than to detect faults. Many errors can be prevented by designing thorough requirements prior to code development. The requirements should be stated in terms of imperatives—statements that command that something must occur. Weak phrases that can be interpreted in more than one way should be eliminated from a requirements document. There should be no incomplete requirements of the "to be determined" (TBD) or "to be supplied" (TBS) type. In addition, the requirements should be structured much like a good modular quality program. Developing a high-quality requirements document is worth the effort.

A software team that begins with a good set of requirements can still introduce errors, of course. General rules for software code include:

- Make it modular.
- Keep it simple—excess complexity is difficult to debug and maintain. (Sometimes referred to as the Shirley Temple rule: "Don't be too cute.")
- Provide copious documentation and comments.

The testing protocol must test every requirement at least once. Adequate resources must be provided to allow for complete testing.

In addition to a product's service life cycle, every product has a market life cycle with these stages: concept/design, introduction, growth, maturity, decline. The reliability engineer can make useful contributions at each of these stages.

The principal way that reliability engineers can influence the service life cycle is to provide data during the *concept/design* stage to help assure that components and assemblies will perform throughout their designed life. These data may be obtained through testing proposed designs, from standard tables, or by using components whose life cycle information is available.

During the *introduction* and *growth* stages the reliability engineer will begin to rely more on warranty data from the field, as discussed in Section I.A.7 of Chapter 1.

In the product *maturity* and *decline* stages the reliability engineer may be called on to study the use of the product in different environments or applications in an effort to extend its market life.

5. DESIGN EVALUATION

> Use validation, verification, and other review techniques to assess the reliability of a product's design at various lifecycle stages. (Analyze)
>
> **Body of Knowledge I.B.5**

The reliability engineer performs evaluation functions in each step of the design process.

Concept

As the earliest design parameters are established, preliminary estimates of reliability requirements should be made.

Design Team Effort

As the more formal design phase is initiated, the reliability engineer should be prepared to provide the team with guidance and judgment regarding various options. Data on the reliability of proposed components and the implications for product reliability should be documented for the team at each design refinement. The allocation of reliability requirements to various subsystems and components should also be studied. The underlying mind-set must be to formally study every potential failure and establish cost-effective ways of preventing them. The two general approaches to failure prevention are *fault tolerance* and *fault avoidance*. A fault-tolerant approach requires design of redundant systems so that a fault does not result in a failure. Dual master cylinders in an automotive braking system is an example of a fault-tolerant system. The fault-avoidance approach requires designing the product with components that are sufficiently reliable to guarantee the minimum product reliability. This may be accomplished, for example, by using heavier structural pieces, more-reliable components, derating, and other techniques. Reliability growth during the design process should be documented. The customer's life cycle cost for the product can be minimized by establishing optimum reliability levels and implementing systems for obtaining them.

Design Review

Reliability engineers should provide the design team with data from testing of the final version of the product, assuring that the design meets the reliability requirements. This is often referred to as *validation*.

Preproduction

Alternative manufacturing processes and parameters should be studied for their impact on reliability in order to meet or exceed design requirements.

Production

A testing program should be established to assure that production output meets reliability requirements. This is often referred to as *verification*.

Post-Production

Once a product is released for production, a system for follow-up should be in place so that any failures, in-house or in the field, can be studied. One approach to doing this is known as *failure reporting, analysis, and corrective action system* (FRACAS). A system that permits traceability of individual components helps establish sources of failure due to design, production, service, customer misuse, and so on. The thrust should be not to fix the *blame* but to fix the *problem*.

Thorough, documented design requirements help assure customer satisfaction. The purpose of design evaluation is to verify that the product meets the requirements at each design stage. The aspects of the requirements relating to product reliability mandate special attention because the issue is not merely "does it work?" but rather "how long will it work and under what conditions?" A strong program of testing and documentation can help avoid disappointed customers.

There are four types of evaluation:

1. *Environmental stress screening.* The unit is exposed to the most severe design environmental stresses. In some cases accelerated life testing may be used (see Chapter 11 for details). The purpose is to identify weak components.

2. *Reliability development/growth tests.* A series of tests conducted periodically from design through production phases to demonstrate the impact of corrective actions on reliability.

3. *Reliability qualification tests* (also called *reliability demonstration tests*). Conducted on a sample from production to determine whether production units meet reliability requirements. These tests serve as a basis for production approval.

4. *Product reliability acceptance tests.* A periodic test during production to determine whether the output continues to meet reliability requirements.

A product's life cycle from the producer's viewpoint may be expanded to include these stages:

1. *Concept/design.* This is the stage at which reliability testing and analysis can have the most impact, especially on lowering the failure rate during the useful life of the product. Reliability engineers can aid in the assessment of design validation (that is, fitness for use in the intended operating conditions) by accelerated life tests of components and subassemblies.

2. *Introduction.* At this stage the reliability program gathers and analyzes data on failure rate and proposes measures to reduce it. During this phase, restudy of installation and usage procedures is appropriate. In addition, reliability engineers can aid in the assessment of design validation (that is, compliance with explicit specifications).

3. *Growth.* As the market for the product grows and improvements are implemented, careful traceability becomes critical. It is important to know if these type X failures occurred on products that were produced after improvement Y was in place.

4. *Maturity.* Market growth has plateaued and the enterprise searches for new markets and applications. New environments or functions require a new assessment of reliability.

5. *Decline.* As the market for the product decreases, the costs involved with continued maintenance support must be analyzed. Typically, the support cost per item in use will grow unless carefully monitored.

6. SYSTEMS ENGINEERING AND INTEGRATION

> Describe how these processes are used to create requirements and prioritize design and development activities. (Understand)
>
> **Body of Knowledge I.B.6**

Systems engineering may be defined as a branch of engineering that integrates the activities of other disciplines to produce a product or service. It is generally recognized as having six steps:

1. *Problem statement.* Begins with an investigation of customer needs. This is different than customer requirements because there may be needs that are not understood or articulated by customers. Was there a huge outcry for iPads before they were produced? The problem should include a list of mandatory features and a list of "nice but not mandatory" features. Design requirements take shape during this phase.

   Some prioritization of these requirements may be possible also. For example "operates at 60 hertz" might be a higher-priority requirement than "switchable between 50 hertz and 60 hertz." The design team studies the customer needs in the market niche for which the product is intended as these requirements are formed and prioritized. The next two steps often uncover additional features that can be included in these requirements.

2. *Study alternatives.* Sometimes called *divergent* thinking, this step opens the box to innovative and creative options, which are then studied for feasibility. One or more of the alternatives is selected to proceed to step 3. Steps 2 and 3 often are in a continuous loop as alternatives are studied and modeled. During these phases it is necessary to solidify the design requirements and to prioritize them. Care must be exercised to avoid "mission creep"—the tendency to add too many new features.

3. *Model or simulate the system.* Using physical or software approaches, the system is operated in an often primitive way, sometimes at a slower speed, in an effort to identify problems and opportunities.

4. *Integrate the various elements.* Interface the various elements of the system so the entire entity operates as one.

5. *Run the system.* Study the outputs to verify that they meet all the mandatory output requirements stated in step 1.

6. *Study performance.* Evaluate the entire system, measuring its performance against alternatives and the requirements stated in step 1.

Each of these steps is subject to constant reevaluation, so the series of steps is more like a series of interlocking and repeating loops. The objective is to build a system that not only has useful and productive elements but that also blends these elements into a smoothly operating unit.

Chapter 3

C. Ethics, Safety, and Liability

1. ETHICAL ISSUES

Identify appropriate ethical behaviors for a reliability engineer in various situations. (Evaluate)

Body of Knowledge I.C.1

The ASQ Code of Ethics (found in Appendix B) provides useful guidelines. Some particularly significant excerpts are given below. In the following paragraphs, quotes from the Code of Ethics are shown in italics.

[I] will do whatever I can to promote the reliability and safety of all products that come within my jurisdiction. This indicates that the reliability engineer's responsibilities are not limited to crunching numbers and producing good analyses but include the promotion of product reliability and safety.

Example: A design team has decided on a more hazardous configuration against the recommendation of the reliability engineer. What should the reliability engineer do? The engineer must answer the question "Have I done whatever I can to promote the reliability and safety of all products?" If the answer is "no," then the code of ethics requires further action. If the design team decided on a more hazardous configuration against the recommendation of the reliability engineer, that decision should be brought to the attention of the reliability engineer's manager and director for their review.

Will be dignified and modest in explaining my work and merit. This phrase requires that all who subscribe to this code of ethics recognize that their efforts should be expended on objective analysis of facts and not on self-promotion.

Will preface any public statements that I may issue by clearly indicating on whose behalf they are made. Engineers are frequently called on to apply their expertise to issues not directly related to their employer. These opportunities vary from service on a committee in a professional organization to providing advice on public works projects. When it is necessary to issue a statement in this capacity, the code of ethics requires a disclaimer separating one's views from those of the employer.

On the other side of the coin, when the engineer is asked to speak for the employer, the statement should make that fact clear as well.

Will inform each client or employer of any business connections, interests, or affiliations which might influence my judgment or impair the equitable character of my services. Professionals of all types make value judgments as part of their responsibilities. This section of the code of ethics requires a conscious search to identify any connections that might bias conclusions. In some situations, especially public service, any connection that could even be perceived as a conflict of interest should be divulged.

Will indicate to my employer or client the adverse consequences to be expected if my professional judgment is overruled. The reliability engineer is required to present both good news and bad news scenarios when making recommendations. This equips the decision maker with options, complete with the likely outcomes of each. If hypothesis tests were used to reach conclusions, the significance level should be disclosed. For sampling reports, the confidence level and margin of error should be included. (See Chapter 5 for a discussion of these concepts.)

Will not disclose information concerning the business affairs or technical processes of any present or former employer or client without his consent. This clause says that even in the absence of a confidentiality agreement, the individual is honor bound to act as if one is in place. As a practical matter, it may be advisable to have a signed statement from the former employer or client releasing the information.

Will take care that credit for the work of others is given to those whom it is due. This clause requires action on the part of the person preparing or presenting a report. Rather than leaving the report uncredited, which might imply that the credit is due the presenter, the "take care" phrase requires an acknowledgment of those involved. If a team is due credit, the team members should usually be named.

The entire ASQ Code of Ethics should be studied and used as a basis for action by all in this field.

2. ROLES AND RESPONSIBILITIES

Describe the roles and responsibilities of a reliability engineer in relation to product safety and liability. (Understand)

Body of Knowledge I.C.2

Producing a product that is safe must be a top priority for every organization. The responsibility of the reliability engineer in meeting this priority includes the following:

1. Collecting and analyzing data regarding failures and failure rates
2. Presenting those data and analyses in an understandable format

3. Making sure that the key decision makers have an understanding of the analyses

In discharging these responsibilities, the following are among the additional items that must be considered:

1. Could the failure of the product cause some chain of events with safety/liability implications? This should be done in conjunction with the safety organization, since the safety organization is responsible for defining risk (human injury or death, or accidents).

   Example: The product is installed as part of a system in which the failure of the product was not contemplated in system design.

2. What aspects of the product could possibly cause safety/liability hazards even though the product hasn't failed?

   Example: During normal maintenance the product must be partially disassembled, which may expose energized electrical conductors.

3. What misuse of the product might cause safety/liability issues?

   Examples: The product, when stacked more than three high for shipping, can cause damage to nearby items. If the product is exposed to temperatures below –15 °F, the seals will fail. If the product is not installed within one degree of level, it presents possible hazards. If the pH of the solvent used in the product is below 3.2, the product will develop hazardous leaks. When used on a windy day, the product functions correctly but endangers downwind organisms.

   In any of these situations the reliability engineer must work with the safety organization to perform a safety hazard analysis.

4. Can the final disposition of the product present safety/liability issues?

   Example: The product, when crushed for recycling, releases gases that produce a reaction in some people.

5. Can the malfunction of other parts of the system cause safety/liability issues for the product?

   Example: When exposed to fluid pressures outside its operating range, the product will act unpredictably.

6. What is the impact of government regulation, current or contemplated, on safety/liability issues?

   Example: Several states are contemplating legislation that will declare some types of metallurgical content of a component hazardous.

7. Does the product design compromise the reliability of components?

   Example: An electronic component has an acceptable reliability based on a minimum level of air circulation, but its enclosure is not properly ventilated.

8. Are components purchased from a supplier reliable?

   Example: An item has been sent out for powder coating. The coating has failed, exposing the item to corrosive fumes.

Part of the reliability engineer's function is to keep issues such as these before management. Management also must be made aware of options, including more robust components and redundant design elements, that will reduce or mitigate product failure.

3. SYSTEM SAFETY

> Identify safety-related issues by analyzing customer feedback, design data, field data, and other information. Use risk management tools (e.g., hazard analysis, FMEA, FTA, risk matrix) to identify and prioritize safety concerns, and identify steps that will minimize the misuse of products and processes. (Analyze)
>
> **Body of Knowledge I.C.3**

A typical system safety program has three key elements:

1. *Identification of safety hazards.* The reliability engineer working with the safety organization to achieve a safe product must be innovative and diligent in the discovery of all possible ways that any failure, combination of failures, or other combination of circumstances would present a safety hazard to personnel. Warranty data and other forms of customer feedback should be analyzed. Product testing reports should include safety issues that may have emerged. All available product data should be searched for occurrence of safety hazards. In addition, if databases for similar products are available, these should also be studied.

2. *Risk analysis.* The standard analysis techniques—failure mode and effects analysis (FMEA), failure mode, effects, and criticality analysis (FMECA), production reliability acceptance test (PRAT), fault tree analysis (FTA), success tree analysis (STA), failure reporting, analysis, and corrective action system (FRACAS)—are discussed in Chapter 17. The risk matrix illustrated in Section I.B.3 of Chapter 2 is another approach to quantifying risk. These techniques can be used to estimate the risk associated with various events. It then becomes possible to establish a prioritized list that will provide guidance as the root causes are attacked and resolved.

3. *Correction and prevention.* Chapter 16 discusses preventive and corrective action in general. In the case of safety hazards, there is the additional urgency to avoid harm to personnel. In cases in which flaws in products or processes permit hazardous conditions to occur, engineering change requests (ECR) should be initiated. It is always necessary to consider human error. From one perspective, human error occurs because no system is in place to prevent it. In other words, the onus is on the product/process design community to reduce or eliminate errors. Procedures for preventing/mitigating human error are variously called *idiotproofing, mistake-proofing, poka-yoke,* and *zero quality control* (ZQC). Human error tends to fall into the following categories: misunderstanding, misidentification, inexperience, inattention, and lack of standards. The ideal way to deal with human mistakes is to incorporate design elements that will prevent them from occurring or prevent their occurrence from causing defects. Principal types of mistake-proofing techniques include:

   - *Physical barriers to errors.* (The round shaft won't fit through the square hole.)
   - *Visual reminders.* (A photograph of correct and incorrect results is better than a note on a print or a paragraph of text.)
   - *Use of automated equipment.* (The conveyor will stop if the microswitch detects an error.)
   - *Standardizing.* (The operator takes the same action on a family of parts.)

Often, the best way to prevent reoccurrence of human error is to approach a person who has made the error with the question "How can we design a system that will make that error impossible to occur?"

Activity must be continuous in the three elements of the system safety program. As the higher-priority hazards are corrected/prevented, action can focus on the lower-priority items.

Part II

Probability and Statistics for Reliability

Chapter 4

A. Basic Concepts

1. STATISTICAL TERMS

> Define and use terms such as population, parameter, statistic, sample, the central limit theorem, etc., and compute their values. (Apply)
>
> **Body of Knowledge II.A.1**

The *probability* that a particular event occurs is a number between 0 and 1 inclusive. For example, if a lot consisting of 100 parts has four defectives, we would say that the probability of randomly drawing a defective is .04, or 4%.

The word *random* implies that each part has an equal chance of being selected. If the lot had no defectives, the probability would be 0 or zero %. If the lot had 100 defectives, the probability would be 1 or 100%.

Statisticians use the word *mean* in place of the word *average.* In the case of discrete values, the mean is also called the *expected value* or *expectation.* For the set of values $x_1, x_2, \ldots, x_n$ the formula for the mean is

$$\frac{\sum x_i}{n}.$$

The mean is called a *measure of central tendency.* The symbol most often used for the mean is $\bar{x}$, pronounced "x-bar." The *median* and *mode* are other measures of central tendency. The median is the middle number when the numbers are sorted by size. For example, the median of a sorted set of eleven values is the sixth from the end. If there is an even number of values, the median is the mean of the two middle values. The mode is the number in the list that appears most frequently.

There are three frequently used *measures of dispersion.* The *range* of a set of numbers is obtained by subtracting the smallest value from the largest value. One of the disadvantages of using the range as a measure of dispersion is that it uses only two of the values in the data set, the largest and the smallest. If the data set is large, the range does not make use of much of the information contained in the data. For this and other reasons, the *standard deviation* is frequently used to measure dispersion. The standard deviation can be found by entering the values of the data set into a calculator that has a standard deviation key. See the calculator manual for appropriate steps.

Statistics texts sometimes refer to the *sample variance,* which has the rather ugly formula

$$\text{Sample variance} = \frac{\Sigma(x-\bar{x})^2}{n-1}.$$

One disadvantage of the variance is that, as the formula indicates, it is measured in units that are the square of the units of the original data set. That is, if the x-values are in inches, the variance is in square inches. If the x-values are in degrees Celsius, the variance is in square degrees Celsius, whatever that may be. For many applications it is useful to use a measure of dispersion that is in the same units as the original data. For this reason, the preferred measure of dispersion is the square root of the variance, which is called the *sample standard deviation.* Its formula is

$$\text{Sample standard deviation} = s = \sqrt{\frac{\Sigma(x-\bar{x})^2}{n-1}}.$$

The sample standard deviation is used to estimate the standard deviation of the data set by using a sample from that data set. In some situations it may be possible to use the entire data set rather than a sample. Statisticians refer to the entire data set as a *population,* and its standard deviation is called the *population standard deviation,* symbolized by the lowercase Greek sigma, σ. It is common to use capital N to refer to the number of values in the population. The only difference in the formula is that the divisor in the fraction is N rather than $n - 1$:

$$\text{Population standard deviation} = \sqrt{\frac{\Sigma(x-\mu)^2}{N}}$$

where μ = Population mean.

When using the standard deviation function on a calculator, care should be taken to use the appropriate key. Unfortunately, there is not universal labeling among calculator manufacturers. Some label the sample standard deviation key s_{n-1} and the population key s_n, while others use S_x and s_x. Try entering the values 2, 7, 9, 2 in a calculator and verify that the sample standard deviation rounds to 3.6 and the population standard deviation rounds to 3.1. One of the uses of the standard deviation is to compare the amount of dispersion for two data sets. It also has applications in statistical inference.

To review the notation, in a statistical study the *population* is defined as the collection of all individuals, items, or data under consideration. The part of the population from which information is collected is called the *sample.* A *random sample* is chosen by selecting units for a sample in such a manner that all combinations of units under consideration have an equal or ascertainable chance of being selected as the sample. For example, if 1500 citizens are randomly selected from the United States and their heights are measured, the *population* would be all U.S. citizens, and the *sample,* in this case, a *random sample,* would be the 1500 who were selected. If the mean of those 1500 heights is 64.29, the conclusion is that the *sample mean* is 64.29.

The value 64.29 is called a *statistic,* which is defined as a descriptive measure of a sample. The next step is to infer the mean height of the population, which is likely to be around 64.29. The actual population mean is called a *parameter,* which is defined as a descriptive measure of a population. So, it can be said that a statistic is an estimated value of a parameter.

Different symbols are used to denote parameters and statistics. Parameters are usually denoted by Greek letters, and statistics are usually denoted by Latin letters. For instance, the Greek letter mu (μ) is used for the population mean and the Latin letter x with a bar above it ($\bar{x}$), pronounced "x-bar," is used for the sample mean.

Censoring, a sampling topic important to reliability engineering, is discussed in the first section of Chapter 2.

The Central Limit Theorem

A frequent question is the validity of $\bar{x}$ control charts when the population is not normal. The control chart operates under the assumption that the region bounded by ±3 contains 99.7 percent of the points from a stable process. This assumes that the distribution of the points is normal. An important statistical principle called the *central limit theorem* comes to the rescue. It states that:

> *The distribution of sample averages is approximately normal even if the population from which the sample is drawn is not normally distributed. The approximation improves as the sample size increases.*

Because the $\bar{x}$ chart plots averages, the central limit theorem says that normality is (approximately) guaranteed. This theorem should be kept in mind when selecting the sample size for $\bar{x}$ charts. However, a sample size of less than five is appropriate only if the population is normal.

The central limit theorem supports three-sigma control limits for $\bar{x}$ charts, but the principal reason for plotting averages rather than individual values is that the average is more sensitive to process shifts than the individual value. In other words, given a shift in the process average, the $\bar{x}$ chart is more likely to detect it than the individuals chart.

2. BASIC PROBABILITY CONCEPTS

> Use basic probability concepts (e.g., independence, mutually exclusive, conditional probability) and compute expected values. (Apply)
>
> **Body of Knowledge II.A.2**

Complementation Rule

The probability that an event A will not occur is given by the formula

1 – (the probability that A does occur).

Stated symbolically, P(not A) = 1 – P(A). Some texts use other symbols for "not A," including –A, ~A, and sometimes $\bar{A}$.

Special Addition Rule

Suppose a card is randomly selected from a standard 52-card deck. What is the probability that the card is a club? Since there are 13 clubs, P(♣) = 13/52 = .25. What is the probability that the card is either a club or a spade? Since there are 26 cards that are either clubs or spades, P(♣ or ♠) = 26/52 = .5. Therefore, it appears that P(♣ or ♠) = P(♣) + P(♠), which, generalized, becomes the special addition rule:

$$P(A \text{ or } B) = P(A) + P(B)$$

Caveat: Use only if events A and B can not occur simultaneously. The general addition rule described next is safer to use because it is always valid.

The General Addition Rule

What is the probability of selecting either a king or a club? Using the special addition rule:

$$P(K \text{ or } \clubsuit) = P(K) + P(\clubsuit) = \frac{4}{52} + \frac{13}{52} = \frac{17}{52}$$

This is incorrect, because there are only 16 cards that are either kings or clubs (the 13 clubs plus K♠, K♦, and K♥). The reason that the special addition rule doesn't work here is that the two events (drawing a king and drawing a club) can occur simultaneously. We'll denote the probability that events A and B both occur as P(A & B). This leads to the general addition rule:

$$P(A \text{ or } B) = P(A) + P(B) - P(A \,\&\, B)$$

The special addition rule has the advantage of being somewhat simpler, but its disadvantage is that it is not valid when A and B can occur simultaneously. The general addition rule, although more complex, is always valid. For the above example,

$$P(K \,\&\, \clubsuit) = \frac{1}{52}$$

since only one card is both a K and a club. To complete the example:

$$P(K \text{ or } \clubsuit) = P(K) + P(\clubsuit) - P(K \,\&\, \clubsuit) = \frac{4}{52} + \frac{13}{52} - \frac{1}{52} = \frac{16}{52}$$

Two events that can't occur simultaneously are called *mutually exclusive* or *disjoint*. So, the caveat for the special addition rule is sometimes stated, "Use only if events A and B are mutually exclusive" or "Use only if events A and B are disjoint."

Contingency Tables

Suppose each part in a lot is one of four colors (red, yellow, green, blue) and one of three sizes (small, medium, large). A tool that displays these attributes is the contingency table:

	Red	Yellow	Green	Blue
Small	16	21	14	19
Medium	12	11	19	15
Large	18	12	21	14

Each part belongs in exactly one column, and each part belongs in exactly one row. So, each part belongs in exactly one of the twelve cells. When columns and rows are totaled, the table becomes:

	Red	Yellow	Green	Blue	Totals
Small	16	21	14	19	70
Medium	12	11	19	15	57
Large	18	12	21	14	65
Totals	46	44	54	48	192

Note that 192 can be computed in two ways. If one of the 192 parts is randomly selected, find the probability that the part is red:

Solution:

$$P(\text{red}) = \frac{46}{192} \approx .240$$

Find the probability that the part is small.

Solution:

$$P(\text{small}) = \frac{70}{192} \approx .365$$

Find the probability that the part is red and small.

Solution: Since there are 16 parts that are both red and small,

$$P(\text{red \& small}) = \frac{16}{192} \approx .083$$

Find the probability that the part is red or small.

Solution: Since it is possible for a part to be both red and small simultaneously, the general addition rule must be used:

$$P(\text{red or small}) = P(\text{red}) + P(\text{small}) - P(\text{red \& small}) = \frac{46}{192} + \frac{70}{192} - \frac{16}{192} \approx .521$$

Find the probability that the part is red or yellow.

Solution: Since no part can be both red and yellow simultaneously, the special addition rule can be used:

$$P(\text{red or yellow}) = P(\text{red}) + P(\text{yellow}) = \frac{46}{192} + \frac{44}{192} \approx .469$$

Notice that the general addition rule could have been used:

$$P(\text{red or yellow}) = P(\text{red}) + P(\text{yellow}) - P(\text{red \& yellow})$$

$$= \frac{46}{192} + \frac{44}{192} - 0 \approx .469$$

Conditional Probability

Continuing with the contingency table, suppose the selected part is known to be green. With this knowledge, what is the probability that the part is large?

Solution: Since the part is located in the green column of the table, it is one of the 54 green parts. So, the denominator in the probability fraction is 54. Since 21 of those 54 parts are large,

$$P(\text{large, given that it is green}) = \frac{21}{54} \approx .389.$$

This is referred to as *conditional probability*. It is denoted P(large|green) and pronounced "The probability that the part is large given that it is green." It is useful to remember that the category to the right of the | in the conditional probability symbol points to the denominator in the probability fraction. Find the following probabilities:

$P(\text{small}|\text{red})$ Solution: $P(\text{small}|\text{red}) = \frac{16}{46} \approx .348$

$P(\text{red}|\text{small})$ Solution: $P(\text{red}|\text{small}) = \frac{16}{70} \approx .229$

$P(\text{red}|\text{green})$ Solution: $P(\text{red}|\text{green}) = \frac{0}{54} = 0$

A formal definition for conditional probability is

$$P(B|A) = \frac{P(A \,\&\, B)}{P(A)}$$

Verifying that this formula is valid in each of the above examples will aid in understanding this concept.

General Multiplication Rule

Multiplying both sides of the conditional probability formula by P(A),

$$P(A \ \& \ B) = P(A) \times P(B|A)$$

This is called the *general multiplication rule.* It is useful to verify that this formula is valid using examples from the contingency table.

Independence and the Special Multiplication Rule

Consider the contingency table:

	X	Y	Z	Totals
F	17	18	14	49
G	18	11	16	45
H	25	13	18	56
Totals	60	42	48	150

$$P(G|X) = \frac{18}{60} = .300$$

and

$$P(G) = \frac{45}{150} = .300$$

so

$$P(G|X) = P(G).$$

The events G and X are called *statistically independent* or just *independent.* Knowing that a part is of type X does not affect the probability that it is of type G. Intuitively, two events are called independent if the occurrence of one does not affect the probability that the other occurs. The formal definition of independence of events A and B is

$$P(B|A) = P(B).$$

Making this substitution in the general multiplication rule produces the special multiplication rule:

$$P(A \ \& \ B) = P(A) \times P(B)$$

Caveat: Use only if A and B are independent.

EXAMPLE 4.1

A box holds 129 parts, of which six are defective. A part is randomly drawn from the box and placed in a fixture. A second part is then drawn from the box. What is the probability that the second part is defective? This is referred to as *drawing without replacement.* In other words, the probabilities associated with successive draws depend on the outcome of previous draws. Use the symbol D1 to denote the event that the first part is defective and G1 to denote the event that the first part is good, and so on. There are two mutually exclusive events that can result in a defective part on the second draw: good on first draw and defective on second, or else defective on first and defective on second. Symbolically, these two events are (G1 and D2) or else (D1 and D2). The first step is to find the probability for each of these events. So, if we wish to know the probability of the first part drawn being good *and* the second part drawn being defective, by the general multiplication rule

$$P(G_1 \& D_1) = P(G_1) \times P(D_2|G_1) = \frac{123}{129} \times \frac{6}{128} \approx 0.045$$

Also, by the general multiplication rule

$$P(D_1 \& D_2) = P(D_1) \times P(D_2|D_1) = \frac{6}{129} \times \frac{5}{128} \approx 0.002$$

Since the two events (G_1 & D_2) and (D_1 & D_2) are mutually exclusive, it is appropriate to use the special addition rule:

$$P(D_2) \approx 0.045 + 0.002 = 0.047$$

When drawing two parts, what is the probability that one will be good and one defective? Drawing one good and one defective can occur in two mutually exclusive ways:

$$P(\text{one good and one defective}) = P(G_1 \& D_2 \text{ or } G_2 \& D_1) = P(G_1 \& D_2) + P(G_2 \& D_1)$$

$$P(G_1 \& D_2) = P(G_1) \times P(D_2|G_1) = \frac{123}{129} \times \frac{6}{128} \approx 0.045$$

$$P(G_2 \& D_1) = P(D_1) \times P(G_2|D_1) = \frac{6}{129} \times \frac{123}{128} \approx 0.045$$

So,

$$P(\text{one good and one defective}) \approx 0.045 + 0.045 = 0.090.$$

SUMMARY OF KEY PROBABILITY RULES

For events A and B:

Special addition rule: P(A or B) = P(A) + P(B) [Use only if A and B are mutually exclusive]

General addition rule: P(A or B) = P(A) + P(B) – P(A & B) [Always true]

Continued

Continued

Special multiplication rule: P(A & B) = P(A) × P(B) [Use only if A and B are independent]

General multiplication rule: P(A & B) = P(A) × P(B|A) [Always true]

Conditional probability: P(B|A) = P(A & B) ÷ P(A)

Mutually exclusive (or disjoint):

1. A and B are mutually exclusive if they can't occur simultaneously.
2. A and B are mutually exclusive if P(A & B) = 0.
3. A and B are mutually exclusive if P(A or B) = P(A) + P(B).

Independence:

1. A and B are independent events if the occurrence of one does not change the probability that the other occurs.
2. A and B are independent events if P(B|A) = P(B).
3. A and B are independent events if P(A & B) = P(A) × P(B).

EXAMPLE 4.2

A box of 20 parts has two defectives. The quality technician inspects the box by randomly selecting two parts. What is the probability that both parts selected are defective? The general formula for this type of problem is:

$$P = \frac{\text{Number of ways an event can occur}}{\text{Number of possible outcomes}}$$

The "event" in this case is selecting two defectives, so "number of ways an event can occur" refers to the number of ways two defective parts could be selected. There is only one way to do this because there are only two defective parts; therefore, the numerator in the fraction is 1. The denominator in the fraction is the number of possible outcomes. This refers to the number of different ways of selecting two parts from the box.

This is also called *the number of combinations of two objects from a collection of 20 objects*. The formula is:

Number of combinations of r objects from a collection of n objects = ${}_nC_r$ where

$${}_nC_r = \frac{n!}{r!(n-r)!}$$

Note: Another symbol for number of combinations is $\binom{n}{r}$

Continued

Continued

In this formula the exclamation mark is pronounced "factorial," so $n!$ is pronounced "n factorial." The value of 6! is $6 \times 5 \times 4 \times 3 \times 2 \times 1 = 720$. The value of $n!$ is the result of multiplying the first n positive whole numbers. Most scientific calculators have a factorial key, typically labeled x! To calculate 6! by using this key, press 6 followed by the x! key. Returning to the previous example, the denominator in the fraction is the number of possible combinations of two objects from a collection of 20 objects. Substituting into this formula:

$$_{20}C_2 = \frac{20!}{2!(20-2)!} = \frac{20!}{2!18!} = 190$$

Returning to the example, the probability is $1/190 \approx .005$.

EXAMPLE 4.3

A box of 20 parts has three defectives. The quality technician inspects the box by randomly selecting two parts. What is the probability that both parts selected are defective?

The bottom term of the fraction remains the same as in the previous example. The top term is the number of combinations of two objects from a collection of three objects:

$$\binom{n}{r} = \frac{n!}{n!(n-r)!} = \binom{3}{2} = \frac{3!}{2!(3-2)!} = \frac{6}{2!1!} = \frac{6}{2} = 3$$

To see that this makes sense, name the three defectives A, B, and C. The number of different two-letter combinations of these three letters is AB, AC, BC. Note that AB is not a different combination than BA, because it has the same two letters. If two defectives are selected, the order in which they are selected is not significant. The answer to the probability problem has a 3 as its top term:

$$P = \frac{3}{190} \approx 0.016$$

An important thing to remember: combinations are used when order is not significant.

Combinations

Note: Calculators have an upper limit to the value that can use the x! key. If a problem requires a higher factorial, use the statistical function in a spreadsheet program such as Microsoft Excel.

Permutations

With combinations, the order of the objects doesn't matter. Permutations are very similar except that the order does matter.

EXAMPLE 4.4

A box has 20 parts labeled A through T. Two parts are randomly selected. What is the probability that the two parts are A and T in that order? Note that selecting A and then T is different from selecting T and then A. The general formula applies:

$$P = \frac{\text{Number of ways an event can occur}}{\text{Number of possible outcomes}}$$

The bottom term of the fraction is the number of orderings or *permutations* of two objects from a collection of 20 objects. The general formula is:

$$\text{Number of permutations of } r \text{ objects from a collection of } n \text{ objects} = {}_nP_r = \frac{n!}{(n-r)!}$$

In this example

$${}_{20}P_2 = \frac{20!}{(20-2)!} = 380$$

Of these 380 possible permutations, only one is AT, so the top term in the fraction is 1. The answer to the probability problem is

$$P = \frac{1}{380} \approx 0.003$$

EXAMPLE 4.5

A team with seven members wants to select a task force of three people to collect data for the next team meeting. How many different three-person task forces could be formed? This is not a permutations problem, because the order in which people are selected doesn't matter. In other words, the task force consisting of Barb, Bill, and Bob is the same task force as the one consisting of Bill, Barb, and Bob. Therefore, the combinations formula will be used to calculate the number of possible combinations of three objects from a collection of seven objects:

$${}_7C_3 = \frac{7!}{(7-3)!3!} = 35$$

Thirty-five different task forces could be formed.

EXAMPLE 4.6

A team with seven members wants to select a cabinet consisting of a chairman, a facilitator, and a scribe. How many ways can the three-person cabinet be formed? Here, the order is important because the cabinet consisting of Barb, Bill, and Bob will have Barb as chairman, Bill as facilitator, and Bob as scribe, while the cabinet consisting of Bill, Barb, and Bob has Bill as chairman, Barb as facilitator, and Bob as scribe. The appropriate formula is the one for permutations of three objects from a collection of seven objects.

$${}_7P_3 = \frac{7!}{(7-3)!} = 210$$

3. DISCRETE AND CONTINUOUS PROBABILITY DISTRIBUTIONS

> Compare and contrast various distributions (binomial, Poisson, exponential, Weibull, normal, lognormal, etc.) and their functions (e.g., cumulative distribution functions (CDFs), probability density functions (PDFs), hazard functions), and relate them to the bathtub curve. (Analyze)
>
> **Body of Knowledge II.A.3**

Example 4.7 introduces the basic concepts of probability distributions.

Distributions based on random variables that can take on only integer values, or isolated and distinct values, are called *discrete distributions.* Distributions based on random variables that can take on an infinite number of values in a finite interval are called *continuous distributions.* The distribution in the previous example was discrete. Other discrete distributions are presented in the next section.

EXAMPLE 4.7

A piece from a wood finishing process has the following specification: no bubbles with diameter larger than 0.5 mm, and a maximum of 10 bubbles with diameter between 0.05 and 0.5 mm inclusive.

A batch of 50 pieces is inspected for number of bubbles with diameters between 0.05 and 0.5 mm with the following results:

Number of bubbles $0.05 \leq f \leq 0.5$ mm x	0	1	2	3	4	≥ 5
Frequency f	11	15	16	6	2	0
Relative frequency p	0.22	0.30	0.32	0.12	0.04	0.0

That is, there were 11 pieces with no bubbles with the stated diameters, and there were 15 pieces with one bubble with the stated diameter, and so forth. There were no pieces with five or more bubbles. Relative frequency is labeled p because if a piece is selected at random from the batch, this is the probability that it will have the stated number of bubbles. For example, the probability that the piece will have three bubbles is 0.12. The number of bubbles is a *variable*, and the number of bubbles on a randomly selected piece is called a *random variable*. The first and third rows of this table constitute what is called a *probability distribution*, and a histogram of these data, as shown in Figure 4.1, is called a *probability histogram*.

Continued

Continued

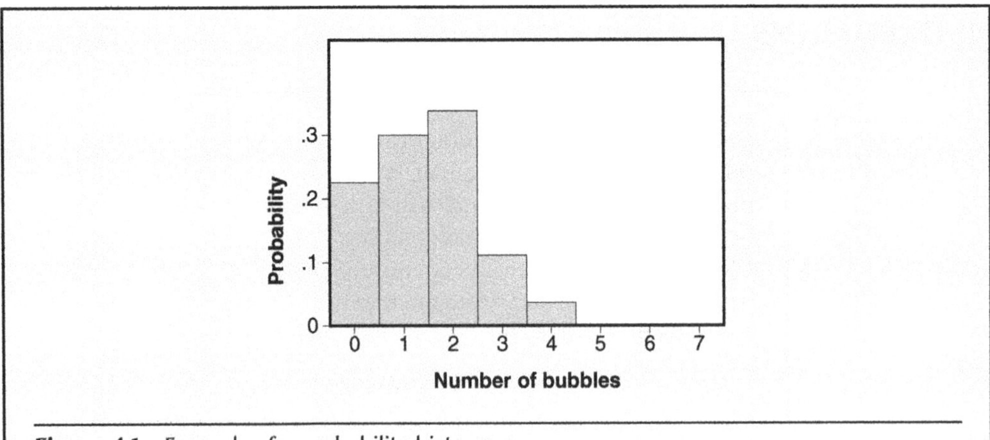

Figure 4.1 Example of a probability histogram.

Discrete Distributions

Binomial Distribution. The *binomial distribution* is a discrete distribution whose random variable can take on one of only two possible values. In reliability applications, the two categories might be *operable* and *failed.*

The function that defines the distribution is called the binomial formula:

$$P(X = x) = \frac{n!}{(n-x)!\,x!} p^x (1-p)^{n-x}$$

where

n = Sample size

x = Number of failures

p = Proportion of the population that has failed

$P(X = x)$ = The probability that the sample has x failures

$a! = a(a-1)(a-2)\ldots(1)$ For example, $5! = 5 \times 4 \times 3 \times 2 \times 1$

EXAMPLE 4.8

Suppose 25% of a very large population of parts has failed. If six parts are selected at random, find the probability that none of the six has failed.

Solution:

In this case, $n = 6$, $p = .25$, $x = 0$. Substituting into the binomial formula,

$$P(X = 0) = \frac{6!}{6!0!}(.25)^0(.75)^6 \approx 0.18$$

EXAMPLE 4.9

Find the binomial distribution for $p = .25$ and $n = 6$, and draw the associated histogram.

As shown in the previous example, $P(X = 0) \approx .18$

$$P(X=1)=\frac{6!}{5!1!}.25^1.75^5 \approx .36$$

$$P(X=2)=\frac{6!}{4!2!}.25^2.75^4 \approx .30$$

$$P(X=3)=\frac{6!}{3!3!}.25^3.75^3 \approx .13$$

$$P(X=4)=\frac{6!}{2!4!}.25^4.75^2 \approx .03$$

$$P(X=5)=\frac{6!}{1!5!}.25^5.75^1 \approx .004$$

$$P(X=6)=\frac{6!}{6!0!}.25^6.75^0 \approx .0002$$

Figure 4.2 shows the complete distribution using $x = 0, 1, \ldots, 6$.

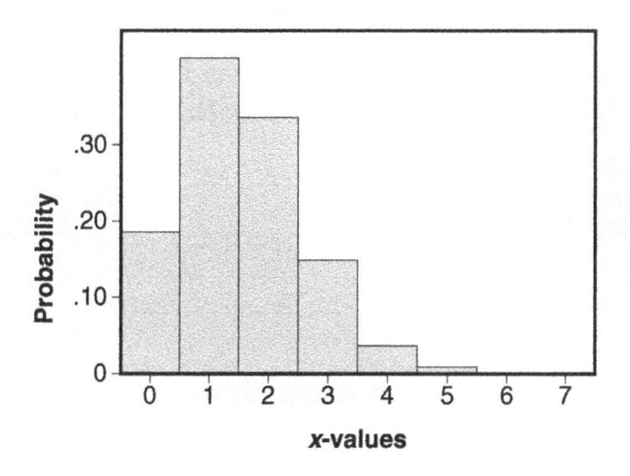

Figure 4.2 Binomial probability distribution and histogram for $p = .25$ and $n = 6$.

Part II.A.3

The binomial distribution consists of the set of possible x-values and their associated probabilities.

The *probability density function* (PDF) is the expression that generates the distribution. In this case it is

$$P(X=x)=\frac{6!}{(6-x)!x!}.25^x.75^{(6-x)}.$$

The *cumulative distribution function* (CDF) $F(x)$ is defined as the sum of the probabilities up to and including the x-value. More precisely, the CDF is defined as

$$F(x) = P(X \le x) = \sum_{t \le x} P(X = t).$$

In Example 4.9, the CDF could be used to answer the question "What is the probability that the sample includes two or fewer failed items?" as follows:

$$F(2) = \sum_{t \le 2} f(t) = P(X = 0) + P(X = 1) + P(X = 2) \approx .18 + .36 + .30 \approx .84$$

The mean and standard deviation of a binomial distribution are given by the formulas

$$\mu = np$$

$$\sigma = \sqrt{np(1-p)}$$

In Example 4.9

$$\mu = 6(.25) = 1.5$$

and

$$\sigma = \sqrt{1.5(1-.25)} \approx 1.06$$

Poisson Distribution. The *Poisson distribution* is a discrete probability distribution that may be used to find the probability that an event will occur a specified number of times. The PDF formula is

$$P(X = x) = e^{-\lambda} \frac{\lambda^x}{x!}$$

where

x = A whole number and λ = A real number.

Since the random variable can take on any whole number, the probability distribution technically extends indefinitely. From Example 4.10 a partial list:

$P(X = 0) \approx .005$, $P(X = 1) \approx .024$, $P(X = 2) \approx .066$, $P(X = 3) \approx .119$, $P(X = 4) \approx .16$, $P(X = 5) \approx .173$, $P(X = 6) \approx .16$, $P(X = 7) \approx .120$, $P(X = 8) \approx .081$, $P(X = 9) \approx .049$

EXAMPLE 4.10

Records indicate that the number of customers that arrive at a bank drive-up window between 1:00 p.m. and 2:00 p.m. has a Poisson distribution with $\lambda = 5.4$. Find the probability that exactly six people arrive.

Solution:

$$P(X = 6) = e^{-5.4} \frac{5.4^6}{6!} \approx .16$$

The CDF for the Poisson distribution is given by

$$\sum_{t \le x} P(X = t) = \sum_{t \le x} e^{-\lambda} \frac{\lambda^t}{t!}.$$

In the previous example the CDF could be used to calculate the probability that at most four people arrive at the drive-up window:

$$\sum_{t \le 4} P(X \le t) = P(X = 0) + P(X = 1) + P(X = 2) + P(X = 3) + P(X = 4)$$

$$\approx .005 + .024 + .066 + .119 + .160 \approx .374$$

The mean and standard deviation of the Poisson distribution are

$$\mu = \lambda$$

$$\sigma = \sqrt{\lambda}.$$

In Example 4.10 $\mu = 5.4$ and $\sigma \approx 2.32$.

Continuous Distributions

Exponential Distribution. The *exponential distribution* is a continuous distribution that is frequently used to model time to failure for products when the failure rate is constant. The PDF is

$$f(t) = \lambda e^{-\lambda t}$$

where

λ = Constant failure rate

t = Time (or some other measure of product use, such as cycles, miles, rounds fired, and so on)

A PDF graph is shown in Example 4.11.

The CDF for the exponential distribution is

$$P(x \le a) = F(a) = \int_0^a \lambda e^{-\lambda t}\, dt = 1 - e^{-\lambda t}.$$

The CDF can be used to determine the probability of failure during the first t hours. The probability that a unit is still operating at t hours is

P(operating at time t) = (1 – probability it has failed by time t) = $e^{-\lambda t}$.

P(operating at time t) is called *reliability at time t*, or R(t), so when the failure rate is constant,

$$R(t) = e^{-\lambda t}.$$

EXAMPLE 4.11

Find the value of the PDF at 1000 hours given that the failure rate is .00053 failures per hour.

$$f(1000) = .00053e^{-.00053 \times 1000} \approx .00031$$

This says that the probability of failure at 1000 hours is about .00031.

A sketch of the PDF in this example is shown in Figure 4.3. The mean and standard deviation of the PDF are $\mu = 1/\lambda$ and $\sigma = 1/\lambda$.

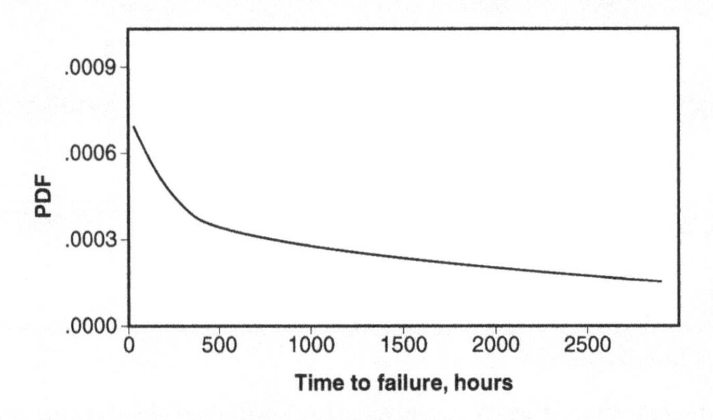

Figure 4.3 Probability density function for Example 4.11.

EXAMPLE 4.12

Find reliability at 1000 hours if $\lambda = 0.00053$ failures/hour.

Solution:

$$R(1000) = e^{-0.00053 \times 1000} \approx .59,$$

which indicates that approximately 59% of the units are still operating after 1000 hours, or the probability that a particular unit is still operating after 1000 hours is .59.

By definition, the mean time to failure (MTTF) is $1/\lambda$. Therefore, the reliability at MTTF is

$$R(\text{MTTF}) = R(1/\lambda) = e^{-\lambda \times 1/\lambda} = e^{-1} \approx .368.$$

This shows that at MTTF only about 37 percent of the products are operating, or that the probability that a particular unit is still operating after MTTF is about .37. For repairable items, MTTF can be replaced by mean time between failures (MTBF) in the discussion in this paragraph.

Weibull Distribution. The PDF for the Weibull distribution is defined as

$$f(t)=\frac{\beta}{\eta}\left(\frac{t}{\eta}\right)^{\beta-1}e^{-\left(\frac{t}{\eta}\right)^{\beta}}$$

where

$t \geq 0$, shape parameter $\beta \geq 0$, scale parameter ≥ 0

Various shapes are possible by selection of different values for β. If $\beta = 1$, the Weibull reduces to the exponential, and if $\beta \approx 3.44$, the curve approximates the normal distribution. See Figure 4.4.

The hazard function is given by

$$h(t)=\frac{\beta}{\eta^{\beta}}(t)^{\beta-1}.$$

The CDF for the Weibull is

$$F(t)=1-e^{-\left(\frac{1}{\eta}\right)^{\beta}}.$$

Again,

$$R(t)=1-F(t)=e^{-\left(\frac{t}{\eta}\right)^{\beta}}.$$

The two curved portions of the bathtub curve (Figure 2.3) are sometimes modeled by using the Weibull with appropriate shape parameters.

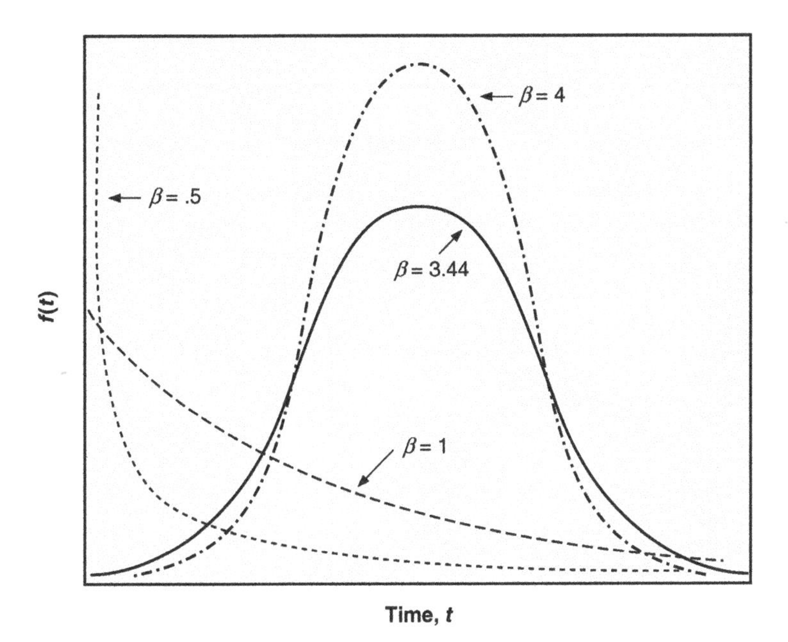

Figure 4.4 Plot of various members of the Weibull distribution.

EXAMPLE 4.13

A product's time to failure has a Weibull distribution with $\beta = .72$ and $\eta = 10{,}000$. Find reliability at 200 hours.

Solution:

$$R(200) = e^{-\left(\frac{200}{10{,}000}\right)^{.72}} \approx .94$$

indicating that about 94 percent of the products are operating after 200 hours.

Normal Distribution. The *normal distribution* is considered the most important distribution in both the theory and practice of statistics. Its PDF is

$$f(x) = \frac{e^{\frac{-(x-\mu)^2}{2\sigma^2}}}{\sigma\sqrt{2\pi}}$$

where

μ and σ are the mean and standard deviation, respectively.

In reliability applications, μ is the mean. Changes in this value cause the center of the distribution to be moved left or right along the *x*-axis. As the standard deviation decreases, the distribution becomes narrower, centered around the mean.

EXAMPLE 4.14

Assuming the following times to failure are sampled from a population that is normally distributed, find the PDF for the distribution.

42.3 45.6 49.5 53.6 54.8

Solution:

The population mean and standard deviation are estimated from the sample mean and sample deviation

$$\hat{\mu} = 49.16$$

and

$$\hat{\sigma} = 5.28$$

Substituting into the generic PDF formula:

$$f(x) = \frac{e^{\frac{-(x-49.16)^2}{55.76}}}{13.23}$$

When units have an increasing failure rate, such as during the wear-out phase, the times to failure are sometimes normally distributed, although it is more common to see a Weibull distribution here.

The standard normal curve has $\mu = 0$ and $\sigma = 1$, so its PDF is

$$f(x) = \frac{e^{-\frac{x^2}{2}}}{\sqrt{2\pi}} \approx .3989e^{-\frac{x^2}{2}}.$$

Applications requiring cumulative calculations are handled by finding the area under this curve. This is typically accomplished using statistical functions embedded in spreadsheet software (for example, NORMDIST in Excel) or using a standard normal table such as that found in Appendix E. For example, the area under the standard normal curve to the right of $z = 1.28$ standard deviations is found in the row labeled 1.2 and the column labeled 0.08. So, the area to the right of $z = 1.28$ is 0.1003, and since the total area under the standard normal curve is 1, it can be said that 10.03 percent of the area under the curve lies to the right of 1.28.

The standard normal table can be used to find specific areas under a nonstandard normal curve if the mean and standard deviation are known, as the following examples show.

EXAMPLE 4.15

The diameters of a batch of turned shafts are normally distributed with $\mu = 2.015$ and $\sigma = 0.0053$.

a. What percent of the batch has a diameter greater than the upper specification of 2.025?

b. What percent of the batch has a diameter smaller than the lower specification of 2.000?

Solution:

a. Find z, the number of standard deviations from the upper specification to the mean. The formula:

$$z = \frac{x - \mu}{\sigma} = \frac{2.025 - 2.015}{0.0053} \approx 1.89$$

From the standard normal table in the row labeled 1.8 and the column labeled 0.09, the value is 0.0294, which indicates that 2.94 percent of the batch exceeds the upper specification.

b. For the lower specification:

$$z = \frac{x - \mu}{\sigma} = \frac{2.000 - 2.015}{0.0053} \approx -2.83$$

The area to the left of –2.83 is the same as the area to the right of +2.83 because of the symmetry of the normal curve. From the standard normal table, 0.23 percent of the batch violates the lower specification.

EXAMPLE 4.16

The time to failure for a product is normally distributed with $\mu = 200$ hours and $\sigma = 1.84$. Find the probability of failure between $t = 201$ hours and $t = 202$ hours. That is, find

$$P(201 \le x \le 202).$$

Solution:

For 202 hours $z = \dfrac{202 - 200}{1.84} \approx 1.09$

For 201 hours $z = \dfrac{201 - 200}{1.84} \approx 0.54$

The area to the left of 1.09 from the standard normal table is 0.8621.

The area to the left of 0.54 from the standard normal table is 0.7054.

To find the area between these two values, subtract:

$$P(x \le 1.09) = .8621$$

Subtracting $$P(x \le .54) = .7054$$

$$P(201 \le x \le 202) = .1567$$

15.67 percent of the units will fail between 201 and 202 hours.

Lognormal Distribution. If the natural logarithm (ln) of a random variable is normally distributed, the variable follows the *lognormal distribution*. The PDF is

$$f(x) = \frac{e^{-0.5\left(\frac{x' - \mu_{x'}}{\sigma_{x'}}\right)^2}}{x\sigma_{x'}\sqrt{2\pi}}$$

where

$x' = \ln x$

$\mu_{x'}$ = Mean of the x' values

$\sigma_{x'}$ = Standard deviation of the x' values

The lognormal distribution has been found to be a good mathematical model for times to failure for some electronic and mechanical products, including transistors, bearings, and electrical insulation. It is sometimes a good model for times to repair a unit after a failure.

The mean and standard deviation of the lognormal distribution are given by

$$\mu = e^{\mu_{x'} + .5\sigma_{x'}^2}$$

$$\sigma = \sqrt{\left(e^{2\mu_x + \sigma_{x'}^2}\right)\left(e^{\sigma_{x'}^2} - 1\right)}$$

EXAMPLE 4.17

The times to failure for a product are known to be lognormally distributed. The times to failure in cycles of a sample of six parts are 850, 925, 1250, 1550, 1800, and 2750 cycles. Find the PDF for the distribution.

Solution:

Find the ln of each of the given values:

x	$x' = \ln x$
850	6.75
925	6.83
1250	7.13
1550	7.35
1800	7.50
2750	7.92

Find the mean and standard deviation of the values in the second column:

$$\bar{x}' = 7.25$$

$$\sigma_{x'} = .44$$

Substituting into the PDF formula:

$$f(x) = \frac{e^{-.5\left(\frac{x'-7.25}{.44}\right)^2}}{x(.44)\sqrt{2\pi}}$$

Statistical software packages such as Minitab and JMP can be used to find the distribution that best fits a given data set.

Descriptive Characteristics of Distributions

Software programs often calculate values of two characteristics of distributions.

Skewness is defined as

$$\text{Skewness} = \frac{n}{(n-1)(n-2)} \sum \left(\frac{x_j - \bar{x}}{s} \right)^3$$

where

n = Sample size

s = Sample standard deviation

x_j = Sample values

If the skewness value is 0, the distribution is symmetric about its mode. If skewness < 0, the distribution is left-skewed; that is, the histogram extends farther to the left of the mode than it does to the right. If skewness > 0, the distribution is right-skewed.

Kurtosis is a measure of flatness and is defined as

$$\text{kurtosis} = \frac{n(n+1)}{(n-1)(n-2)(n-3)}\sum\left(\frac{x_j - \bar{x}}{s}\right)^4 - \frac{3(n-1)^2}{(n-2)(n-3)}$$

where

n = Sample size

s = Sample standard deviation

x_j = Sample values

A normal distribution has kurtosis = 0. If kurtosis < 0, the distribution is flatter than a normal distribution, and if kurtosis > 0, the distribution is more peaked than a normal distribution.

4. POISSON PROCESS MODELS

> Define and describe homogeneous and non-homogeneous Poisson process models (HPP and NHPP). (Understand)
>
> **Body of Knowledge II.A.4**

As we have seen, when a system is operating on the flat part of the bathtub curve, the failure rate is constant, that is, $\lambda(t) = \lambda$. This model for failure is referred to as the *homogeneous Poisson process* (HPP). It is the most widely used model in reliability engineering.

If the failure rate is a nonconstant function of t, the resulting model is called a *non-homogeneous process model* (NHPM). This model has been found to be useful for predicting failure rates in software products. The case in which $\lambda(t) = \alpha t^{-\beta}$ $\alpha > 0$, $\beta < 1$ is referred to as the *power law model*, which has been found to be quite useful for many applications, particularly failure times for repairable systems. This model is also known as the *Duane model* and the AMSAA (for *Army materiel systems analysis activity*) model.

Note that when $\lambda = 0$, the power law model becomes the HPP.

5. NON-PARAMETRIC STATISTICAL METHODS

> Apply non-parametric statistical methods, including median, Kaplan-Meier, Mann-Whitney, etc., in various situations. (Apply)
>
> Body of Knowledge II.A.5

When using a hypothesis test when the underlying distribution of the data is not known, it is necessary to use nonparametric methods.

Median Test

This test is used to determine whether there is sufficient evidence to conclude that k medians from samples of size n_j are equal. This test permits the use of unequal sample sizes.

Procedure (see steps in hypothesis testing in Chapter 5):

1. Conditions: no conditions on the population distributions are required.
2. The null hypothesis H_0: $\tilde{\mu}_1 = \tilde{\mu}_2 = \ldots = \tilde{\mu}_k$

   The alternative hypothesis H_a: $\tilde{\mu}_1 \neq \tilde{\mu}_j$ for at least one pair (i, j)
3. Determine α
4. Indentify the critical value using the χ^2 table in Appendix I. The critical value is defined as $\chi^2_{\alpha,k-1}$ where α = level of significance. This is a right-tail test, so the reject region consists of the values of the test statistics ≥ the critical value.
5. Construct a table with column 1 showing the number of observations above the overall median and column 3 showing the number of readings below the overall median for each sample. Half the observations that are equal to the overall median should be counted in the "above" column and half in the "below" column. Columns 2 and 4 show the expected number above and below the overall medians. The expected number will be half the sample size. *Observed* frequencies, which will be denoted as O-values O_A and O_B, refer to the observed frequencies above and below the overall median.

   E_A and E_B refer to the expected frequency above and below the overall median.

   For each row, calculate

$$\frac{(O_A - E_A)^2}{E_A} + \frac{(O_B - E_B)^2}{E_B}.$$

The test statistic value is the sum of these values.

6. Reject H_0 if the test statistic is in the reject region.
7. State the conclusion in terms of the original problem.

EXAMPLE 4.18

Failure times of a component from four suppliers are being compared. Random samples from the four populations are collected and operated to failure. Are at least two of the population medians different at the 0.05 significance level?

The data:

Supplier 1: 10, 8, 10, 12, 13, 7, 9, 11, 9, 7

Supplier 2: 7, 9, 12, 7, 8, 7, 11, 10

Supplier 3: 8, 7, 8, 6, 9, 7, 6, 7

Supplier 4: 14, 13, 12, 13, 11, 10, 6

Overall median: $\tilde{x} = 9$

Procedure:

1. Conditions are met
2. H_0: $\tilde{\mu}_1 = \tilde{\mu}_2 = \tilde{\mu}_3 = \tilde{\mu}_4$
   H_a: $\tilde{\mu}_1 \neq \tilde{\mu}_j$ for at least one pair (i, j)
   From Appendix I, the critical value $\chi^2_{.05,3} = 7.815$
3. $\alpha = 0.05$
4. Reject region: test statistic ≥ 7.815
5. Construct the table

Column #	1	2	3	4	5	6	7
	# above $\tilde{x}$		# below $\tilde{x}$		$\frac{(O_A - E_A)^2}{E_A^2}$	$\frac{(O_B - E_B)^2}{E_B^2}$	$\frac{(O_A - E_A)^2}{E_A^2} + \frac{(O_B - E_B)^2}{E_B^2}$
Supplier	Obs.	Exp.	Obs.	Exp.			
1	6	5	4	5	0.2	0.2	0.4
2	3.5	4	4.5	4	0.0625	0.0625	0.063
3	.5	4	7.5	4	3.0625	3.0625	6.125
4	6	3.5	1	3.5	1.7857	1.7857	3.571
							Test statistic = Σ = 10.159

6. Since the test statistic ≥ 7.815 the null hypothesis can be rejected.
7. The conclusion is that at least two of the population medians are different at the 0.05 significance level.

Part II.A.5

Kaplan-Meier Estimate for Reliability

This procedure provides a nonparametric procedure for estimating reliability at failure times. Censored failure data are permitted, and exact failure times are required. The calculation table is constructed as follows:

Column 1: Failure times, t_i, in ascending order.

Column 2: Number of units on test at beginning of time period.

Column 3: Number of units censored during time period. These are units that have been taken off test for any reason other than failure.

Column 4: Number of units on test at end of period (column 2 – column 3 – 1).

Column 5: For the first line of the table use (column 4)/(column 2 – column 3).

For successive columns use:

(Column 4)/(column 2 – column 3) (column 5 entry from previous line)

Column 5 is the estimate of reliability at time t_i.

EXAMPLE 4.19

Fifty items are placed in test fixtures. Failure times in hours are 38, 52, 68, 70, 85, and 98 hours, respectively. Test fixtures fail at 40, 62, and 80 hours into the test. Two non-failed units are removed for examination at 75 hours.

1	2	3	4	5
Failure times	**# units at beg. of period**	**# units censored**	**# units @ end**	
38	50	0	49	$\hat{R}(38)=\frac{49}{50}$
52	49	1	47	$\hat{R}(50)=\frac{49}{50}\times\frac{47}{48}$
68	47	1	45	$\hat{R}(68)=\frac{49}{50}\times\frac{47}{48}\times\frac{45}{46}$
70	45	0	44	$\hat{R}(70)=\frac{49}{50}\times\frac{47}{48}\times\frac{45}{46}\times\frac{44}{45}$
85	44	3	40	$\hat{R}(85)=\frac{49}{50}\times\frac{47}{48}\times\frac{45}{46}\times\frac{44}{45}\times\frac{40}{41}$
98	40	0	39	$\hat{R}(85)=\frac{49}{50}\times\frac{47}{48}\times\frac{45}{46}\times\frac{44}{45}\times\frac{40}{41}\times\frac{39}{40}$

Mann-Whitney Test for Means of Two Populations

This nonparametric test can be used to compare means of two populations.

Procedure (see steps in hypothesis testing in Chapter 5):

1. Conditions: The two populations have the same shape.
2. H_0: $\mu_1 = \mu_2$ H_a: $\mu_1 \neq \mu_2$ (two-tailed test), or

   $\mu_1 < \mu_2$ (left-tailed test), or

   $\mu_1 > \mu_2$ (right-tailed test)

   (Use the sample with the fewest values as sample 1.)
3. Determine α
4. Find the critical values from Appendix L and define reject regions.

   Use Table 1 for one-tailed tests with $\alpha = 0.025$ and two-tailed tests with $\alpha = 0.05$.

   Use Table 2 for one-tailed tests with $\alpha = 0.05$ and two-tailed tests with $\alpha = 0.10$.
5. Assign an overall rank (considering all the values in both samples) to each value. If two or more values are tied, each is given the mean of the ranks they would have had if there were no ties. The test statistic M is the sum of the overall ranks for the sample with smaller sample size.
6. Reject H_0 if M is in a reject region.
7. State the conclusion in terms of the original problem.

EXAMPLE 4.20

Samples from two suppliers are tested until all have failed. At the 0.05 significance level, do the populations from which the samples were drawn have different means? Assume the two populations have the same shape.

Failure times supplier A	Failure times supplier B
260	230
265	241
266	265
283	280
290	284
292	288
	290
	294
	297

Continued

Continued

Solution:

1. The condition is met per the problem statement (the two populations have the same shape).
2. $H_0: \mu_1 = \mu_2 \quad H_a: \mu_1 \neq \mu_2$
3. $\alpha = 0.05$
4. From Appendix L, critical values are 31 and 65. Reject regions are $M \geq 65$ or $M \leq 31$
5. The ranks:

Failure times supplier A	Overall rank	Failure times supplier B	Overall rank
260	3	230	1
265	4.5*	241	2
266	6	265	4.5*
283	8	280	7
290	11.5	284	9
292	13	288	10
		290	11.5
		294	14
		297	15

*There are two 265's in ranks 4 and 5, so each gets the mean rank.

$M = 46$ (the sum of supplier A ranks)

6. Since M is not in a reject region, do not reject H_0.
7. At the 0.05 significance level, the data do not show that the two populations have different means.

6. SAMPLE SIZE DETERMINATION

Use various theories, tables, and formulas to determine appropriate sample sizes for statistical and reliability testing. (Apply)

Body of Knowledge II.A.6

Methods for calculating the width of a confidence interval are discussed in Chapter 5. In some cases the width of the interval is specified in advance. It is then necessary to calculate the sample size that will meet that specification. Table 4.1 lists formulas for this task. The formulas use E to denote *margin of error,* which is half the width of the confidence interval. The confidence level is $1 - \alpha$.

Table 4.1 Formulas for sample size

Confidence interval for	Sample size n Round up to the nearest whole number
μ, population mean	$\left(\frac{z_{\alpha/2}\sigma}{E}\right)^2$
p, population proportion	$\left(\frac{z_{\alpha/2}}{E}\right)^2 \hat{p}(1-\hat{p})$ $\hat{p}$ denotes an estimate for p
p, population proportion	$0.25\left(\frac{z_{\alpha/2}}{E}\right)^2$ Use when no estimate for p is available
$\mu_1 - \mu_2$, the difference between two population means	$n_1 = n_2 = \frac{(z_{\alpha/2})^2(\sigma_1^2+\sigma_2^2)}{E^2}$
$p_1 - p_2$, the difference between two population proportions	$n_1 = n_2 = 0.5\left(\frac{z_{\alpha/2}}{E}\right)^2$

EXAMPLE 4.21

Determine the sample size required to construct a 95% confidence interval for the mean if the CI is to be 0.020 wide given that $\sigma = 0.028$

In this case the first formula is appropriate because a confidence interval for the mean is sought. The value of $z_{\alpha/2}$ is found in the normal table. For 95% confidence, $\alpha = 0.05$, $\alpha/2 = 0.025$, and $z_{\alpha/2} = 1.96$. The margin of error E = 0.010.

$$n = \left(\frac{z_{\alpha/2}\sigma}{E}\right)^2 = \left(\frac{1.96 \times 0.028}{0.010}\right)^2 \approx 30.1$$

So, use a sample size $n = 31$ (always round up).

The justification for the use of the normal table is the central limit theorem. As discussed in the first section of this chapter, this theorem states that the means are normally distributed.

Suppose the sample of 31 was selected with due care for randomness, and the sample mean $\bar{x} = 25.6$. The 95% confidence interval would be 25.6 ± 0.01. That is, there is 95% confidence that the population mean is between 25.59 and 25.61.

EXAMPLE 4.22

Determine the sample size to construct a 95% confidence interval that is 0.06 wide for the proportion of defectives in a batch, assuming the historic proportion has been 5%.

The second formula in Table 4.1 deals with a proportion:

$$n=\left(\frac{z_{\alpha/2}}{E}\right)^2 \hat{p}(1-\hat{p})=\left(\frac{1.96}{0.03}\right)^2 (0.05)(0.95)\approx 202.75$$

so, use sample size $n = 203$

Although the justification is not provided here, statistical theory states that the normal table is appropriate. If a sample of 203 is polled and the sample proportion is $p_o = .042$, the 95% confidence interval is $.042 \pm .03$.

EXAMPLE 4.23

Determine the sample size needed to construct a 95% confidence interval that is 0.06 wide for the proportion of defectives in a batch when no estimate is available.

The third formula in Table 4.1 is appropriate in this case:

$$n=0.25\left(\frac{z_{\alpha/2}}{E}\right)^2=0.25\left(\frac{1.96}{0.03}\right)^2\approx 1067.1$$

so, use sample size $n = 1068$.

This illustrates why it is best to use an estimate if available.

EXAMPLE 4.24

A vendor claims that product A has an average thickness at least 2mm larger than that of product B. How many of each product type should be sampled to construct a 95% confidence interval to check the vendor's claim? Assume the two populations have standard deviations of 3.3 mm and 3.8 mm, respectively.

The fourth formula in Table 4.1 deals with the difference between population means.

$$n=\frac{(z_{\alpha/2})^2(\sigma_1^2+\sigma_2^2)}{E^2}=\frac{1.96^2(3.3^2+3.8^2)}{2^2}\approx 24.3$$

so, use sample size $n = 25$.

If samples of size 25 are selected from each product type, and $\bar{x}_A = 45.6$ and $\bar{x}_B = 37.6$, then the difference of the sample means is $\bar{x}_A - \bar{x}_B = 8$, and the confidence interval for the difference is 8 ± 2, that is, there is 95% confidence that the difference is between 6 mm and 10 mm. Therefore, there is 95% confidence that the vendor's claim is valid.

EXAMPLE 4.25

A reliability engineer has two competing designs that are stress-tested for 100 hours. Let p_A = the proportion of design *A* that fails, and p_B = the proportion of design *B* that fails. It is necessary to calculate the 95% confidence interval with width = 0.4 for the difference $p_A - p_B$.

Using the fifth formula in Table 4.1:

$$n_1 = n_2 = 0.5\left(\frac{z_{\alpha/2}}{E}\right)^2 = 0.5\left(\frac{1.96}{0.2}\right)^2 \approx 48.02$$

So, use $n_1 = n_2 = 49$.

The results of testing 49 items of each design are $p_{A_0} = 0.29$ and $p_{B_0} = 0.02$

So, $p_{A_0} - p_{A_0} = 0.27$ and the confidence interval is 0.27 ± 0.2.

For example, there is 95% confidence that the difference between the two proportions is between 0.07 and 0.47, which means that design B is significantly better than design A.

7. STATISTICAL PROCESS CONTROL (SPC) AND PROCESS CAPABILITY

> Define and describe SPC and process capability studies (C_p, C_{pk}, etc.), their control charts, and how they are all related to reliability. (Understand)
>
> **Body of Knowledge II.A.7**

The central tool of *statistical process control* (SPC) is the *control chart*, whose purpose is to provide an early signal when a process changes. Historic data are used to calculate the mean *m* and standard deviation *s* of the characteristic to be watched. The control chart is a graph with an *upper control limit* drawn at $\mu + 3\sigma$ and a *lower control limit* drawn at $\mu - 3\sigma$. (The lower limit may be omitted if it is not meaningful, such as a negative value for number of defects.) During the operation of the process, values are periodically plotted on the control chart. Points that plot outside the control limits are considered signals of *process change.* Other tests are also used to detect process change, such as seven successive points below (or above) the process mean. In general, events occurring that are very unlikely are considered evidence of process change.

An alternate way of wording these ideas is to say that when the process does not exhibit any of the process change signals, it is called *in control,* and otherwise the process is *out of control.* Some authors state that an in-control process is experiencing *common cause variation,* while an out-of-control process exhibits *special cause variation.* In this terminology, the purpose of a control chart is to detect the presence of special causes.

A number of events are very unlikely to occur unless the process has changed (that is, a special cause is present) and thus serve as statistical indicators of process change. The lists of rules that reflect these statistical indicators vary somewhat from textbook to textbook, but two of the most widely used lists of rules are the eight rules used by the software package Minitab and those listed by the Automotive Industry Action Group (AIAG) in its SPC manual. The eight Minitab rules are:

1. One point more than three sigma from the centerline (either side)
2. Nine points in a row on the same side of the centerline
3. Six points in a row, all increasing or decreasing
4. Fourteen points in a row, alternating up and down
5. Two out of three points more than two sigma from the centerline (same side)
6. Four out of five points more than one sigma from the centerline (same side)
7. Fifteen points in a row within one sigma of the centerline (either side)
8. Eight points in a row more than one sigma from the centerline (either side)

The second edition of the AIAG SPC manual lists a Summary of Typical Special Cause Criteria that is identical to the Minitab list except rule 2, which says:

2. Seven points in a row on one side of the centerline

The AIAG manual emphasizes that ". . . the decision as to which criteria to use depends on the process being studied/controlled." It may be useful to generate additional tests for particular situations. If, for instance, an increase of values represents a safety hazard, it would not be necessary to wait for the specified number of successively increasing points to take action. The $\pm 3s$ location for the control limits is somewhat arbitrary and could conceivably be adjusted based on the economic trade-off between the costs of not taking action when a special cause has occurred and taking action when a false special cause signal has occurred.

Various process characteristics may be plotted, including physical measurements such as dimensions, weights, hardness, and so on. It is often useful to plot measurements of inputs such as pressure or voltage, or measurements of raw materials. The two main categories of control charts are *attribute* and *variables.*

Attribute Charts

Attribute charts are used for count data. In attribute control charts, if every item is in one of two categories, such as good or bad, "nonconforming items" are counted. The p- and np-charts are used for plotting nonconforming items. For example, if a leak test is performed, and items that leak are rejected, the p- or np-chart would be appropriate. If the samples to be tested are of the same size, the *np*-chart may be used. If the sample size varies, the *p*-chart should be used.

If each item may have several flaws, "nonconformities" are counted. The *c*- and *u*-charts are used for plotting nonconformities. For example, nonconformities on a pane of glass include bubbles, scratches, chips, inclusions, waves, and dips. Assume that none of these requires the pane to be rejected, although there may be specification limits on some of them such that a scratch of a certain depth may result in rejection of the pane. If the samples to be inspected are of the same size, the *c*-chart may be used. If the sample size varies, the *u*-chart should be used. These control charts are illustrated in the following examples.

A test for the presence of the "Rh–" factor in 13 samples of donated blood has the following results. These data are plotted on a *p*-chart in Figure 4.5.

	Test number												
	1	2	3	4	5	6	7	8	9	10	11	12	13
No. of units of blood	125	111	133	120	118	137	108	110	124	128	144	138	132
No. of Rh– units	14	18	13	17	15	15	16	11	14	13	14	17	16

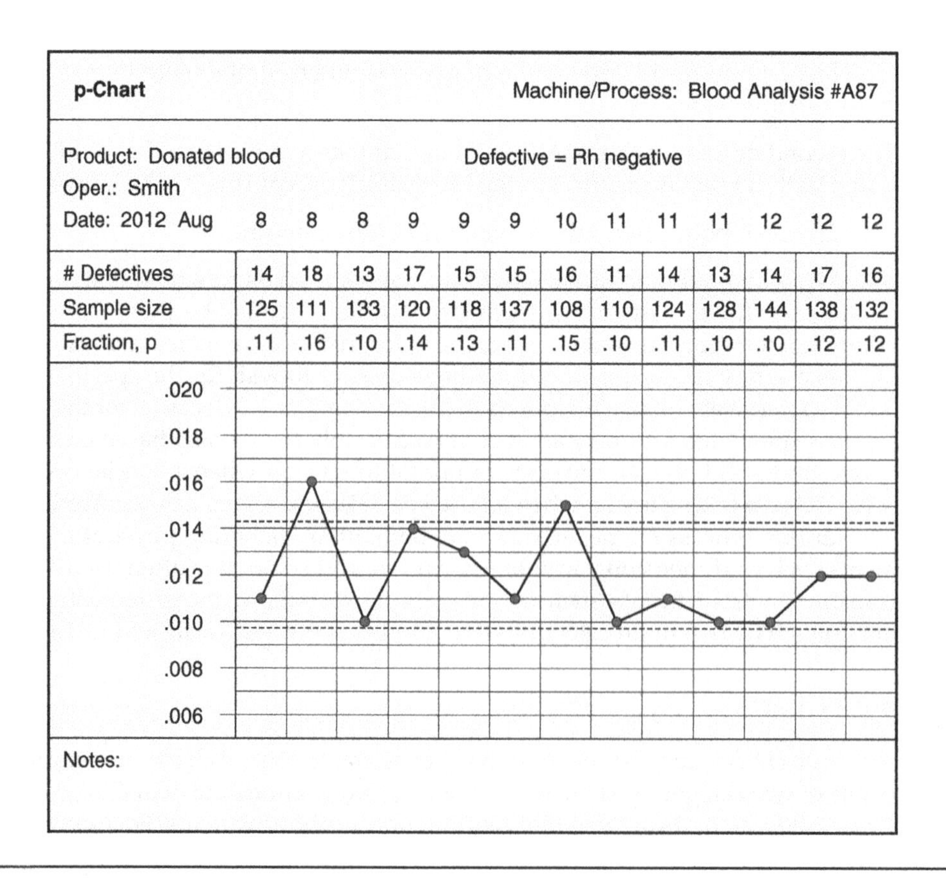

Figure 4.5 *p*-chart example.

Note that the p-chart in Figure 4.5 has two points that are outside the control limits. These points indicate that the process was "out of statistical control," which is sometimes referred to as "out of control." It means that there is a very low probability that these points came from the same distribution as the one used to calculate the control limits. It is therefore very probable that the distribution has changed. These "out of control" points are a statistical signal that the process needs attention of some type. People familiar with the process need to decide how to react to various points that are outside the control limits. In the situation in this example an unusually high number of units of blood test Rh–. This could indicate a different population of donors or possibly a malfunction of the testing equipment or procedure. Control limit formulas for the p-chart are listed in Appendix C and are repeated here for convenience:

$$\bar{p} \pm \sqrt{\frac{\bar{p}(1-\bar{p})}{\bar{n}}}$$

where

$\bar{p}$ = Average value of p

$\bar{n}$ = Average sample size

If defectives are being counted and the sample size remains constant, the np-chart can be used.

Example: Packages containing 1000 light bulbs are randomly selected, and all 1000 bulbs are light-tested. The np-chart is shown in Figure 4.6. Note that on March 25 the point is outside the control limits. This means there is a high probability that the process was different on that day than on the days that were used to construct the control limits. In this case the process was different in a good way. It would be advisable to pay attention to the process to see what went right and to see if the conditions could be incorporated into the standard way of running the process. Notice the operator note at the bottom of the chart.

The control limits for the np-chart are given by the formulas:

$$n\bar{p} \pm 3\sqrt{n\bar{p}(1-\bar{p})}$$

where

n = Sample size

$\bar{p}$ = Average number of defectives per sample

The u- and c-charts are used when defects rather than defectives are being counted. If the sample size varies, the u-chart is used. If the sample size is constant, the c-chart may be used. An example of a u-chart is shown in Figure 4.7. A c-chart would look much like the np-chart illustrated in Figure 4.6 and is not shown here.

To decide which attribute chart to use:

- For defectives, use p or np:

  Use p for varying sample size.

  Use np for constant sample size.

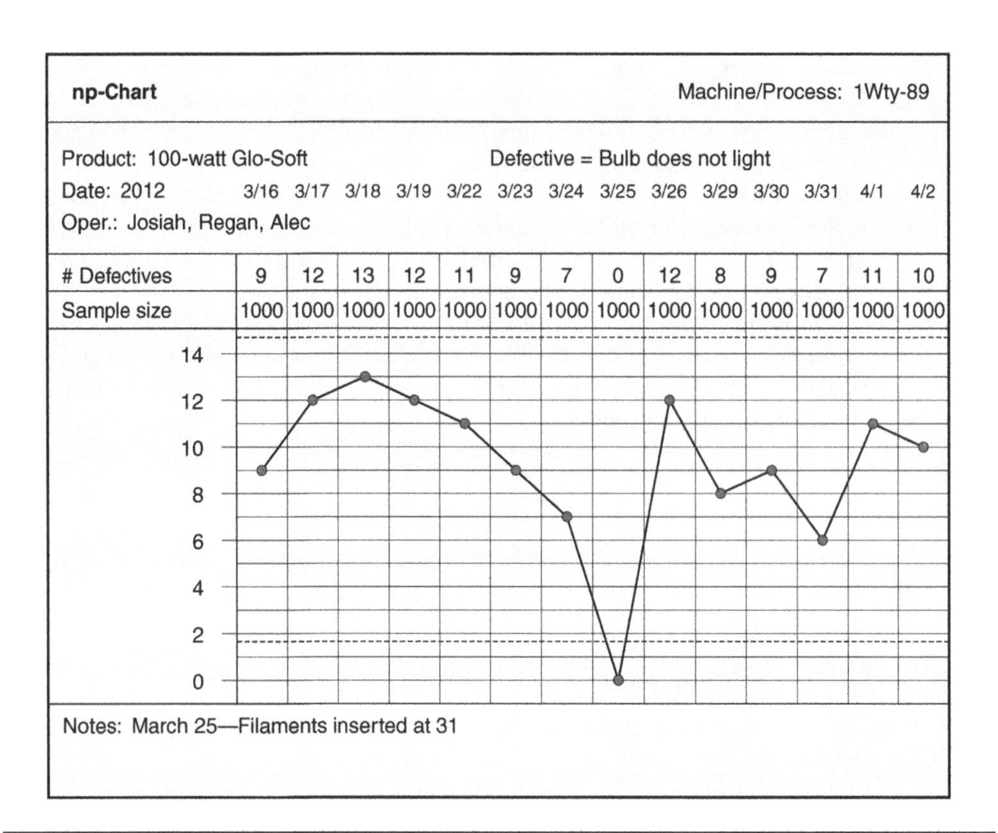

Figure 4.6 Example of an *np* control chart.

- For defects, use u or c:

  Use u for varying sample size.

  Use c for constant sample size.

$$\text{The control limit formula for the } u\text{-chart is } \bar{u} \pm 3\sqrt{\frac{\bar{u}}{\bar{n}}}$$

where

$$\bar{u} = \text{Average defect fraction} = \frac{\sum \text{number of defects}}{\sum \text{sample sizes}}$$

$\bar{n}$ = Average sample size

Using the data in Figure 4.7, $\bar{u} \approx .518$ and $\bar{n} \approx 9.786$, the control limits would be:

$$\text{UCL} = .518 + 3\sqrt{.053} \approx 1.21$$

$$\text{LCL} = .518 - 3\sqrt{.053}$$

Since the latter value is negative there is no LCL.

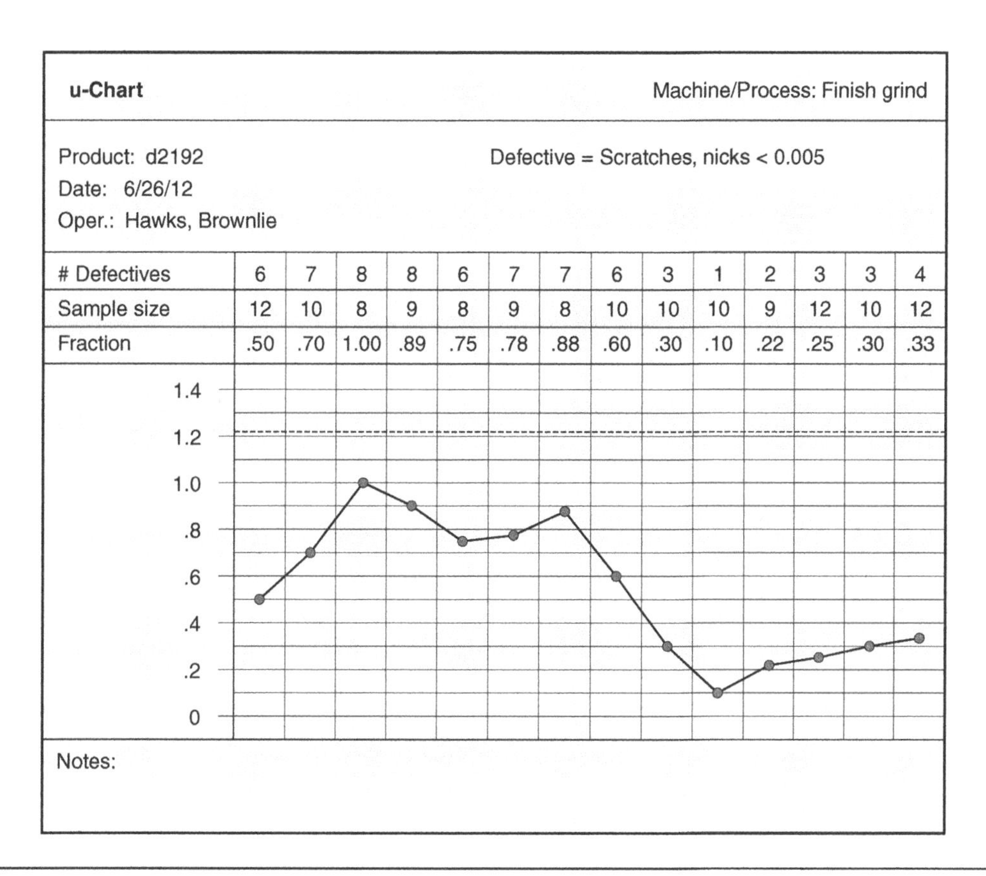

# Defectives	6	7	8	8	6	7	7	6	3	1	2	3	3	4
Sample size	12	10	8	9	8	9	8	10	10	10	9	12	10	12
Fraction	.50	.70	1.00	.89	.75	.78	.88	.60	.30	.10	.22	.25	.30	.33

Figure 4.7 Example of a u control chart.

Variables Charts

Variables charts are used when measurements on some continuous scale are to be plotted. A continuous scale has an infinite number of possible values between each pair of values. Examples include length, weight, light intensity, pH, and percentage carbon. For instance, in measuring length there are an infinite number of values between 1.250 and 1.251, values such as 1.2503, 1.2508, and so on. Common variables control charts are the $\overline{X}$ and R chart, ImR (individuals and moving range) chart, and the median chart.

An example of an $\overline{X}$ and R control chart is shown in Figure 4.8. Values are calculated by averaging the sample of readings taken at a particular time, and the range values are found by subtracting the low value from the high value for each sample.

Control limits are typically set at three standard deviations above and below the average. The standard deviation is messy to calculate, so the standard way to locate control limits is to use formulas using control limit constants. These formulas are summarized in Appendix C, and the constants are listed in Appendix D.

For the $\overline{X}$ and R chart, Appendix C shows the following formulas:

Averages chart: $\overline{\overline{x}} \pm A_2\overline{R}$

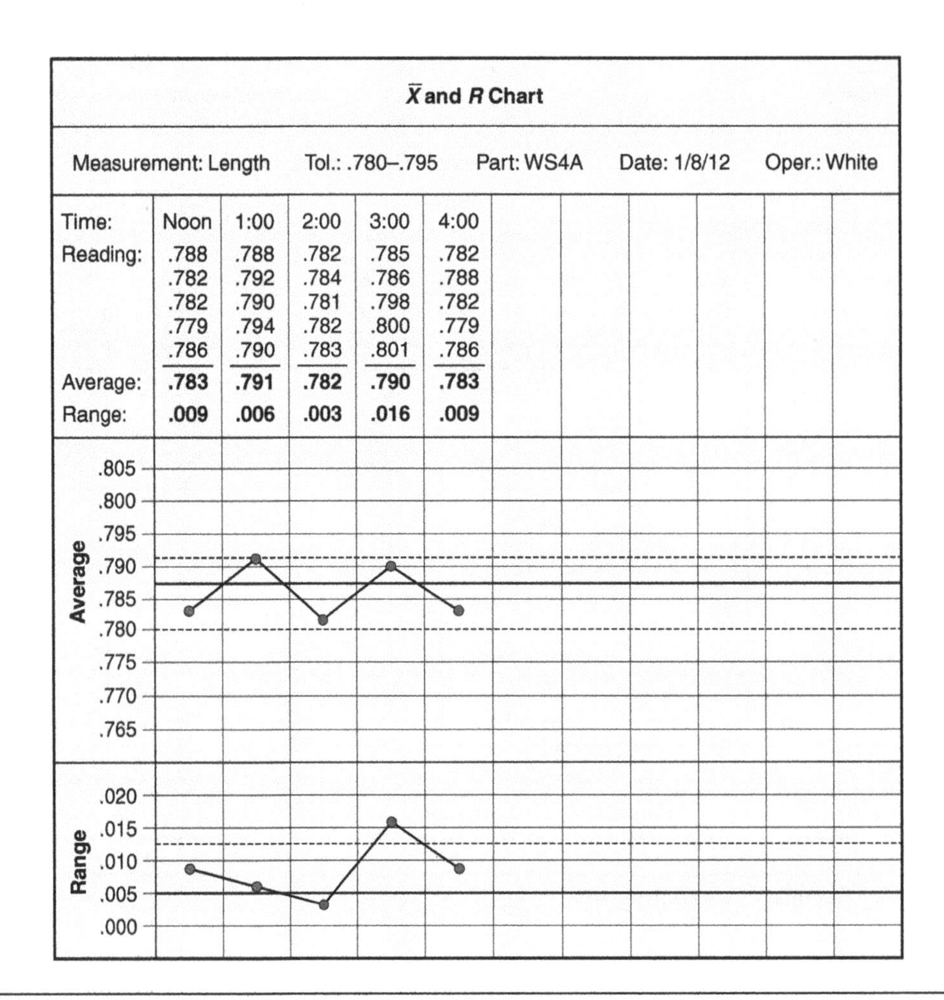

$\bar{X}$ and R Chart

Measurement: Length | Tol.: .780–.795 | Part: WS4A | Date: 1/8/12 | Oper.: White

Time:	Noon	1:00	2:00	3:00	4:00
Reading:	.788	.788	.782	.785	.782
	.782	.792	.784	.786	.788
	.782	.790	.781	.798	.782
	.779	.794	.782	.800	.779
	.786	.790	.783	.801	.786
Average:	**.783**	**.791**	**.782**	**.790**	**.783**
Range:	**.009**	**.006**	**.003**	**.016**	**.009**

Figure 4.8 Example of an $\bar{X}$ and R control chart.

Range chart: LCL = $D_3\bar{R}$ UCL = $D_4\bar{R}$

For these formulas,

$\bar{\bar{x}}$ = The average of all the average values (that is, the process mean)

$\bar{R}$ = The average of all the range values

A_2, D_3, and D_4 are constants from Appendix E and are dependent on sample size.

Example: Suppose that data have been collected from the process depicted in Figure 4.8 using samples of size five. The averages and ranges of these samples are calculated with the following results:

From Appendix D, using sample size n = 5: A_2 = .577, D_3 is undefined, and D_4 = 2.114.

Using the formulas shown above, the control limits would be

Average chart: $.786 \pm .577 \times .010 \approx .786 \pm .006 = .792$ and .780

Range chart: LCL isn't defined, UCL = $2.114 \times .010 = .021$

These control limits have been drawn on the chart shown in Figure 4.8.

An important issue here is that the control limits are not arbitrary but rather are calculated using data from the process. How much data is needed? The more, the better. Some textbooks say that at least 25 samples should be used. The sample size used when collecting the data dictates the sample size used in the control chart.

The $\bar{X}$ and s chart is another variables control chart. It works a lot like the $\bar{X}$ and R chart, but instead of calculating and plotting the range of each sample, the standard deviation of each sample is calculated and plotted. Not surprisingly, different control limit formulas and constants are used:

Averages chart: $\bar{\bar{x}} \pm A_3\bar{s}$

Standard deviation chart: $LCL = B_3\bar{R}$ $UCL = B_4\bar{R}$

For these formulas,

$\bar{\bar{x}}$ = The average of all the average values (that is, the process mean)

$\bar{s}$ = The average of all the sample standard deviation values

A_3, B_3, and B_4 are constants from Appendix E and are dependent on sample size.

The charts are constructed and interpreted in the same manner as the $\bar{X}$ and R charts.

The *median chart* is another variables control chart. An example of the median chart is shown in Figure 4.9. In this chart, all the readings in the sample are plotted, and the medians of the samples are connected with a broken line.

One advantage of this chart is that no calculation by the operator is required. The control limit formula is: $\bar{\tilde{x}} \pm A_2\bar{R}$

where

$\bar{\tilde{x}}$ = The average of the medians

A_2 is a special constant for median charts found in Appendix D.

One disadvantage of this chart is that it does not capture the signal when a range is too high. Some authors suggest constructing a paper or plastic mask with width equal to the UCL for the range. If this mask can't cover all the plotted points for a particular sample, the range is above the UCL for range. An example of this sort of mask is shown as a shaded rectangle on Figure 4.9. The formula for UCL of the range is the same as that for the $\bar{X}$ and R chart.

The *individuals and moving range* (ImR) chart is another variables control chart. It is used when the sample size is 1. An example of this chart is shown in Figure 4.10. Note that the moving range is the absolute value of the difference between the current reading and the previous reading. This means that the first reading will have no moving range.

The control limit formulas for the ImR (also called the XmR) chart are:

Individuals: $\bar{x} \pm E_2\bar{R}$

Moving range: $UCL = D_4\bar{R}$ $LCL = D_3\bar{R}$

The sample size is the width of the moving window. In the example, the moving window is two readings wide. *Note*: The moving range points are what statisticians

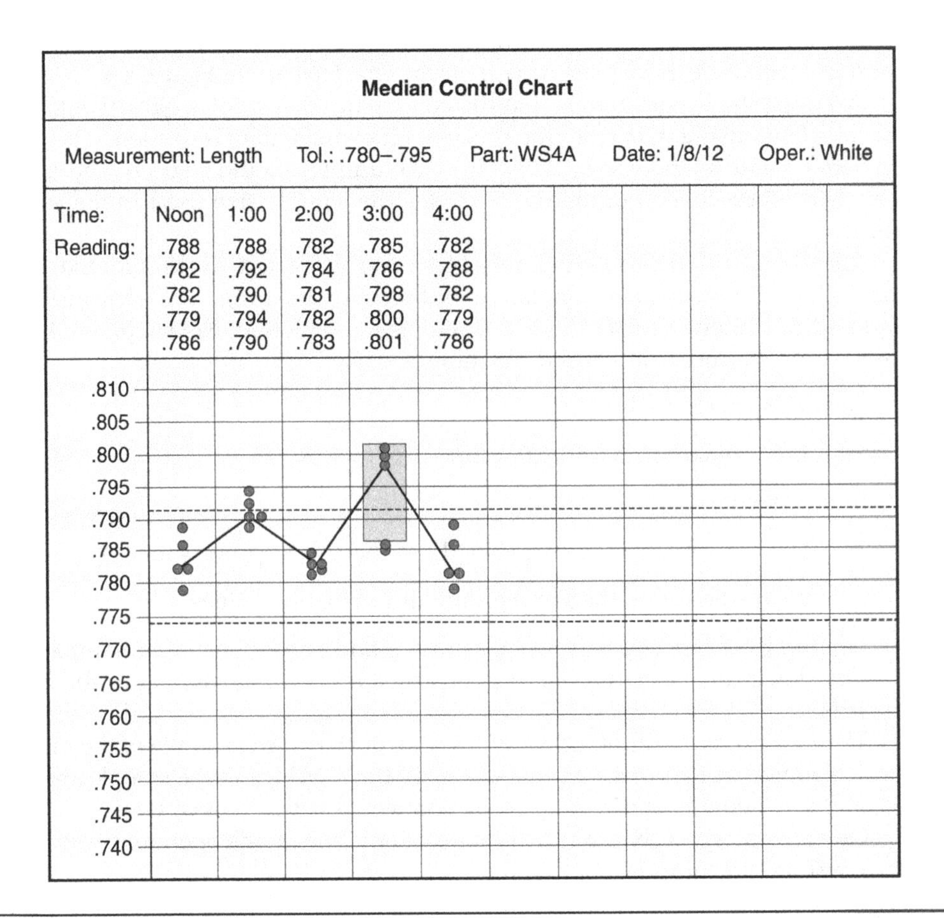

Figure 4.9 Example of median control chart.

call *correlated.* This means that the only points on the MR chart that signal a process change are those outside the control limits.

Reliability engineers are interested in process control because the reliability of a product is dependent on the process that produced it. It may be useful to specify the use of SPC for monitoring those characteristics that are known to impact the life cycle characteristics of the item.

Capability Analysis

The use of control charts is referred to as the *online* or *real-time* application of SPC. The data from control charts and other sources can also be analyzed later, or *off-line,* for further process insights. Long-term trends may be discovered, for instance. These data can be used to calculate the extent to which the process meets specifications. This use of data is known as *capability analysis.* The two principal *capability indices* are denoted C_p and C_{pk} and are defined as follows:

$$C_p = \frac{USL - LSL}{6\sigma}$$

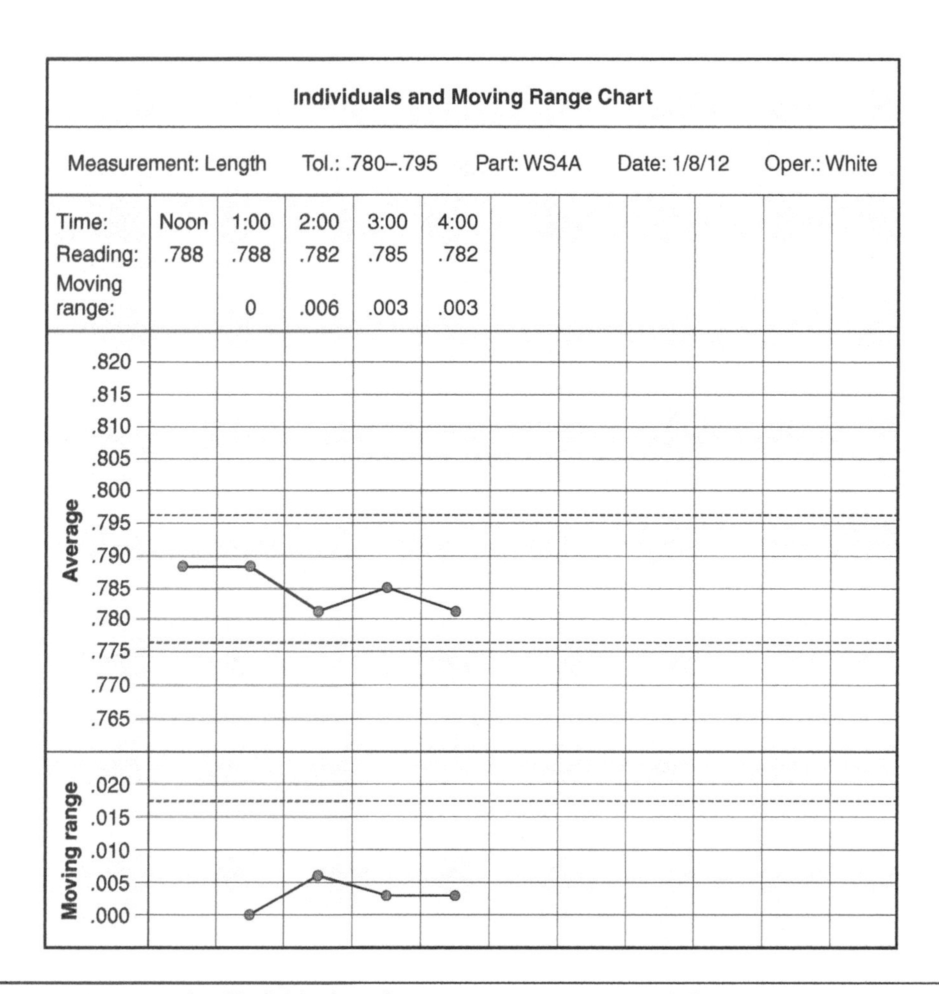

Figure 4.10 Example of individuals and moving range (ImR) control chart.

where

USL and LSL are upper and lower specification limits, respectively

σ = Process standard deviation

$$C_{pk} = \frac{\min[Z_U, Z_L]}{3}$$

where

$$Z_U = \frac{USL - \bar{\bar{x}}}{\sigma} \text{ and } Z_L = \frac{\bar{\bar{x}} - LSL}{\sigma}$$

where $\bar{\bar{x}}$ is the process mean.

For example, if a product has specification limits of 1.000 to 1.010, and the process has a process mean of $\bar{\bar{x}} = 1.003$ and standard deviation $\sigma = 0.002$, then

$$C_p = \frac{0.010}{0.012} \approx 0.83 \text{ and } Z_U = 3.5, Z_L = 1.5, \text{so } C_{pk} = \frac{1.5}{3} = 0.5$$

In general, C_p shows how good C_{pk} could be if the process were centered.

Historically, a process was considered capable if both C_p and C_{pk} were greater than 1. More recently, stricter requirements have been used in some cases. Note that since the formula for C_p doesn't use the process mean, it doesn't detect a non-centered process.

Chapter 5
B. Statistical Inference

1. POINT ESTIMATES OF PARAMETERS

> Obtain point estimates of model parameters using probability plots, maximum likelihood methods, etc. Analyze the efficiency and bias of the estimators. (Evaluate)
>
> **Body of Knowledge II.B.1**

Suppose an estimate is needed for the mean coating thickness of a population of 1000 circuit boards received from a supplier. Rather than measure all 1000 boards, a random sample of 40 is selected for measurement. The mean of the 40 values is 0.003 inches. Based on this sample, the estimate for the mean coating thickness for the entire lot of 1000 boards is 0.003 inches. This value is called the *point estimate*. In this case the sample mean is an estimator of the population mean. Stated in other words, a statistic, in this case the sample mean, is used to estimate the parameter, in this case the population mean (keep in mind that *statistic* refers to a value obtained from a sample and *parameter* is a value from the population). An estimator is called *unbiased* if the mean of all possible values is equal to the parameter being estimated. The sample mean is an unbiased estimator for the population mean as a result of the central limit theorem. An example of a *biased* estimator is the sample standard deviation s. That is, $\mu_s \neq \sigma$.

One estimator for a parameter is called more *efficient* than another if it requires fewer samples to obtain an equally good approximation. If two estimators A and B are unbiased, A is defined to be more efficient than B if it has a smaller variance. The efficiency E of A relative to B is defined as

$$E = \frac{\sigma_A^2}{\sigma_B^2}.$$

In control chart calculations the population standard deviation is often calculated from the range of the sample. This method is unbiased. However, as sample size gets larger it is better to use the sample standard deviation because the relative efficiency of the range method decreases. That is, it takes more samples to obtain an equally good estimate for *s*. For sample sizes smaller than $n = 6$, however, the relative efficiency is greater than 0.95.

Maximum Likelihood Estimates

In some cases a distribution has a known probability density function (PDF) type with an unknown parameter *Y* for which an estimator is needed. The values of the elements of the distribution depend on the PDF type and the value of the parameter *Y*. That is, the PDF can be written $f(x,Y)$.

Let $x_1, x_2, \ldots, x_n$ be a random sample from the distribution. The likelihood function is defined as the probability that these *n* values will be the ones selected. The probability that the first value is selected is $f(x_1,Y)$, the probability that the second is selected is $f(x_2,Y)$, and so on. So the probability that this sample will be drawn is the product of these values:

$$L(Y) = f(x_1,Y)\,f(x_2,Y)\ldots.f(x_n,Y)$$

This product is known as the likelihood function. In order to find the value of *Y* that will maximize L(*Y*) we will set its derivative equal to zero and solve the resulting equation for *Y*, that is, solve $L'(Y) = 0$. This will produce the value of *Y* that maximizes the probability that the randomly selected numbers will have values $x_1, x_2, \ldots, x_n$.

EXAMPLE 5.1

Suppose the PDF is the Bernoulli distribution:

$$f(x) = p^x(1-p)^{1-x} \quad x = 0 \text{ or } 1 \quad 0 \le p \le 1$$

Given a random sample $x_1, x_2, \ldots, x_n$ we want a value of *Y* that is a good estimate of *p*. In other words, the question is "What value of *p* would have the highest likelihood of producing this set of random values?" The answer is found by differentiating the likelihood function:

$$\begin{aligned} L(p) &= f(x_1,p)f(x_2,p)\ldots f(x_n,p) \\ &= p^{x_1}(1-p)^{1-x_1}p^{x_2}(1-p)^{1-x_2}\ldots p^{x_n}(1-p)^{1-x_n} \\ &= p^{x_1+x_2+\ldots+x_n}(1-p)^{1-x_1+1-x_2+\ldots+1-x_n} \\ &= p^{\sum x_i}(1-p)^{n-\sum x_i} \end{aligned}$$

Setting the derivative of this expression equal to zero and solving leads (eventually) to the conclusion that

$$p = \frac{\sum x_i}{n} = \bar{x}.$$

So the value of *p* that maximizes the likelihood that this sample will be drawn is the mean of the sample values.

Point Estimates

A point estimate of a population parameter can be made using the values of a sample from that population. One population parameter often estimated is the mean. It is also sometimes desirable to estimate the population standard deviation.

Consistency

It is important for estimators to be *consistent*. Consider a series of estimators for some parameter E and denote the elements of the series $E_1, E_2, \ldots, E_n$. The estimator is called consistent if E_n approaches E as n gets larger.

The Estimate of the Mean

The population mean μ can be estimated using the sample mean $\bar{x}$. If the sample consists of n values, $x_1, x_2, \ldots, x_n$, the estimate of the population mean can be calculated as

$$\hat{\mu} = \bar{x} = \frac{\sum x(i)}{n}.$$

The Estimate of the Standard Deviation

The population standard deviation (σ) can be estimated using the sample standard deviation (s). If the sample contains n values, $x_1, x_2, \ldots, x_n$, the estimate of the population standard deviation can be calculated as

$$\hat{\sigma} = s = \sqrt{\frac{\sum (x_i - \bar{x})^2}{n-1}}.$$

EXAMPLE 5.2

A sample of five units from a population is tested to failure.

Unit number 1 failed at 162 hours.
Unit number 2 failed at 157 hours.
Unit number 3 failed at 146 hours.
Unit number 4 failed at 173 hours.
Unit number 5 failed at 155 hours.
Total = 793 hours.

The point estimate of the mean of the failure times of the population can be calculated as

$$\hat{\mu} = \bar{x} = \frac{793}{5} = 158.6 \text{ hours}$$

EXAMPLE 5.3 (INTERVAL CENSORED DATA)

Assume the units in Example 5.2 were checked every five hours for failure.
The exact time of failure is not known.

Unit 1 failed between 160 and 165 hours.
Unit 2 failed between 155 and 160 hours.
Unit 3 failed between 145 and 150 hours.
Unit 4 failed between 170 and 175 hours.
Unit 5 failed between 150 and 155 hours.

Totals = 780 and 805 hours.

An interval estimate of the mean can be calculated. The interval end points can be obtained using the two totals:

$$\text{Upper estimate: } \hat{\mu} = \frac{805}{5} = 161 \text{ hours}$$

$$\text{Lower estimate: } \hat{\mu} = \frac{780}{5} = 156 \text{ hours.}$$

The interval estimate for the mean of the population is between 156 and 161 hours

EXAMPLE 5.4

The point estimate of the population standard deviation using the data in Example 5.2 can be calculated as

$$\hat{\sigma} = \sqrt{\frac{(162-158.6)^2 + (157-158.6)^2 + \ldots + (155-158.6)^2}{4}}$$

$$\hat{\sigma} = 9.915$$

Any scientific calculator will perform the calculations for estimating the mean and standard deviation using sample data. Be sure to use the function key with $n - 1$ in the denominator when calculating the estimate for standard deviation.

Note that the denominator in the estimate of the population standard deviation is one less than the sample size.

The point estimates of the population mean and the population standard deviation are distribution free. This means they are valid regardless of the population distribution. What meaning these values have, however, is very much dependent on the distribution of the population.

2. STATISTICAL INTERVAL ESTIMATES

> Compute confidence intervals, tolerance intervals, etc., and draw conclusions from the results. (Evaluate)
>
> **Body of Knowledge II.B**

Confidence Intervals

A statistical confidence interval incorporating a given risk (α) can be placed about the point estimate of any population parameter made from a sample taken from the population. The statement can then be made that the true parameter of the population is within that interval with a confidence of $1 - \alpha$. The risk denoted as α is known as the *significance*.

Usually, the mean is the parameter of interest. However, confidence intervals can be set about other estimates, such as the standard deviation. The central limit theorem can be used to set a confidence interval about the estimate of the mean regardless of the population distribution if the sample size is large. It can also be used with small samples if the population distribution does not differ greatly from the normal distribution. In general, the criterion is that the population is unimodal (one peak) and is somewhat symmetrical. This is not always the case for reliability data. If the test is to determine failure times, and if the units are designed not to fail, there is usually not enough time or enough units available to get a significant number of failures. Also, some of the distributions used to model reliability data, especially the exponential, are significantly different from a normal distribution. If the population distribution is known (or assumed), exact confidence limit values can be calculated.

The Normal Distribution

Confidence interval for the mean: standard deviation known.

If the population distribution is normal and the standard deviation (σ) is known, the confidence interval for the mean is:

$$\bar{x} \pm \frac{Z_{\alpha/2}}{\sqrt{n}}\sigma.$$

α = Probability that the population mean is not in the interval (called α risk)

$1 - \alpha$ = The confidence level

$Z_{\alpha/2}$ = The value from the Z-table (area under the normal curve) with an area of $\alpha/2$ in the tail of the distribution

The upper confidence limit on the mean is $\bar{x}+\frac{Z_{\alpha/2}}{\sqrt{n}}\sigma$.

The lower confidence limit on the mean is $\bar{x}-\frac{Z_{\alpha/2}}{\sqrt{n}}\sigma$.

Minimal error will be incurred using the above formulas, even if the standard deviation must be estimated, if the sample size is at least 30 ($n \geq 30$).

When a confidence interval for the mean of a population is calculated, the margin of error is given by the formula

$$E=\frac{Z_{\alpha/2}}{\sqrt{n}}\sigma$$

To determine the sample size necessary to obtain a given margin of error, solve for n rounded to a whole number:

$$n=\left(\frac{Z_{\alpha/2}\sigma}{E}\right)^2$$

EXAMPLE 5.5

Calculate the 0.90 confidence limits for the mean of the coating thickness of the circuit boards described at the beginning of this chapter.

The coating thickness of 40 circuit boards is measured in order to make an estimate of the coating thickness of the entire population of 1000 circuit boards. The coating process is known to have a standard deviation of 0.0005 inches.

$\bar{X} = 0.003$ inches

$\sigma = 0.0005$

$\alpha = 0.10$ (the risk)

$1 - \alpha = 0.90$ (the confidence)

$Z_{(\alpha/2)} = Z_{(0.05)} = 1.645$ (from the normal tables in Appendix E)

Since the standard deviation is known, the Z-distribution can be used to calculate the confidence limits.

$$\text{Confidence interval} = \bar{x}\pm\frac{Z_{\alpha/2}}{\sqrt{n}}\sigma$$

$$\text{Confidence interval} = 0.003\pm\frac{1.645}{\sqrt{40}}(0.0005)=0.003\pm0.00013$$

Upper confidence limit = 0.003 + 0.00013 = 0.00313 inches

Lower confidence limit = 0.003 – 0.00013 = 0.00287 inches

With a confidence of 0.90, the true mean of the population is between 0.00287 and 0.00313 inches.

EXAMPLE 5.6

For Example 5.5, calculate the required sample size to obtain a 0.99 percent confidence interval ($\alpha = 0.01$) with the same margin of error:

$$E = \frac{Z_{\alpha/2}}{\sqrt{n}}\sigma = \frac{(1.645)(0.0005)}{\sqrt{40}} = 0.00013$$

$$Z_{\alpha/2} = Z_{0.05} = 2.575$$

$$n = \left(\frac{Z_{\alpha/2}}{E}\sigma\right)^2 = \left[\frac{(2.575)(0.0005)}{0.00013}\right]^2 = 98 \text{ (rounded value)}$$

To reduce the margin of error or to increase the confidence level, increase the sample size.

Confidence interval for the mean: standard deviation unknown.

If the standard deviation of the population is not known and must be estimated from the sample data, the t-distribution is used to set the confidence limits. The t-distribution is similar to the Z-distribution in that it is symmetrical about the mean. The t value is larger than the corresponding Z value because of the uncertainty involved in the estimate of σ. The smaller the sample size, the more the uncertainty. The value of t depends on the sample size. Statisticians define a value called the degrees of freedom (ν). For the t-distribution the degrees of freedom are $n - 1$. A table of t values is in Appendix J. It should be stressed that the use of the t-table is valid only under the assumption that the population distribution is normal. However, the formulas work fairly well for moderate sample sizes and nearly normal populations. Statisticians say the procedure is robust to the normality assumption:

$$\text{Confidence interval for the mean} = \bar{x} \pm \frac{t_{\alpha/2,\nu}}{\sqrt{n}} s.$$

s is the estimate of the population standard deviation.

$t_{(\alpha/2,\upsilon)}$ is the value from the t-table for a risk of α and $\nu = n - 1$ degrees of freedom.

$$\text{Upper confidence limit} = \bar{x} + \frac{t_{\alpha/2,\nu}}{\sqrt{n}} s$$

$$\text{Lower confidence limit} = \bar{x} - \frac{t_{\alpha/2,\nu}}{\sqrt{n}} s$$

Note: The difference in the use of the Z-distribution and the t-distribution to calculate confidence limits on the mean is of significance to the statistician. The difference may be of little practical importance to the practicing reliability engineer. Questions on the CRE exam will be stated to test for the understanding of this

EXAMPLE 5.7

Assume that the sample in Example 5.2 is from a population that can be approximated by a normal distribution. Find the 0.90 confidence interval for the mean:

$n = 5$

$\bar{X} = 158.6$ hours

$s = 9.915$ hours

The standard deviation is estimated from the sample; therefore, the *t*-distribution is used to calculate the confidence limits:

$$t_{(\alpha/2=0.05,\nu=n-1=4)} = 2.132$$

$$\text{Upper confidence limit} = 158.6 + \frac{(9.915)(2.132)}{\sqrt{5}} = 168.05$$

$$\text{Lower confidence limit} = 158.6 - \frac{(9.915)(2.132)}{\sqrt{5}} = 149.15$$

The true mean of the population is between these limits with a confidence of 0.90.

$149.15 < \mu < 168.05$ with a confidence of 0.90.

difference. The student is encouraged to pay close attention to the difference in the use of the two distributions and to the way the questions are stated.

Whenever sampling is used to estimate the value of population parameters, it is possible to calculate an interval of risk. When journalists report the results of sampling polls, they add a margin of error, which is another format for a confidence interval, usually with the default confidence level of 95 percent. For instance, suppose the results state that 43 percent of the respondents answered "A" and 46 percent responded "B" with a ± 3.5 percent margin of error. This is equivalent to saying that there is 95 percent confidence that the percent of the population that would have answered "A" is between 39.5 percent and 46.5 percent, while the percent that would have answered "B" is between 42.5 percent and 49.5 percent. Since these intervals overlap, one can't be 95 percent confident which answer is most popular, a so-called statistical dead heat.

Some statistical software packages will produce curves with associated confidence intervals. Figure 5.1 displays a plot from Minitab showing 95 percent confidence interval curves for the survival plot of a variable named "Start."

Confidence Interval on the Standard Deviation

The *chi square* (χ^2) *distribution* is used to calculate confidence limits on the standard deviation of the normal distribution. The chi square distribution is not symmetrical; therefore, both the upper and lower values must be found from the tables. The chi square values are also defined using degrees of freedom. The chi square value is dependent on the sample size. Values of the chi square distribution are given in Appendix I.

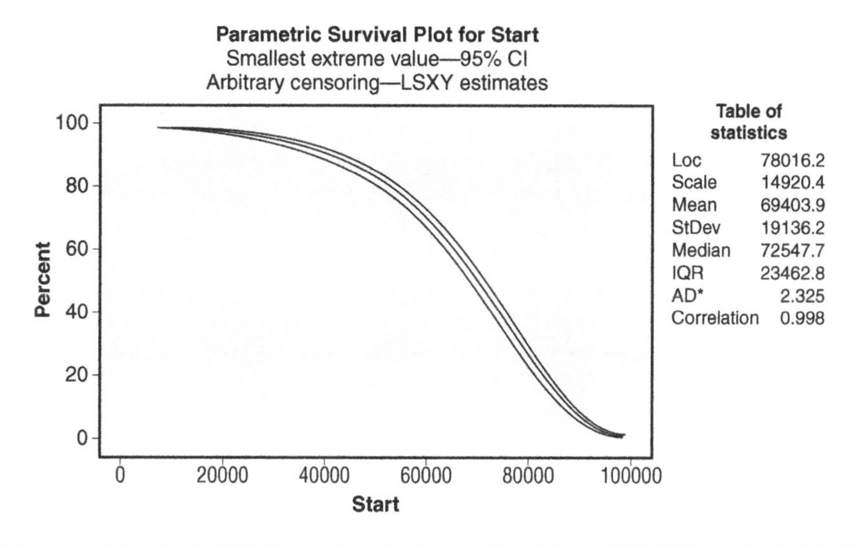

Figure 5.1 Plot of a curve with confidence interval.

The confidence interval for the population standard deviation is given by

$$\sqrt{\frac{n-1}{\chi^2_{(\alpha/2,\nu=n-1)}}}s < \sigma < \sqrt{\frac{n-1}{\chi^2_{(1-\alpha/2,\nu=n-1)}}}s$$

χ^2 is the value from the chi square table using $n - 1$ degrees of freedom.

EXAMPLE 5.8

Calculate the 0.90 confidence interval for the population standard deviation using the data from Example 5.2.

$n = 5$

$\hat{\sigma} = s = 9.915$ hours

From the chi square tables:

$\chi^2_{(0.05,4)} = 9.488$

$\chi^2_{(0.95,4)} = 0.711$

Lower confidence value = $\sqrt{\frac{4}{9.488}}(9.915) = 6.44$

Upper confidence value = $\sqrt{\frac{4}{0.711}}(9.915) = 23.5$

$6.44 < \sigma < 23.5$ with a confidence of 0.90

Statistical Tolerance Interval

Statistical tolerance intervals, sometimes called *statistical tolerance limits,* are used to determine confidence values on the area under the normal curve when using a mean and a standard deviation estimated from a sample. Sample data from the population are analyzed to obtain $\overline{X}$, an estimate of the population mean, and *s*, an estimate of the population standard deviation. The two-sided tolerance interval is $\overline{X} \pm Ks$, where K is a value from the two-sided tolerance table and is dependent of the sample size, the desired confidence level, and the proportion of the area to be included within the interval. If the one-sided limit is desired, the tolerance value is $\overline{X} - Ks$ where K is a value from the one-sided tolerance table. Tables for one-sided and two-sided statistical tolerance factors are given in Appendix K.

EXAMPLE 5.9

Fifteen units from a population assumed to be normal are tested to failure. The sample mean and the sample standard deviation are calculated:

$n = 15$

$\overline{X} = 1076$ hours

$s = 45.6$ hours

a. What is the estimate of the reliability for a mission time of 1000 hours?

The reliability $[R(t_1)]$ is the area under the failure distribution to the right of t_1.

$$Z = \frac{(1000 - 1067)}{45.6} = -1.47$$

$R(t = 1000) = 0.93$ (using the normal table in Appendix E).

b. How many hours can be accumulated on a unit and maintain a reliability of at least 0.99?

It is necessary to find Z from the normal tables such that 99% of the area is to the right of that value:

$Z = -2.33$

$t = 1067 - (2.33)(45.6) = 961$ hours

c. With a confidence of 0.95, how many hours can be accumulated on a unit and maintain a reliability of at least 0.99?

From the one-sided tolerance interval table with a confidence of 0.95 and a sample size of 15:

$K = 3.520$

$t = \overline{X} - Ks = 1067 - (3.520)(45.6) = 906$ hours

d. What time interval will contain 99% of the failures?

It is necessary to find Z values from the normal tables such that 0.5% of the area is in each tail of the distribution:

Continued

Continued

$|Z| = 2.57$

$t_{(Upper)} = 1067 + (2.57)(45.6) = 1184$ hours

$t_{(Lower)} = 1067 - (2.57)(45.6) = 950$ hours.

99% of the failures are expected to occur between 950 and 1184 hours of use.

e. With a confidence of 0.95, what interval will contain 99% of the failures?

From the two-sided tolerance table with a confidence of 0.95 and a sample size of 15:

$K = 3.878$

$t_{(Upper)} = \bar{X} + Ks = 1067 + (3.878)(45.6) = 1244$ hours

$t_{(Lower)} = \bar{X} - Ks = 1067 - (3.878)(45.6) = 890$ hours

99% of the failures will occur between 890 and 1244 hours of use with a confidence of 0.95.

Exponential Distribution

The exponential distribution is expressed as

$$f(t) = \frac{1}{\theta} e^{\frac{-t}{\theta}}.$$

The mean of the exponential distribution is equal to θ, and is referred to as the mean time to failure (MTTF) if the units are not repairable, and the mean time between failures (MTBF) if the units are repairable.

The exponential distribution is unique in that its failure rate (λ) is constant. The failure rate is equal to the reciprocal of the mean:

$$\lambda = \frac{1}{\theta}.$$

Because the failure rate is constant, the probability of failure is not dependent on the age of the unit, but only on the length of the mission time. Also, because of the constant failure rate property, the mean can be estimated using censored data. It is not necessary for each unit in the sample to fail in order to make an estimate. A test can be terminated on a given failure, referred to as *failure censored*, or at a given time, referred to as *time censored*. To make an estimate of the mean, it is only necessary to know the total time (T) accumulated on the units and the number of failures (r). The total time is accumulated using the units that fail as well as the units that do not fail. The estimate of the mean is given as

$$\hat{\theta} = \frac{T}{r}.$$

EXAMPLE 5.10

Ten units from a distribution assumed to be exponential are placed on a life test. Four of the units failed at the times shown. The units were not replaced when they failed. The test ended at 5000 hours (time censored) with six units still operating. Estimate the population mean:

$t(1) = 1150$ hours

$t(2) = 2150$ hours

$t(3) = 3350$ hours

$t(4) = 4950$ hours

Total of the failure times $\Sigma t(i) = 11{,}600$ hours

Number of units on test $n = 10$

Number of failures $r = 4$

The total accumulated test time (T):

$T = \Sigma t(i) + (n - r)(5500) = 11{,}600 + (6)(5000) = 41{,}600$ hours

$$\hat{\theta} = \frac{T}{r} = \frac{41{,}600}{4} = 10{,}400 \text{ hours.}$$

This also gives an estimate of failure rate (λ).

$$\hat{\lambda} = \frac{1}{\hat{\theta}} = \frac{1}{10{,}400} = 96 \times 10^{-6} \text{ failures per hour (96 failures per million hours).}$$

Confidence Interval for the Mean of the Exponential Distribution

The chi square (χ^2) distribution is used to calculate the confidence limits on the mean of the exponential distribution.

For the two-sided $1 - \alpha$ confidence interval, the lower limit is given by

$$\theta_L = \frac{2T}{\chi^2_{(\alpha/2, \nu = 2r+2)}}$$

if the test is time censored (ended at a given time), and

$$\theta_L = \frac{2T}{\chi^2_{(\alpha/2, \nu = 2r)}}$$

if the test is failure censored (ended on a given failure).

The difference in the equations is in the number of degrees of freedom for the χ^2 term. If the test is time censored, the number of degrees of freedom is $2r + 2$, and if the test is failure censored, the number of degrees of freedom is $2r$.

The upper limit for both time-censored and failure-censored tests is

$$\theta_U = \frac{2T}{\chi^2_{(1-\alpha/2, \nu=2r)}}.$$

EXAMPLE 5.11

Calculate the 0.90 confidence interval for the data in Example 5.10:

The test is time censored

$T = 41{,}600$ hours

$r = 4$

$\alpha = 0.10$

$\hat{\theta} = 10{,}400$ hours

The lower confidence limit is $\theta_L = \dfrac{2T}{\chi^2_{(\alpha/2, \nu=2r+2)}}$

From the chi square tables (Appendix I),

$$\chi^2_{(\alpha/2, \nu=2r+2)} = \chi^2_{(0.05,10)} = 18.307$$

$$\theta_L = \frac{2(41{,}600)}{18.307} = 4545 \text{ hours.}$$

The upper confidence limit is $\theta_U = \dfrac{2T}{\chi^2_{(1-\alpha/2, \nu=2r)}}$

From the chi square tables,

$$\chi^2_{(1-\alpha/2, \nu=2r)} = \chi^2_{(0.95,8)} = 2.733$$

$$\theta_U = \frac{2(41{,}600)}{2.744} = 30{,}320 \text{ hours.}$$

$$4545 \text{ hours} < \theta < 30{,}320 \text{ hours.}$$

The true mean of the population is between 4545 and 30,320 hours with a confidence of 0.90.

To set the one-sided confidence limit with a confidence of 0.90:

$$\theta_L = \frac{2T}{\chi^2_{(\alpha, \nu=2r+2)}}$$

$$\chi^2_{(\alpha, \nu=2r=2)} = \chi^2_{(0.10,10)} = 15.987$$

$$\theta_L = \frac{2(41{,}600)}{15.987} = 5200 \text{ hours.}$$

$$\theta > 5200 \text{ hours.}$$

The true mean of the population is greater than 5200 hours with a confidence of 0.90.

For the one-sided confidence limit, the full α risk is placed in the lower tail:

$$\theta_L = \frac{2T}{\chi^2_{(\alpha, \nu=2r+2)}}$$

if the test is time censored, and

$$\theta_L = \frac{2T}{\chi^2_{(\alpha, \nu=2r)}}$$

if the test is failure censored.

Binomial Distribution

The binomial is a discrete distribution. This means the distribution is only defined for integer values. The binomial distribution gives the probability of exactly x success outcomes out of n possible outcomes if the probability (p) of success remains the same for each trial.

The PDF for the binomial is

$$P(x) = C_n^x p^x (1-p)^{n-x}$$

C_n^x is the number of combinations that can be made from n items taken x at a time.

$$C_n^x = \frac{n!}{x!(n-x)!}$$

When the binomial is used to model reliability data, x could be the number of units that fail, and is denoted as r. If x becomes the number of units that do not fail, it is denoted as $n - r$. The probability of success on each trial is p, which is the mean of the distribution.

EXAMPLE 5.12

Find the number of combinations that can be made using 10 items if you take them four at a time:

$$C_4^{10} = \frac{10!}{4!6!} = 210$$

This problem can be worked on most scientific calculators using the factorial key or using the combinations key to find the solution directly.

EXAMPLE 5.13

It is known that units have a 0.92 probably of surviving a given test. Twenty units are subjected to the test.

a. What is the probability that exactly 18 units will survive (not fail)?

$p = 0.92$ (Probability the unit does not fail)

$n = 20$

$r = 2 \quad n - r = 18$ (Number of units that do not fail)

$$P(18) = C_{18}^{20}(0.92)^{18}(1-0.92)^{2} = 0.2711$$

This problem could be worked by finding the probability that exactly two of the units fail and then subtracting from one:

$P = 0.08$ (Probability the unit fails)

$n = 20$

$r = 2$ (Number of units that fail)

$$P(18) = 1 - C_{2}^{20}(0.08)^{2}(0.92)^{18} = 0.2711$$

b. What is the probability that at least 18 units will survive the test?

The probabilities are mutually exclusive and therefore can be added:

$$P(n-r \geq 18) = P(18) + P(19) + P(20)$$
$$= C_{18}^{20}(0.92)^{18}(0.08)^{2} + C_{19}^{20}(0.92)^{19}(0.08) + C_{20}^{20}(0.92)^{20}$$
$$= 0.2711 + 0.3282 + 0.1887$$
$$= 0.788$$

Estimating the Binomial Mean from a Sample

An estimate of p can be made using data from a test. The data must be recorded as unit success or unit failure. This is known as attribute data, as the exact times of the failures are not known. If n units are tested and r units fail ($n - r$ units do not), the estimate of the probability of unit success, p, is the number of successes divided by the number tested. This probability is the reliability of the unit for a mission equivalent to the test:

$$\hat{p} = \frac{n-r}{n}.$$

EXAMPLE 5.14

Thirty sensors are tested to determine if they will detect a given force. Two of the sensors failed to detect the force. What is the estimate of the reliability of the sensor to detect the force?

$$n = 30$$

$$r = 2$$

$$\hat{p} = \hat{R} = \frac{(30-2)}{30} = 0.933$$

Confidence Interval for the Mean of the Binomial Distribution

Reliability test data are usually very limited. Sample sizes are small, and if the units have high reliability, there will be few failures. The exact relationships should be used to calculate the confidence limit for the mean. Excessive error can be introduced using the normal approximation. To calculate the confidence limit for the mean of the binomial, it is necessary to use the *F*-distribution. The *F* value is the ratio of two chi square values, and has two sets of degrees of freedom.

The lower $1 - \alpha$ confidence limit is given as

$$R_L = \frac{n-r}{(n-r)+(r+1)F_{\alpha,\nu_1,\nu_2}}.$$

F_{α,ν_1,ν_2} is a value from the *F*-distribution table with a risk of α and ν_1 degrees of freedom for the numerator and ν_2 degrees of freedom for the denominator:

$$\nu_1 = 2(r+1)$$

$$\nu_2 = 2(n-r)$$

F-tables are in Appendix F, Appendix G, and Appendix H.

The *F*-tables can be confusing to read. It is important that the student studying for and taking the CRE exam have tables they can read quickly and correctly.

Success Testing

If there are no failures during a test, a point estimate of the mean can not be made. A lower confidence value for the mean can be calculated.

The lower $1 - \alpha$ confidence value for a no-failure test is given as

$$R_L = (\alpha)^{\frac{1}{n}}.$$

EXAMPLE 5.15

Calculate the lower 0.90 confidence limit on the mean estimated in Example 5.14:

$n = 30$

$r = 2$

$n - r = 28$

$\alpha = 0.10$

$F_{\alpha,\nu_1,\nu_2} = F_{0.10,6,56} = 1.88$

When the number of degrees of freedom exceeds 30, the F-tables require interpolation.

The lower 0.90 confidence limit is

$$R_L = \frac{28}{[28+3(1.88)]} = 0.832$$

$$R > 0.832$$

The reliability is greater than 0.832 with a confidence of 0.90.

For graphs that can be used to make this calculation, see Appendix B of Ireson, Coombs, and Moss, *The Handbook of Reliability Engineering and Management.*

EXAMPLE 5.16

Calculate the lower 0.90 confidence limit for the reliability for the sensors in Example 5.14 if the 30 units were tested with no failures:

$$R_L = (0.10)^{\frac{1}{30}} = 0.926$$

$$R > 0.926$$

The reliability of the sensor is greater than 0.926 with a confidence of 0.90.

A frequently asked question is "What sample size is required to obtain a given reliability with a given confidence?" If the test can be run with no failures, the required sample size can be found from the above equation.

Solve for n:

$$n = \frac{\log(\alpha)}{\log(R_L)}$$

n is usually rounded to the next higher integer.

Part II.B.2

EXAMPLE 5.17

How many no-failure tests are required to demonstrate a reliability of 0.95 with a confidence of 0.90?

$$n = \frac{\log(0.10)}{\log(0.95)} = 45$$

Note: The logarithms used here can be either base 10 or base e. Base 10 logarithms are denoted as log. Base e logarithms, called natural logarithms, are usually denoted as ln.

3. HYPOTHESIS TESTING (PARAMETRIC AND NON-PARAMETRIC)

Apply hypothesis testing for parameters such as means, variance, proportions, and distribution parameters. Interpret significance levels and Type I and Type II errors for accepting/rejecting the null hypothesis. (Evaluate)

Body of Knowledge II.B.3

The *hypothesis test*, another tool used in inferential statistics, is closely related to confidence intervals. A few key terms used in hypothesis tests are listed below.

Terminology

Null Hypothesis, H_0. This is the hypothesis that there is no difference (null) between the population of the sample and the specified population (or between the populations associated with each sample). The null hypothesis can never be proved true, but it can be shown (with specified risks of error) to be untrue, that is, that a difference exists between the populations.

For example, given a random sample from a population, a typical null hypothesis would be that the population mean is equal to 10. This statement is denoted H_0: $\mu = 10$.

Alternative Hypothesis, H_a. This is a hypothesis that is accepted if the null hypothesis (H_0) is rejected.

Consider the null hypothesis that the statistical model for a population is a normal distribution. The alternative hypothesis to this null hypothesis is that the statistical model of the population is *not* a normal distribution.

Note 1: The alternative hypothesis is a statement that contradicts the null hypothesis. The corresponding test statistic is used to decide between the null and alternative hypotheses.

Note 2: The alternative hypothesis can also be denoted H_1, H_A, or H_a with no clear preference as long as the symbolism parallels the null hypothesis notation.

One-Tailed Test. A hypothesis test that involves only one of the tails of a distribution. For example, we wish to reject the null hypothesis H_0 only if the population mean is *larger* than 10: H_a: $\mu > 10$. This is a right-tailed test. A one-tailed test is either right-tailed or left-tailed, depending on the direction of the inequality of the alternative hypothesis.

Two-Tailed Test. A hypothesis test that involves two tails of a distribution. For example, we wish to reject the null hypothesis H_0 if the population mean is not equal to 10. H_a: $\mu \neq 10$.

Test Statistic. A statistic calculated using data from a sample. It is used to determine whether the null hypothesis will be rejected.

Rejection Region. The numerical values of the test statistic for which the null hypothesis will be rejected.

Critical Value(s). The numerical value(s) of the test statistic that determine the rejection region.

Steps in Hypothesis Testing

Textbooks tend to treat hypothesis tests as somewhat more formal procedures. Many list seven or eight steps to be followed for each type of test. Although not all books agree on the steps themselves, this list is fairly generic:

1. Determine that the conditions or assumptions required for the test are met.

2. State the null and alternative hypotheses (H_0 and H_a) and determine whether it is a one-tailed or two-tailed test.

3. Determine the α value. This is similar to the use of α in confidence intervals. In hypothesis testing jargon, the value of α is referred to as the *significance level.*

4. Determine the critical values. These are typically found in a table such as the Z, *t*, or χ^2 tables. Use these values to define the reject region.

5. Calculate the test statistic. Each hypothesis test type has a formula for the test statistic. Some of the inputs to the formulas come from the sample data.

6. Determine whether the null hypothesis should be rejected. If the value of the test statistic is in the reject region, then the null hypothesis is rejected and the alternative hypothesis is accepted. If the value of the test statistic does not fall in the reject region, the null hypothesis is not rejected.

7. State the conclusion in terms of the original problem.

Hypothesis Tests for Means

The hypothesis test usually studied first is the one-sample z-test for population mean. Its steps are:

1. Conditions:
   a. Normal population or large sample ($n \geq 30$)
   b. σ known
2. H_0: $\mu = \mu_0$ and H_a: $\mu \neq \mu_0$ or $\mu < \mu_0$ or $\mu > \mu_0$

   This is a two-tailed test when H_a has the $\neq$ sign. It is a left-tailed test when H_a has the < sign, and a right-tailed test when H_a has the > sign.
3. Determine the α value.
4. Determine the critical values:
   a. For a two-tailed test, use a z-table to find the value that has an area of $\alpha/2$ to its right. This value and its negative are the two critical values. The reject region is the area to the right of the positive value and the area to the left of the negative value.
   b. For a left-tailed test, use a z-table to find the value that has an area of α to its right. The negative of this value is the critical value. The reject region is the area to the left of the negative value.
   c. For a right-tailed test, use a z-table to find the value that has an area of α to its right. This value is the critical value. The reject region is the area to the right of the positive value.
5. Calculate the test statistic:

$$z = (\bar{x} - \mu_0)\frac{\sqrt{n}}{\sigma}$$

6. If the test statistic is in the reject region, reject H_0. Otherwise, do not reject H_0.
7. State the conclusion in terms of the problem.

In many applications, the population standard deviation is not known. As with the confidence interval, the appropriate distribution is found in the t-table. The procedure is the same as the previous seven steps, except for steps 1, 4, and 5, which will now read:

1. Condition: population is normally distributed or $n \geq 30$.
4. The critical values are obtained from the t-table using degrees of freedom of $n - 1$.
5. The formula for the test statistic is

$$t = (\bar{x} - \mu_0)\frac{\sqrt{n}}{s}$$

where s is the sample standard deviation.

This hypothesis test is referred to as the t-test for one population mean.

It is important to note here that the fact that the null hypothesis is not rejected does not mean it is true. The conclusion is that the probability that it is true is less than 90 percent.

EXAMPLE 5.18

A vendor claims that the average weight of a shipment of parts is 1.84. The customer randomly chooses 64 parts and finds that the sample has an average of 1.88. Suppose that the standard deviation of the population is known to be 0.03. Should the customer reject the lot? Assume the customer wants to be 95 percent confident that the supplier's claim is incorrect before he rejects.

1. Conditions (a) and (b) are met.
2. H_0: $\mu = 1.84$ and H_a: $\mu \neq 1.84$. This is a two-tailed test.
3. From the problem, $\alpha = 0.05$.
4. Critical values are the z-value that has 0.025 to its right and the negative of this value. These values are 1.96 and –1.96. The reject region consists of the area to the right of 1.96 and the area to the left of –1.96.
5. $z = (1.88 - 1.84)\dfrac{\sqrt{64}}{.03} \approx 10.7$
6. Since 10.7 is in the reject region, H_0 is rejected.
7. At the .05 significance level, the data suggest that the vendor's assertion that the average weight is 1.84 is false.

EXAMPLE 5.19

A vendor claims that the average weight of a shipment of parts is 1.84. The customer randomly chooses 64 parts and finds that the sample has an average of 1.88 and standard deviation of .03. Should the customer reject the lot? Assume the customer wants to be 95 percent confident that the supplier's claim is incorrect before he rejects. (This is the same as the last example, except that .03 is the sample standard deviation rather than the population standard deviation.)

1. Condition is met.
2. H_0: $\mu = 1.84$ and H_a: $\mu \neq 1.84$. This is a two-tailed test.

Continued

Continued

3. From the problem, $\alpha = 0.05$.
4. The positive critical value would be in row 63 of the .025 column of the t-table. Since the table has no row 63, it is appropriate to use the more conservative row 60. This value is 2.000. The other critical value is –2.000. The reject region consists of the area to the right of 2.000 and the area to the left of –2.000.
5. $t = (1.88 - 1.84)\frac{\sqrt{64}}{.03} \approx 10.7$
6. Since 10.7 is in the reject region, H_0 is rejected.
7. At the 0.05 significance level, the data suggest that the vendor's assertion that the average weight is 1.84 is false.

EXAMPLE 5.20

A cut-off saw has been producing parts with a mean length of 4.125. A new blade is installed, and we want to know whether the mean has decreased. We select a random sample of 20, measure the length of each part, and find that the average length is 4.123 and the sample standard deviation is 0.008. Assume that the population is normally distributed. Use a significance level of 0.10 to determine whether the mean length has decreased.

Since the population standard deviation is unknown, the t-test will be used.

1. Condition is met.
2. H_0: $\mu = 4.125$ and H_a: $\mu < 4.125$. This is a left-tailed test.
3. From the problem, $\alpha = 0.10$.
4. The positive critical value is in the 19th row of the 0.10 column of the t-table. This value is 1.328. The critical value is –1.328. The reject region consists of the area to the left of –1.328.
5. $t = (4.123 - 4.125)\frac{\sqrt{20}}{.008} \approx -1.1$
6. Since –1.1 is not in the reject region, H_0 is not rejected.
7. At the 0.10 significance level, the data do not indicate that the average length has decreased.

Hypothesis Tests for Means of Two Populations

The next two hypothesis tests are for means of two populations. The procedure for the non-pooled t-test for two population means is:

1. Conditions:

   a. Normal populations or large samples ($n \geq 30$)

   b. Independent samples (that is, each of the pairs of sets of samples is equally likely to be selected)

2. H_0: $\mu_1 = \mu_2$ and H_a: $\mu_1 \neq \mu_2$ or $\mu_1 < \mu_2$ or $\mu_1 > \mu_2$

   This is a two-tailed test when H_a has the $\neq$ sign. It is a left-tailed test when H_a has the $<$ sign and a right-tailed test when H_a has the $>$ sign.

3. Determine the α value.

4. Determine the critical values.

   a. For a two-tailed test, use a t-table to find the value that has an area of $\alpha/2$ to its right. This value and its negative are the two critical values. The reject region is the area to the right of the positive value and the area to the left of the negative value.

   b. For a left-tailed test, use a t-table to find the value that has an area of α to its right. The negative of this value is the critical value. The reject region is the area to the left of the negative value.

   c. For a right-tailed test, use a t-table to find the value that has an area of α to its right. This value is the critical value. The reject region is the area to the right of the positive value. The bad news here is that the degrees of freedom, instead of being $n - 1$, is obtained from the formulas

$$a_1 = \frac{s_1^2}{n_1} \qquad a_2 = \frac{s_2^2}{n_2}$$

   where s_1 is the standard deviation of the sample from population 1, n_1 is the number of elements in the sample from population 1, and s_2 and n_2 are those from population 2, and

$$\text{Degrees of freedom} = \frac{(a_1 + a_2)^2}{\left[\frac{a_1^2}{n_1 - 1} + \frac{a_2^2}{n_2 - 1}\right]}$$

   rounded down to the nearest whole number.

5. Calculate the test statistic:

$$t = (\bar{x}_1 - \bar{x}_2) / \sqrt{a_1 + a_2}$$

6. If the test statistic is in the reject region, reject H_0. Otherwise, do not reject H_0.

7. State the conclusion in terms of the problem.

EXAMPLE 5.21

Two vendors of a valve diaphragm present significantly different cost quotations. The wall thickness is the critical quality characteristic. Use the following data to determine whether the average thickness of the products from vendor 1 is greater than that from vendor 2. Test at the 0.10 significance level. Assume the populations are normally distributed and that the samples are independent.

Wall thickness measurements:

Vendor 1: 86 82 91 88 89 85 88 90 84 87 88 83 84 89

Vendor 2: 79 78 82 85 77 86 84 78 80 82 79 76

Solution:

Analysis of the data on a scientific calculator shows that

$$\bar{x}_1 = 86.7, \bar{x}_2 = 80.5, s_1 = 2.76, s_2 = 3.26, n_1 = 14, \text{ and } n_2 = 12.$$

1. Conditions are met.
2. H_0: $\mu_1 = \mu_2$ and H_a: $\mu_1 > \mu_2$. This is a right-tailed test.
3. $\alpha = 0.10$

$$a_1 = \frac{2.76^2}{14} = .54 \qquad a_2 = \frac{3.26^2}{12} = .89$$

$$df = \frac{(.54 + .89)^2}{\left(\frac{.54^2}{13} + \frac{.89^2}{11}\right)} = \frac{2.04}{.09} \approx 22 \text{ (when rounded down)}$$

4. The critical value is found in the 22nd row of the $t_{.10}$ column of the t-table. This value is 1.321. The reject region is the area to the right of 1.321.
5. $t = \dfrac{(86.7 - 80.5)}{\sqrt{.54 + .89}} = 5.2$
6. Reject H_0 since the value of the test statistic is in the reject region.
7. At the 0.10 significance level, the data indicate that the average wall thickness of the product produced by vendor 1 is larger than the average wall thickness of the product produced by vendor 2.

You may be wondering why step 7 is phrased ". . . the data indicate . . ." rather than something like "the data prove." The 0.10 significance level means that there is a 10 percent chance that a rejected null hypothesis is actually true. Rejecting a true null hypothesis is referred to as a *type I error,* denoted by α, and is sometimes called the *producer's risk* because in lot sampling plans it is the probability that the plan will reject a good lot. Failing to reject a false hypothesis is called a *type II error* and is sometimes referred to as the *consumer's risk.* The probability of type II error is denoted β.

Paired-Comparison Tests

The next hypothesis test is called the *paired* t-*test for two population means.* Each pair in a paired sample consists of a member of one population and that member's corresponding member in the other population. Suppose, for instance, that we want to determine if a gasoline additive increases average mileage in the population consisting of several hundred company cars of various types and vintage. One approach would be to randomly select 10 vehicles and record mileages using gasoline without the additive, and randomly select another 10 vehicles and record mileages using gasoline with the additive. These two samples could be tested using the previous hypothesis test (assuming the populations are normally distributed). If the additive causes a large increase in average mileage, this procedure would likely reject the null hypothesis that the averages are equal. If the additive causes a small increase in average mileage, the test might not detect it because of the large variation between cars. Thus, the test might fail to reject the null hypothesis even though it is false. Statisticians would say the test lacks *sensitivity.*

An alternate approach has probably already occurred to you: choose 10 cars at random, record their mileages using gas without the additive, then use gas with the additive in those same ten cars. This arrangement reduces the sampling variation encountered when two samples of 10 are used. The approach is called the *paired sample* method and provides a very powerful test when it can be used. Sometimes it is impractical to use. Suppose we need to know whether the average effect of a particular drug is different for people with type A blood than for people with type B. One possible approach would be to select 10 type A people and measure the effects of the drug on them, then drain their blood and refill them with type B and again measure the effects of the drug. One might conclude that the drug was fatal for the population having type B blood.

The procedure for the paired t-test for two population means is:

1. Conditions:

    a. Paired sample

    b. Large sample or differences are normally distributed

2. H_0: $\mu_1 = \mu_2$ and H_a: $\mu_1 \neq \mu_2$ or $\mu_1 < \mu_2$ or $\mu_1 > \mu_2$

    This is a two-tailed test when H_a has the $\neq$ sign, a left-tailed test when H_a has the $<$ sign, and a right-tailed test when H_a has the $>$ sign.

3. Determine α.

4. Find the critical value(s) from the t-table using degrees of freedom = $n - 1$.

5. Calculate the test statistic:

    a. Let d_1 be the difference within the first element of the sample.

    b. Let d_2 be the difference within the second element of the sample, and so on.

c. Find the average $\bar{d}$ and standard deviation s_d of these d values.

d. The test statistic is

$$t = \bar{d}\frac{\sqrt{n}}{s_d}.$$

6. If the test statistic is in the reject region, reject H_0. Otherwise, do not reject H_0.

7. State the conclusion in terms of the problem.

EXAMPLE 5.22

For the gasoline additive problem discussed above, suppose the data are:

Vehicle #:	1	2	3	4	5	6	7	8	9	10
mpg with additive:	21	23	20	20	27	18	22	19	36	25
mpg without additive:	20	20	21	18	24	17	22	18	37	20

Do the data indicate that the additive increases average gas mileage at the .05 significance level? Assume that the differences are normally distributed.

1. Conditions are met.
2. H_0: $\mu_1 = \mu_2$ and H_a: $\mu_1 > \mu_2$. This is a right-tailed test.
3. $\alpha = .05$.
4. Since df = 10 – 1, the critical value is in the ninth row of the $t_{0.05}$ column. This value is 1.833.
5. Use the following table to get the data for the test statistic formula:

Vehicle #:	1	2	3	4	5	6	7	8	9	10
mpg with additive:	21	23	20	20	27	18	22	19	36	25
mpg without additive:	20	20	21	18	24	17	22	18	37	20
Difference d	1	3	–1	2	3	1	0	1	–1	5

Using a scientific calculator, $\bar{d} = 1.4$ and $s_d = 1.90$.

The test statistic is

$$t = \frac{1.4\sqrt{10}}{1.90} \approx 2.33.$$

6. Since 2.33 is in the reject region, reject H_0.
7. At the .05 significance level, the data indicate that the average mpg is increased by using the additive.

Hypothesis Test for Two Population Standard Deviations

Many process improvement efforts are designed to reduce variation. This test is used to determine whether the standard deviations of two populations are different:

1. Conditions: the two populations are normally distributed and the samples are independent.
2. H_0: $\sigma_1 = \sigma_2$

   H_a: $\sigma_1 \neq \sigma_2$ (two-tailed), $\sigma_1 < \sigma_2$ (left-tailed), $\sigma_1 > \sigma_2$ (right-tailed)
3. Determine the significance level α.
4. The critical values are obtained from the F-table (Appendixes F, G, and H). They are $F_{1-\alpha/2}$ and $F_{\alpha/2}$ for the two-tailed test, $F_{1-\alpha}$ for the left-tailed test, and $F\alpha$ for the right-tailed test. Use numerator df = $n_1 - 1$ and denominator df = $n_2 - 1$ where n_1, n_2 are sample sizes.
5. The test statistic is

$$F = \frac{s_1^2}{s_2^2}$$

   where s_1 and s_2 are the sample standard deviations.
6. If the test statistic is in the reject region, reject H_0. Otherwise, do not reject H_0.
7. State the conclusion in terms of the problem.

EXAMPLE 5.23

Data from two competing machines include the following statistics:

Machine 1: $n_1 = 21$ $s_1 = 0.0032$

Machine 2: $n_2 = 25$ $s_2 = 0.0028$

Do these data suggest that the standard deviations of the machines are different at the 0.10 significance level? The populations are normal, and the samples have been drawn independently.

1. The conditions are met.
2. H_0: $\sigma_1 = \sigma_2$ H_a: $\sigma_1 \neq \sigma_2$
3. $\alpha = 0.10$

Continued

Continued

4. This is a two-tailed test. The critical values are

$$F_{\alpha/2}\begin{bmatrix}20\\24\end{bmatrix}=F_{.05}\begin{bmatrix}20\\24\end{bmatrix}=\frac{1}{F_{.95}\begin{bmatrix}24\\20\end{bmatrix}}=\frac{1}{2.08}\approx 0.48$$

$$F_{1-\alpha/2}\begin{bmatrix}20\\24\end{bmatrix}=F_{.95}\begin{bmatrix}20\\24\end{bmatrix}=2.03.$$

The reject region is the area to the left of 0.48 and the area to the right of 2.03.

5. The test statistic is

$$F=\frac{s_1^2}{s_2^2}=\frac{0.0032^2}{0.0028^2}\approx 1.31$$

6. Since the test statistic does not lie in the reject region, do not reject the null hypothesis.
7. At the 0.10 significance level, the data do not support the conclusion that the standard deviations of the two machines are different.

THE LEFT TAIL OF THE F-DISTRIBUTION

Note that the F-tables in the Appendixes are limited to $F_{.90}$, $F_{.95}$, and $F_{.99}$. This appears to restrict the user to right-tailed tests. A special property of the F-distribution is used to find the left tail: let F_a with numerator df = n and denominator df = d be denoted by

$$F_\alpha\begin{bmatrix}n\\d\end{bmatrix}.$$

Then the special property may be stated

$$F_\alpha\begin{bmatrix}n\\d\end{bmatrix}=\frac{1}{F_{1-\alpha}\begin{bmatrix}d\\n\end{bmatrix}}.$$

Example: find $F_{0.05}$ with numerator df = 10 and denominator df = 20.

$$F_{.05}\begin{bmatrix}10\\20\end{bmatrix}=\frac{1}{F_{.95}\begin{bmatrix}20\\10\end{bmatrix}}=\frac{1}{2.77}\approx .036$$

Goodness-of-Fit Tests

Chi-square and other goodness-of-fit tests help determine whether a discrete sample has been drawn from a known population. Example 5.24a will illustrate this concept.

EXAMPLE 5.24A

Suppose that all rejected products have exactly one of four types of defectives and that historically they have been distributed as follows:

Paint run	16%
Paint blister	28%
Decal crooked	42%
Door cracked	14%
Total	100%

Data on rejected parts for a randomly selected week in the current year:

Paint run	27
Paint blister	65
Decal crooked	95
Door cracked	21

The question—"Is the distribution of defective types for the selected week different from the historical distribution?" The test that answers this question is rather awkwardly called the χ^2 *goodness-of-fit test.* To get a feel for this test, construct a table that displays the number of defectives that would be expected in each category if the sample exactly followed the historical percentages:

Defective type	Probability	Observed frequency	Expected frequency
Paint run	.16	27	33.28
Paint blister	.28	65	58.24
Decal crooked	.42	95	87.36
Door cracked	.14	21	29.12
Total		208	

The expected frequency for "paint run" is found by calculating 16 percent of 208, for "paint blister" use 28 percent of 208, and so on. The question to be decided is whether the difference between the expected frequencies and observed frequencies is sufficiently large to conclude that the sample comes from a population that has a different distribution. Test this at the 0.05 significance level.

The test statistic is obtained by calculating the value of

$$\frac{(\text{Observed}-\text{Expected})^2}{\text{Expected}}$$

Continued

Part II.B.3

Continued

for each defective type:

Defective type	Probability	Observed frequency	Expected frequency	O − E	$(O - E)^2/E$
Paint run	.16	27	33.28	−6.28	1.19
Paint blister	.28	65	58.24	6.76	.78
Decal crooked	.42	95	87.36	7.64	.67
Door cracked	.14	21	29.12	−8.12	2.26
Total		208			

The null hypothesis is that the distribution hasn't changed. This hypothesis will be rejected if the total of the last column is too large.

The procedure:

1. Conditions:
   a. All expected frequencies are at least 1.
   b. At most, 20 percent of the expected frequencies are less than 5.
2. H_0: The distribution has not changed.

   H_a: The distribution has changed.
3. Determine α, the significance level.
4. Find the critical value in row $k - 1$ in the χ^2_α column of the χ^2 table, where k = number of categories in the distribution. This is always a right-tailed test, so the reject region is the area to the right of this critical value.
5. Calculate the test statistic using the formula

$$\chi^2 = \sum \frac{(O-E)^2}{E}$$

   (the sum of the last column in the table).
6. Reject H_0 if the test statistic is in the reject region. Otherwise do not reject H_0.
7. State the conclusion.

Using the data and calculations of Example 5.24a, follow the seven steps of the procedure to arrive at a reject or not reject decision as shown in Example 5.24b.

EXAMPLE 5.24B

1. The conditions are met.
2. H_0: The distribution of defective types has not changed.

   H_a: The distribution of defective types has changed.
3. $\alpha = 0.05$
4. From row 3 of the $\chi^2_{0.05}$ column, the critical value is 7.815. The reject region is the area to the right of 7.815.
5. $\chi^2 = \frac{(O-E)^2}{E} \approx 4.9$
6. Since the test statistic does not fall in the reject region, do not reject H_0.
7. At the .05 significance level, the data do not indicate that the distribution has changed.

EXAMPLE 5.25A

A vendor claims that at most two percent of a shipment of parts is defective. Receiving inspection chooses a random sample of 500 and finds 15 defectives. At the 0.05 significance level, do these data indicate that the vendor is wrong?

Hypothesis Tests for Proportions

The next hypothesis test is for *one population proportion.* To set the stage for this test, consider the problem posed in Example 5.25a. To solve this problem, the following symbolism and steps will be used:

p = Population proportion

n = Sample size

x = Number of items in the sample with the defined attribute

p' = Sample proportion = x/n

p_0 = The hypothesized proportion

The problem posed in Example 5.25a is solved in Example 5.25b.

The hypothesis test steps:

1. Conditions: $np_0 \geq 5$ and $n(1-p_0) \geq 5$.
2. H_0: $p = p_0$

   H_a: $p \neq p_0$ or $p < p_0$ or $p > p_0$ (two-tailed, left-tailed, and right-tailed, respectively)

3. Decide on α, the significance level.
4. Find the critical values in a standard normal table:

   $\pm z_{\alpha/2}$, $-z_{\alpha}$, z_{α} (two-tailed, left-tailed, and right-tailed, respectively)
5. Calculate the test statistic using the formula

$$z = \frac{p' - p_0}{\sqrt{p_0(1-p_0)/n}}.$$

6. If the test statistic is in the reject region, reject H_0. Otherwise, do not reject H_0.
7. State the conclusion in terms of the problem.

The next hypothesis test is for *two population proportions*. The symbolism to be used is

p_1 & p_2 = Proportions of populations 1 and 2 with the defined attribute

n_1 & n_2 = Sample sizes

x_1 & x_2 = Number of items in the sample with the defined attribute

p_1' & p_2' = Sample proportions = x_1/n_1 and x_2/n_2, respectively

The hypothesis test steps:

1. Conditions: Samples are independent

$$x_1 \geq 5, n_1 - x_1 \geq 5, x_2 \geq 5, n_2 - x_2 \geq 5$$

EXAMPLE 5.25B

$N = 500 \quad x = 15 \quad p' = 15 \div 500 = 0.03 \quad p_0 = 0.02$

1. $np_0 = 500 \times 0.02 = 10$ and $n(1 - p_0) = 500 \times 0.98 = 490$.

   Both values are ≥ 5, so conditions are met.
2. H_0: $p = 0.02$ H_a: $p > 0.02$ (right-tailed test).
3. $\alpha = 0.05$.
4. Critical value = 1.645 from a normal table.
5. $z = \frac{0.03 - 0.02}{\sqrt{0.02 \times 0.98 \div 500}} \approx 1.597.$
6. Do not reject H_0.
7. At the 0.05 significance level, the data do not support a conclusion that the vendor is incorrect in asserting that at most two percent of the shipment is defective.

2. H_0: $p_1 = p_2$

   H_a: $p_1 \neq p_2$ or $p_1 < p_2$ or $p_1 > p_2$ (two-tailed, left-tailed, and right-tailed, respectively)

3. Decide on α, the significance level.

4. Find the critical values in a standard normal table: $\pm z_{\alpha/2}$, $-z_{\alpha}$, z_{α} (two-tailed, left-tailed, and right-tailed respectively).

5. Calculate the test statistic using the formula

$$z = \frac{p'_1 - p'_2}{\sqrt{p'_p(1-p'_p)}\sqrt{(1/n_1)+(1/n_2)}} \quad \text{where } p'_p = \frac{x_1 + x_2}{n_1 + n_2}.$$

6. If the test statistic is in the reject region, reject H_0. Otherwise, do not reject H_0.

7. State the conclusion in terms of the problem.

Although not usually on a standard list of hypothesis tests, *Tukey's quick compact two-sample test* can be useful in comparing failure rates of two populations.

EXAMPLE 5.26

Two machines produce the same parts. A random sample of 1500 parts from machine 1 has 36 defectives, and a random sample of 1680 parts from machine 2 has 39 defectives. Does machine 2 have a lower defective rate? Test at 0.01 significance level.

$$n_1 = 1500 \quad n_2 = 1680$$
$$x_1 = 36 \quad x_2 = 39$$
$$p'_1 = 36/1500 = 0.024 \quad p'_2 = 39/1680 \approx 0.0232$$

1. Conditions are satisfied.

2. H_0: $p_1 = p_2$

   H_a: $p_1 > p_2$ (right-tailed test)

3. $\alpha = 0.01$.

4. From a standard normal table the z-value with 0.01 to its right is 2.33.

5. $p'_p = (x_1 + x_2)/(n_1 + n_2) = (36 + 39)/(1500 + 1680) = 0.0236$

$$z = \frac{0.0240 - 0.0232}{\sqrt{0.0236(0.9764)}\sqrt{0.0006667 + 0.0005952}} \approx 0.148$$

6. Do not reject H_0.

7. At the 0.01 significance level, the data do not support the conclusion that machine 2 has a lower defective rate.

EXAMPLE 5.27

Ten items are randomly selected from population A and 10 from population B. The 20 items are tested to failure with the following results:

Time to failure	948	942	939	930	926	918	917	910	897	895	886	880	870	865	862	850	835	830	821
Population In	B	B	B	B	A	B	B	A	B	B	A	A	B	A	A	B	A	A	A

Lead count = 4 Lag count = 3

In this case, lead count = 4 and lag count = 3, so end count $h = 3 + 4 = 7$ and

$$\alpha \le \frac{h}{2^h} = \frac{7}{2^7} \approx 0.055.$$

So, the conclusion is that the items in population B are more reliable than the items in population A at the 0.055 significance level.

Random samples of size $n \ge 8$ from each population are tested to failure, and the times to failure are sorted from highest to lowest. Define lead count as the number of consecutive items of one type at the top of the sorted list. Define lag count as the number of consecutive items at the bottom of the list from the other population. Define end count h as

$$h = \text{Lead count} + \text{Lag count}.$$

The test shows that the population represented by the lead count is more reliable than the other population, with a significance level of

$$\alpha \le \frac{h}{2^h}.$$

Statistical versus Practical Significance

In some situations it may be possible to detect a statistically significant difference between two populations when there is no practical difference. For example, suppose that a test is devised to determine whether there is a significant difference in the surface finish when a lathe is operated at 400 rpm and 700 rpm. If large sample sizes are used, it may be possible to determine that the 400 rpm population has a tiny but statistically significant surface improvement. However, if both speeds produce surface finishes that are capable of meeting the specifications, the best decision might be to go with the faster speed because of its associated increase in throughput. Thus, the difference between two populations, although statistically significant, must be weighed against other economic and engineering considerations.

Significance Level, Power, Type I and Type II Errors

Since every hypothesis test infers properties of a population based on analysis of a sample, there is some chance that although the analysis is flawless, the conclusion may be incorrect. These *sampling errors* are not errors in the usual sense because they can't be corrected (without using 100 percent sampling with no measurement errors). The two possible types of errors have been named type I and type II.

A *type I error* occurs when a true null hypothesis is rejected. The probability of type I error is denoted α. This is the same α that is used in the formulas for confidence intervals and critical values. Hence, when a hypothesis test is conducted at the .05 significance level, there is a probability of .05 that the hypothesis will be rejected when it shouldn't have been. When using α in the construction of a confidence interval for the mean of a population, the probability that the population mean is not in the interval is α.

A *type II error* occurs when a false null hypothesis is not rejected. The probability of type II error is denoted β.

It is helpful to think of a sampling example. Suppose a lot of 1200 is to be inspected to an acceptable quality level (AQL) of 2.5 percent and that the appropriate sampling plan calls for a sample size of 80 with a reject number of 6, that is, 80 parts are randomly selected and inspected and if six or more are defective, the entire lot is rejected. One of the myths of sampling theory is that any lot rejected by the sampling procedure fails to meet the 2.5 percent AQL requirement. But assume that the lot of 1200 has only 25 defectives, well below the 2.5 percent level. Is it possible that the sample of 80 could include at least six of those 25 defectives? Of course. In fact, the probability that this occurs is α. The null hypothesis here is that the lot is good. The null hypothesis is true, but the sampling plan causes us to reject it erroneously. This type I error is due to sampling error. If we took many samples of size 80, at most only α percent of them should have six or more defectives. It is easy to see why α is sometimes called the *producer's risk* because it is the probability that the lot, although meeting the AQL, will be rejected via sampling error.

Another myth of sampling theory is that a lot passed by the sampling procedure must meet the AQL. Assume that the lot of 1200 has 60 defectives, twice the defective level allowed by the AQL. It is possible that a sample of 80 could have fewer than six defectives. The probability that this will occur is β. β is also referred to as the *consumer's risk* because it is the probability that a lot that fails to meet the AQL will nevertheless be accepted through sampling error. The more powerful sampling plans keep the β value low. Thus the *power* of a sampling plan is defined as $1 - \beta$. The smaller the β, the larger the power.

For a given sample size and lot quality, the values of α and β depend on the accept and reject numbers of the sampling plan. If these numbers are adjusted to reduce α, then β will be increased and vice versa. To reduce both α and β it would be necessary to increase the sample size. Both could be reduced to zero if the sampling plan called for 100 percent sampling and no inspection errors occurred. But then it wouldn't be a sampling plan.

Part III

Reliability in Design and Development

Chapter 6

A. Reliability Design Techniques

1. ENVIRONMENTAL AND USE FACTORS

> Identify environmental and use factors (e.g., temperature, humidity, vibration) and stresses (e.g., severity of service, electrostatic discharge [ESD], throughput) to which a product may be subjected. (Apply)
>
> **Body of Knowledge III.A.1**

The following are environmental and other stress factors that can negatively affect the reliability of a product: temperature, vibration, humidity, a corrosive environment, electrostatic discharge (usually encountered during assembly), RF interference, cyclic stresses, and environments containing salt, dust, chlorine, and other contaminants.

Not all of the factors will have equal effects on various products. In general, temperature will have a much greater effect on a solid-state electronic system than on a mechanical system. Cyclic stresses will result in fatigue of a mechanical system. High humidity or a saline environment will corrode mechanical parts, but could also have an effect on electronic components.

Use factors are those environmental conditions that are normal for the product. For example, a laundry product may be designed to handle four loads per day with a temperature range of 40 °F to 120 °F and pH from 4 to 10. *Stresses* are those conditions that are additional or outside the design ranges. The presence of radiation or extremes of water pressure or voltage are examples of stresses.

2. STRESS-STRENGTH ANALYSIS

> Apply stress-strength analysis method of computing probability of failure, and interpret the results. (Evaluate)
>
> **Body of Knowledge III.A.2**

Failure of a part will occur when a stress exceeds the strength of the part for that stress. Accepted design practice is to design so that the strength is always greater than the expected stress. A good design will incorporate safety factors or safety margins to insure that the strength is always greater than the stress. To use these design techniques, the stress the part will encounter as well as the strength of the part must be viewed as single-point values. To use the stress-strength analysis method, both the stress and the strength are viewed as distributions.

For example, suppose the stress and strength are normally distributed random variables. The mean of the stress distribution is μ_s, and the standard deviation of the stress distribution is σ_s. The mean and standard deviation of the strength distribution are μ_S and σ_S. Good design would dictate that $\mu_S > \mu_s$. The safety factor is equal to μ_S/μ_s and is greater than 1. If viewed as single point values, failure can not occur. However, when viewed as distributions, there can be an interference region in which it is possible for stress to exceed strength (see Figure 6.1).

The *difference distribution* can be used to solve for the probability of failure. The difference distribution is the distribution of strength minus stress. The difference distribution will have a mean $\mu_D = \mu_S - \mu_s$ and standard deviation

$$\sigma_D = \sqrt{\sigma_S^2 + \sigma_s^2}.$$

The area to the left of zero in the difference distribution is the probability of failure as this is the region where stress exceeds strength. The reliability is equal to 1 – (probability of failure), or the area to the right of zero.

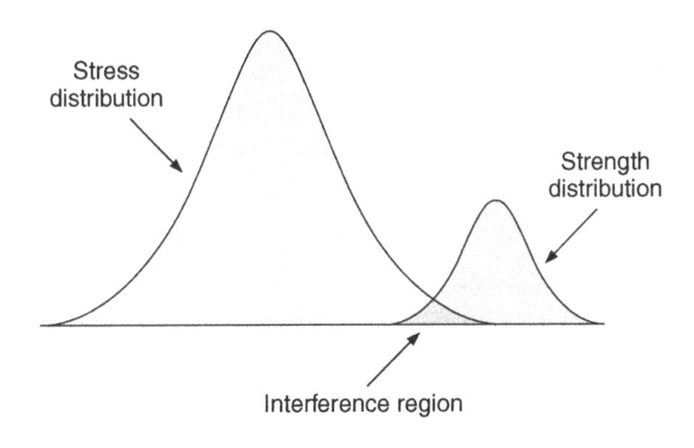

Figure 6.1 Diagram showing stress-strength interference region.

EXAMPLE 6.1

$\sigma_S = 4000$ psi $\quad \sigma_s = 3000$ psi

$\mu_S = 40{,}000$ psi $\quad \mu_s = 30{,}000$ psi

Safety factor $\eta = 40{,}000/30{,}000 = 1.33$

$\mu_D = 40{,}000 - 30{,}000 = 10{,}000$ psi

The probability of failure is the area to the left of zero in the difference distribution. It is the area from the standard normal tables to the left of a *z* value equal to:

$$z = \frac{(0 - \sigma_D)}{\sigma_D} = (-10{,}000/5000) = -2$$

Probability of failure = 0.023

Reliability = 1 – 0.023 = .977

Viewing stress-strength relationships in this manner emphasizes the four basic ways the designer can improve on the reliability.

The designer usually has more control over the strength:

- Increase the mean strength: use different materials or different design.
- Decrease the strength variation: reduce variation in the materials and in the process.

The designer may have some control over the stress:

- Decrease the mean stress: control the loading.
- Decrease the stress variation: limit the use environment.

3. FMEA AND FMECA

Define and distinguish between failure mode and effects analysis and failure mode, effects, and criticality analysis and apply these techniques in products, processes, and designs. (Analyze)

Body of Knowledge III.A.3

FMEA

Failure mode and effects analysis (FMEA) is an engineering technique for system reliability improvement. It is a structured analysis of a system or subsystem to identify potential failures at the component level, the causes of these failures, and the effect these failures will have on the operation of the system. *Modes* are technical events that occur, whereas *effects* refers to the impact of those events.

An example of the heading for an FMEA table is shown in Figure 6.2.

The purpose of FMEA is the anticipation and mitigation of the negative effects of possible failures prior to the time they occur. It is best implemented by a broad-based team with representation from all functions that can be impacted by the results. This typically includes design, production, purchasing, quality, sales, and others as appropriate. Although customers are not always represented on the team, it behooves all team members to think as customers during FMEA procedures. FMEA should be used at every stage from product and process design through service delivery activities.

The steps of FMEA are:

1. Prior to the first team meeting, determine the product to be analyzed and gather associated data.
2. List all possible failure modes. It is very important that this step be given time and resources. If one or more failure modes are overlooked, the remainder of the FMEA becomes less valuable. For each failure mode ask how, where, when, and why the failure would occur and what impact the failure would have.
3. Calculate a risk priority number (RPN) for each mode. Do this by assigning a value from one to 10 for each of these categories:

   S = Severity: a judgment regarding the impact this failure would have

   O = Occurrence: an estimate based on the probability that this failure will occur

   D = Detection: an estimate of the probability that the failure would be detected once it has occurred

   The RPN is the product of the three numbers: rpn = S × O × D
4. Develop a corrective action plan for the risks the team and management personnel deem most significant.
5. Document and report the results.

Item	Potential failure mode	Effects	Sev	Causes	Occ	Current controls	Det	RPN	Recom'ed actions	Responsible and date	Action	Results			
											Actions taken	Sev	Occ	Det	RPN
A29	Pin bends	V16 sags	6	Low carbon	1	Hardness check	8	48	Change spec. to supplier	Jim Hzn. 8/21/12	Spec. chgd.	6	1	2	12

Figure 6.2 An example of the heading for an FMEA table.

Part III.A.3

Each of these steps requires a great deal of serious effort and time. Some guidelines for assigning the numbers required in step 3 follow.

Severity

- The numbers 9 and 10 are reserved for failures that will endanger the safety of individuals, with 10 usually denoting a safety hazard that will occur without warning.
- The numbers 5 through 8 are associated with various levels of dissatisfaction of the customer.
- The numbers 2 though 4 refer to some lack of function.
- The number 1 means the failure would have no impact.

Occurrence

Often, the best estimate can be obtained by looking at data from similar products and processes. Otherwise:

- Use 10 if probability is about .5
- Use 9 if probability is about .3
- Use 8 if probability is about .1
- Use 7 if probability is about .05
- Use 6 if probability is about .01
- Use 5 if probability is about .003
- Use 4 if probability is about .0005
- Use 3 if probability is about .00007
- Use 2 if probability is about .000007
- Use 1 if probability is about .0000007

Detection

- 10 is used for failures that are considered impossible to detect.
- 1 is used for failures that are almost certain to be detected.
- Numbers between 1 and 10 indicate likelihood of detection, from most likely to least likely.

Once the RPN has been calculated for each failure mode, the next step, number 4 on the list, is to determine which ones should have corrective or preventive actions. If several of the RPN values are clearly clumped at the upper end, these will typically be the place to begin. When two failure modes have very similar RPN values, it is necessary to look a little deeper to determine which to work on first. One approach to prioritizing in this situation is to recalculate RPN with the occurrence factor omitted.

There is an inherent mathematical problem with RPN values, so it is important to exercise some judgment when using them. For example, suppose the S, O, and D values for one failure mode are 10, 7, and 4, respectively, giving an RPN of 280, while another failure mode has 5, 8, and 8, respectively, for an RPN of 320. This would mean that the failure mode that would produce moderate customer dissatisfaction, which occurs about 10 percent of the time and is very unlikely to be detected, should have a higher priority than one that causes a safety hazard without warning, occurs five percent of the time, and has a moderate likelihood of being detected. This is, of course, ridiculous. An alternate prioritization plan, avoiding RPN values altogether, is to rank all failure modes by severity, and within each severity value, rank them by occurrence. Then within each occurrence value, rank them by detection.

FMECA

Failure mode, effects, and criticality analysis (FMECA) bases the prioritization scheme on probability of occurrence and severity of the failure mode. These two variables can be used to construct a two-dimensional graph displaying each failure mode as a point. Severity can be plotted on the horizontal axis and probability on the vertical axis. More-severe failure modes will plot further to the right, and more-probable ones further up from the horizontal axis. If a set of failure modes is grouped in the upper right corner, those should be attacked with highest priority. In Example 6.2, failure modes ⑧, ⑤, and ③ appear to be the place to start. The

EXAMPLE 6.2

A team has identified 10 failure modes denoted ① through ⑩ for a product, and have determined severity and probability of occurrence rankings for each (see Figure 6.3).

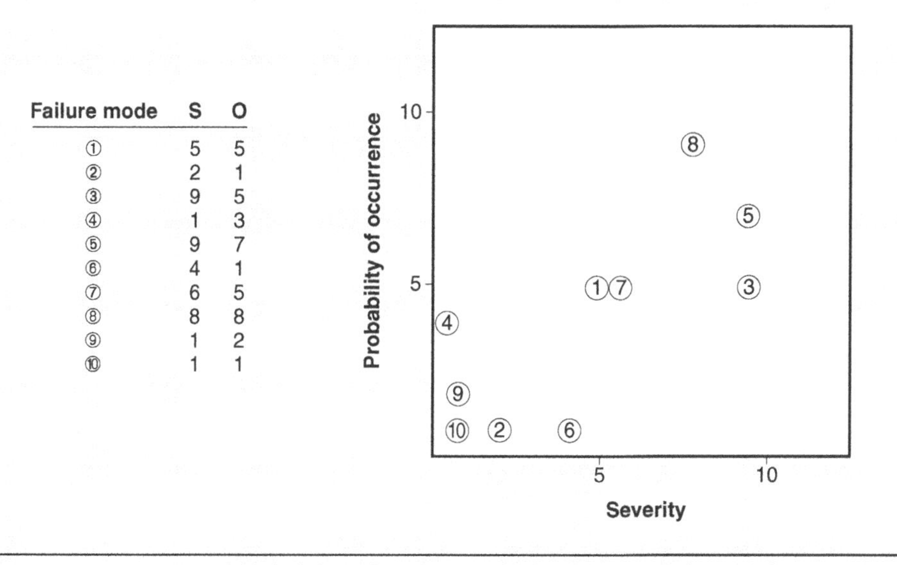

Failure mode	S	O
①	5	5
②	2	1
③	9	5
④	1	3
⑤	9	7
⑥	4	1
⑦	6	5
⑧	8	8
⑨	1	2
⑩	1	1

Figure 6.3 Severity and probability of occurrence rankings for 10 failure modes.

probability of occurrence and the severity scales are typically the same as those defined in the previous section.

Other tools for assessing criticality include flowcharts and block diagrams. These graphical tools display the relationships and dependencies of various subsystems and components. They are especially useful for complex systems. Users of FMECA have the option of including additional concerns such as safety, downtime, preventive maintenance, and stocks of spare parts.

FMEA in Design

Reliability improvement will occur when design changes are incorporated into the system to eliminate a failure, reduce the probability the failure will occur, or reduce the effect the failure has on the operation of the system. A most important factor for the success of the technique is that the FMEA is completed in time to economically make changes to the design using the results of the FMEA activity. The FMEA is meant to be a before-the-fact action, not an after-the-fact exercise.

The FMEA is changed from a qualitative analysis to a quantitative analysis by assigning values to the probability of the failure occurring, to the severity of the effect of the failure on the operation of the system, and to the probability that the system controls will detect and eliminate the failure before the design is complete. The RPN is then calculated as the product of the three assigned values. This gives a numerical ranking to each failure. The failures with the top rankings are selected as possible candidates for design changes and thus for reliability improvement. Special consideration should be given to any failure with a high severity rating regardless of its RPN value. It is recommended that a 10-point scale be used to rank probability of occurrence, severity, and detection (see Table 6.1).

Each failure will then have an RPN between 1 and 1000. High RPN values would result in corrective action to reduce the risk, reduce the severity, or increase the detection probability.

An FMEA can be performed on a product or on a process. The design FMEA (DFMEA) is performed on the design or product. The process FMEA (PFMEA) is performed on the process. The DFMEA is considered a reliability engineering function, while the PFMEA is considered to be a quality engineering function. Both activities will improve the reliability of the product.

The DFMEA will document weaknesses in the design that can cause failures to occur during product use. A change made to the design because of the DFMEA will reduce the failure rate during the useful life period or increase the duration of the useful life by eliminating early-wear failure. The result is improved reliability

Table 6.1 Failure ranking using a 10-point scale.

Level	Probability of occurrence	Severity	Detection probability
High	8–10	8–10	1–3
Moderate	4–7	4–7	4–7
Low	1–3	1–3	8–10

of the product. Product safety can also be improved by elimination of any unsafe conditions that might result from a failure. The PFMEA will uncover the potential of the process to add nonconformity to the product. Process improvement or the addition of process detection controls because of the PFMEA will result in a reduction of product early-life failures. The FMEA team should represent a broad cross-section of technical and nontechnical expertise. The team should include, but not be limited to, representation from product design, product service, manufacturing engineering, quality engineering, reliability engineering, purchasing (to represent suppliers), and marketing (to represent the customer), as well as experts in materials, thermal stresses, vibration, fatigue, and corrosive environments.

In the mid 1990s the major automotive companies created and adopted the Society of Automotive Engineers standard SAE J1739 *Potential Failure Mode and Effects Analysis in Design* for their use and the use of their suppliers. This standard gives specific rankings for probability, severity, and detection. The ranking of a 9 or a 10 for severity is limited to hazardous or life-threatening effects or operation outside of government regulations. The system being totally inoperable has a severity ranking of 8.

The breadth and depth of experience for the design FMEA team is critical. Some members must have a background with similar products and with any equipment that interfaces with the product being designed. If electrical, pneumatic, mechanical, or software linkages are needed, expertise in those fields is also essential.

FMECA in Design

At each step of the design process the FMECA team should provide the design group with a list of potential failure modes and their impact on the satisfaction of customer needs. Compiling such a list can be more difficult than the post-design FMECA due to the incompleteness of the design. On the other hand, the design process typically affords more options for preventive and corrective actions.

The FMECA process places emphasis on the criticality of the failure mode. Again, addressing critical failure modes at the earliest possible stage of design is most cost-effective, as seen in Example 6.3.

EXAMPLE 6.3

A group is charged with producing a design for an operator cab for an existing agricultural tractor. The FMECA team identifies UV deterioration of the seal where the steering column intersects the cab wall as one of the critical failure modes. Possible preventive actions include:

- Protection of the seal from UV exposure
- Using UV resistant seal material
- Designing the steering linkage so the seal isn't needed

It is likely that none of these solutions would be available once the design has been finalized.

4. COMMON MODE FAILURE ANALYSIS

> Describe this type of failure (also known as common cause mode failure) and how it affects design for reliability.
>
> **Body of Knowledge III.A.4**

To avoid confusion it is necessary to distinguish this concept from that of *common cause variation* as applied to control charts. A *common cause failure* (CCF) occurs when two or more components fail at the same time as a result of a shared cause. Recognizing and mitigating common cause failure is a vital part of designing for reliability. These failures often are not exhibited from the standard fault tree analysis.

Clearly, the reliability engineer must be aware of common cause failure and how it may be studied and controlled. *Fault tree analysis* (FTA), as discussed in Section 5 of this chapter, is the primary tool when each fault is studied for additional potential failures. The reliability engineer must take a holistic view. For example, if the fault being studied is a sudden spike in engine compartment temperature, a fault tree analysis that only considers the impact of this event on the braking system would be too narrow. Other functions that could be impaired might include the power steering system, computer motherboard, wiring for lighting, and so on. So a single fault might generate multiple fault trees. For a discussion of how to reduce or eliminate the impact of the fault, see the discussion on fault tolerance (Section 8 of this chapter).

5. FAULT TREE ANALYSIS (FTA) AND SUCCESS TREE ANALYSIS (STA)

> Apply these techniques to develop models that can be used to evaluate undesirable (FTA) and desirable (STA) events. (Analyze)
>
> **Body of Knowledge III.A.5**

AND and OR Gates

Once a failure mode has been identified as one requiring additional study, fault tree analysis can be used. Basic symbols used in FTA are borrowed from the electronic and logic fields. The fundamental symbols are the AND gate and the OR gate. Each of these has at least two inputs and a single output (see Figure 6.4).

The output for the AND gate occurs if and only if all inputs occur. The output for the OR gate occurs if and only if at least one input occurs. Rectangles are

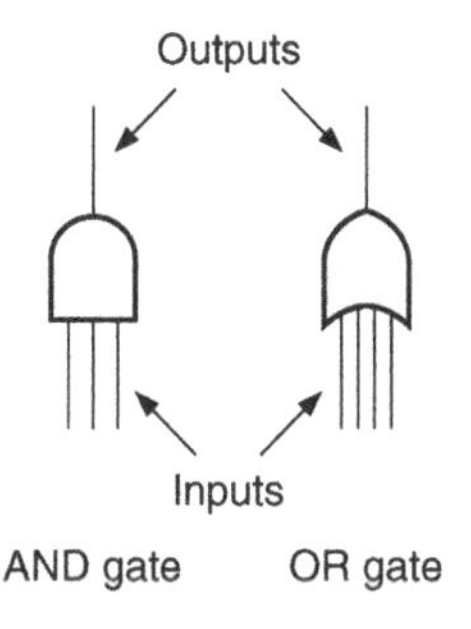

Figure 6.4 AND and OR gate symbols.

typically used for labeling inputs and outputs. The failure mode being studied is sometimes referred to as the "top" or "head" event. An FTA helps the user to consider underlying causes for a failure mode and to study relationships between various failures.

Voting OR Gates

In this gate, the output occurs if and only if k or more of the input events occur, where k is specified, usually on the gate symbol (see Figure 6.6).

Success tree analysis (STA) approaches a system more positively, asking "What must occur for the system to function successfully?"

Fault tree analysis (FTA) is very useful in the early stages of design, especially in situations where the product being designed is complex and/or has interdependencies with other components. FTA provides a diagram of the complexities and helps users visualize possible preventive/corrective actions. In typical use the

EXAMPLE 6.4

The failure mode being studied is the stoppage of agitation in a tank before mixing is complete. This becomes the top event. Further team study indicates that this will occur if any of the following occurs:

- Power loss
- Timer shuts off too soon
- Agitator motor failure
- Agitator power train failure

Power loss will occur if the external power source fails and the backup generator fails. The timer shuts off too soon if it is set incorrectly or it has a mechanical failure. The agitator motor fails if overheated, or a fuse or capacitor fails. The agitator power train fails if both belts A and B break, or clutch or transmission fails. This is symbolized by the FTA diagram shown in Figure 6.5.

Continued

Continued

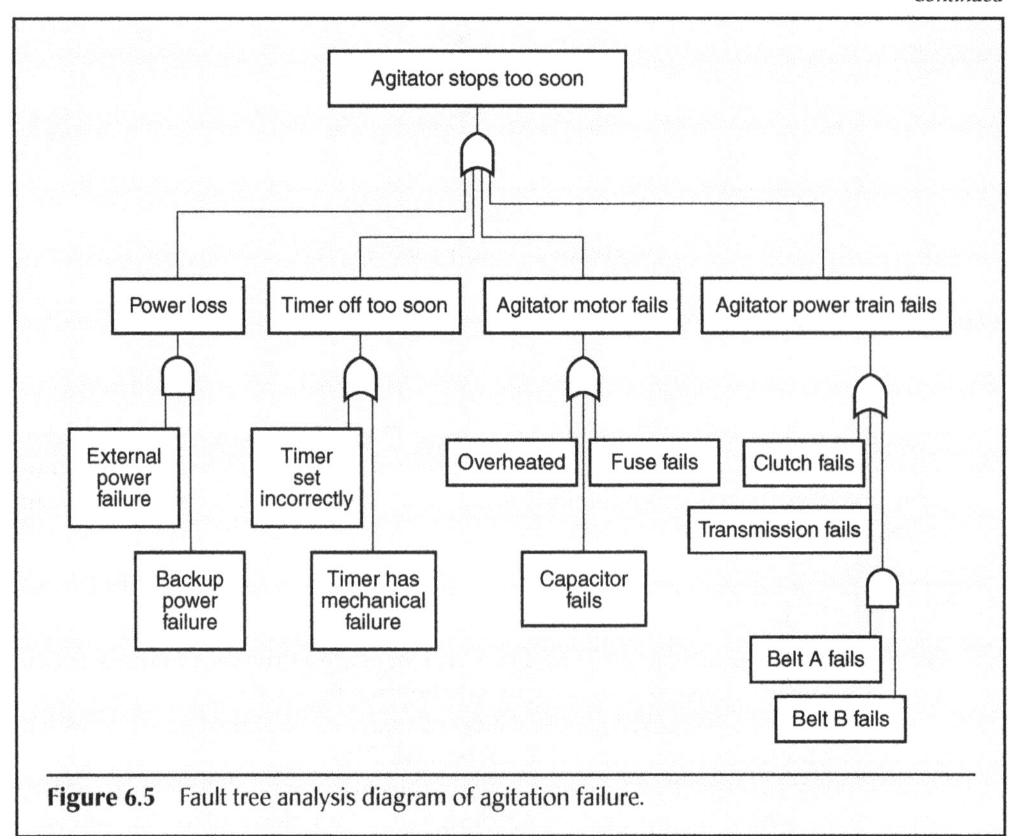

Figure 6.5 Fault tree analysis diagram of agitation failure.

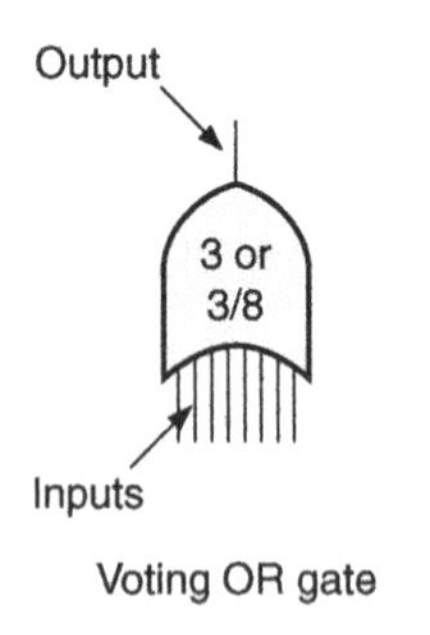

Figure 6.6 Voting OR gate.

analysis begins with a fault or failure that has been singled out for further study and asks, "What condition(s) could cause this failure?"

Further analysis and assignment of probabilities could be done using meteorological records and insulator testing results. At this point the design team could determine whether the wire strength, insulator design, amount of sag, and so on, are appropriate.

EXAMPLE 6.5

An STA diagram for the previous example is shown in Figure 6.7.

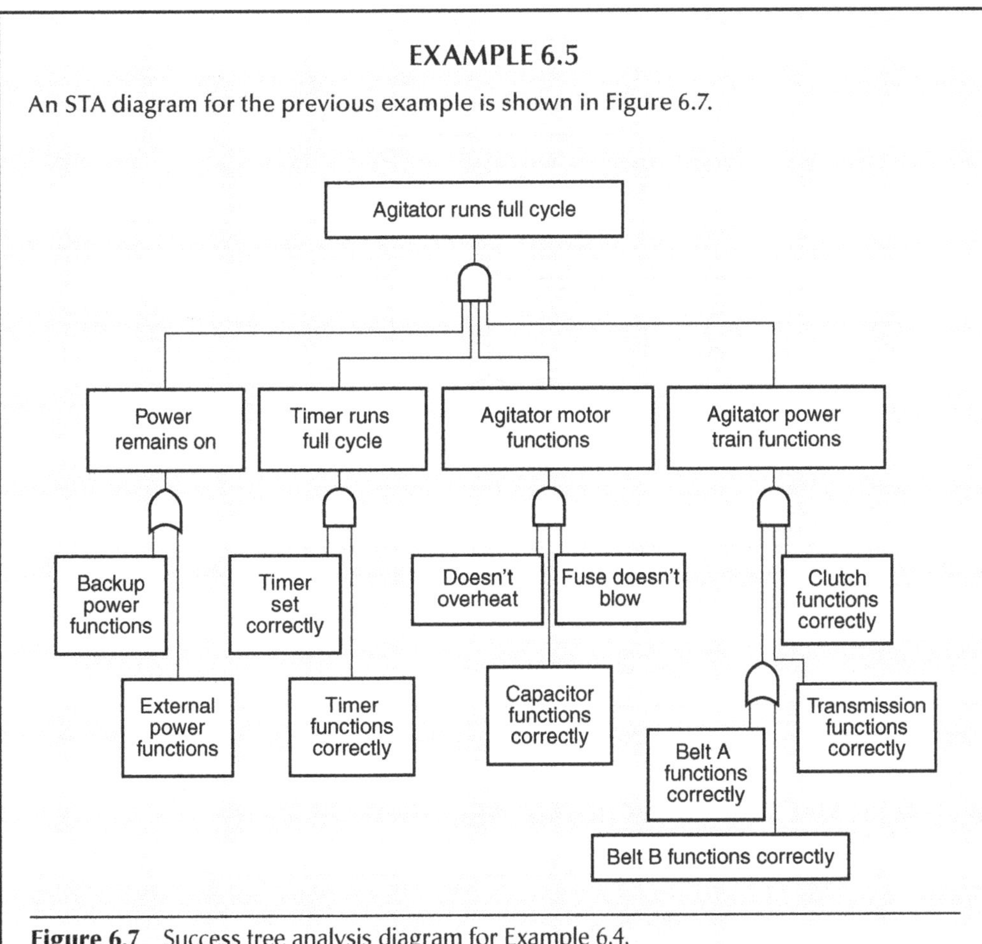

Figure 6.7 Success tree analysis diagram for Example 6.4.

EXAMPLE 6.6

The failure to be studied is specified as a loss of continuity between points A and B on a power transmission line. Suppose the team has determined three events, any one of which would cause the outage:

1. Wind more than 130 mph
2. Wind more than 70 mph with more than one inch of ice buildup on the line
3. Insulator failure

The fault tree could be drawn as shown in Figure 6.8

Continued

Continued

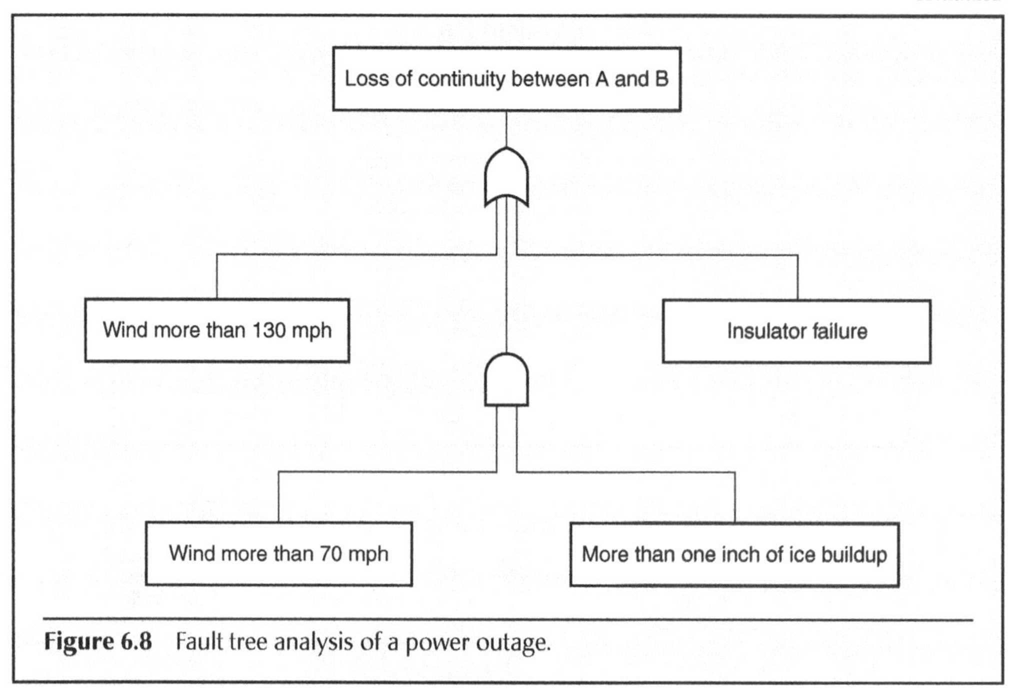

Figure 6.8 Fault tree analysis of a power outage.

6. TOLERANCE AND WORST-CASE ANALYSES

Describe how tolerance and worst-case analyses (e.g., root of sum of squares, extreme value) can be used to characterize variation that affects reliability. (Understand)

Body of Knowledge III.A.6

Reliability testing should include products whose components are at the extremes of their tolerance limits. A machining example will illustrate this issue. If the process is capable and normal, the method shown in solution 2 has obvious advantages.

EXAMPLE 6.7

A set of three spacers called A, B, and C whose lengths have dimensions 1.000 ± .003, 2.000 ± .003, and 3.000 ± .003, respectively, will be assembled on a shaft (see Figure 6.9).

Continued

Continued

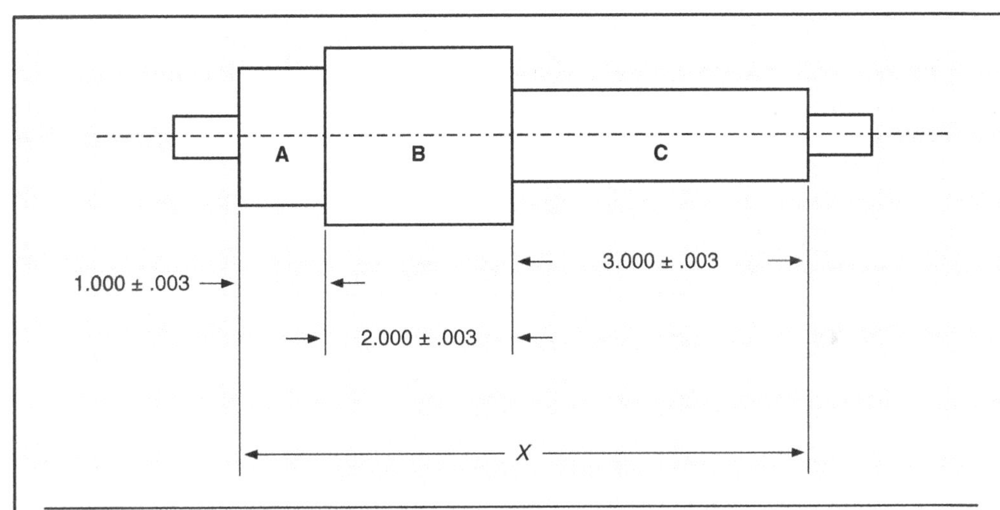

Figure 6.9 Three spacers for assembly on a shaft.

Problem:

What are appropriate extreme values for the assembled overall length *X*? There are two approaches to this problem.

Solution 1: (Worst-Case Scenario)

The conventional and most conservative approach is to obtain the minimum value for dimension *X* by adding the minimums of the three components:

$$X_{\text{Min}} = A_{\text{Min}} + B_{\text{Min}} + C_{\text{Min}} = .997 + 1.997 + 2.997 = 5.991$$

Similarly, the maximum value of *X* is found by

$$X_{\text{Max}} = A_{\text{Max}} + B_{\text{Max}} + C_{\text{Max}} = 1.003 + 2.003 + 3.003 = 6.009$$

This approach indicates that reliability testing should include some components whose assembled length is 5.991 and others whose assembled length is 6.009.

Solution 2: (Statistical Tolerancing)

Suppose the production processes that produce parts A, B, and C each have capability $C_{pk} = 2$ (for six sigma), and the lengths are normally distributed and centered at the nominal dimensions. Then, the standard deviation of each of the three sets of lengths is .0005. Denote the three standard deviations as s_A , s_B, and s_C, and let s_X be the standard deviation of length *X*. The statistical formula that relates these standard deviations is

$$\sigma_X = \sqrt{\sigma_A^2 + \sigma_B^2 + \sigma_C^2}$$

(sometimes referred to as root sum squared).

Substituting,

$$\sigma_X = \sqrt{.0005^2 + .0005^2 + .0005^2} \approx .0009 \text{ and so } 6\sigma_X \approx .0054.$$

Then, the six sigma limits on dimension *X* are 6.0000 ± .0054, or 5.9946 and 6.0054.

Six sigma covers all but 3.4 items per million, so testing component combinations with *X*-values at these 6σ extremes will cover 99.99966 percent of the combinations. Note that this is a considerably smaller testing range that that required in solution 1. Note that this reduction in test requirements is only valid for capable, centered normal processes. However, C_{pk} need not be equal to 2. The analysis could be redone for any C_{pk}.

7. DESIGN OF EXPERIMENTS

> Plan and conduct standard design of experiments (DOE) (e.g., full-factorial, fractional factorial, Latin square design). Implement robust-design approaches (e.g., Taguchi design, parametric design, DOE incorporating noise factors) to improve or optimize design. (Analyze)
>
> **Body of Knowledge III.A.7**

Terminology

This section provides definitions for some important basic terms:

experimental error—The variation in the response variable when levels and factors are held constant.

factor or variable—An assignable cause that may affect the responses, and of which different levels are included in the experiment.

levels—The possible values of a factor in an experimental design.

noise factors—Those factors that aren't controlled in an experiment.

replication—The repetition of the set of all the treatment combinations to be compared in an experiment. Each of the repetitions is called a *replicate.*

response variable—The variable that shows the observed results of an experimental treatment.

treatment—A combination of the levels of each of the factors assigned to an experimental unit.

The objective of a designed experiment is to generate knowledge about a product or process. The experiment seeks to find the effect a set of independent variables has on a set of dependent variables. Mathematically, this relationship can be denoted $y = f(x)$, where x is a list of independent variables and y is the dependent variable. For example, suppose a machine operator who can adjust the feed, speed, and coolant temperature wishes to find the settings that will produce the best surface finish. The feed, speed, and coolant temperature are called *independent variables.* The surface finish is called the *dependent variable* because its value depends on the values of the independent variables. Independent variables may be thought of as input variables, and dependent variables as output variables. There may be additional independent variables, such as the hardness of the material or humidity of the room, that have an effect on the dependent variable. The independent variables that the experimenter controls are called *control factors* or sometimes just *factors*. The other factors, such as hardness or humidity, are called *noise factors*. In this example, the experimental design may specify that the speed will be set at 1300 rev/min for part of the experiment and at 1800 rev/min for the remainder. These values are referred to as the *levels* of the speed factor. The experimenting team decides to test each factor at two levels, as follows:

Feed: (F): .01 and .04 in/rev

Speed (S): 1300 and 1800 rev/min

Coolant temp (C): 100 and 140 °F

They opt for a full-factorial experiment so they can generate the maximum amount of process knowledge. A full-factorial experiment tests all possible combinations of levels and factors, using one run for each combination. The formula for number of runs is

$$n = L^F$$

where

n = Number of runs

L = Number of levels

F = Number of factors

In this situation, $n = 2^3 = 8$ runs. The team develops a data collection sheet listing those eight runs with room for recording five repetitions for each run (see Table 6.2).

The factor–level combinations are also called *treatments*, and the repetition of runs is sometimes called *replication*. In experimenter's jargon, it is said that there

Table 6.2 A 2^3 full-factorial data collection sheet.

Run #	Feed	Speed	C temp	1	2	3	4	5
1	.01	1300	100					
2	.01	1300	140					
3	.01	1800	100					
4	.01	1800	140					
5	.04	1300	100					
6	.04	1300	140					
7	.04	1800	100					
8	.04	1800	140					

Table 6.3 A 2^3 full-factorial data collection sheet with data entered.

Run #	Feed	Speed	C temp	1	2	3	4	5
1	.01	1300	100	10.1	10.0	10.2	9.8	9.9
2	.01	1300	140	3.0	4.0	3.0	5.0	5.0
3	.01	1800	100	6.5	7.0	5.3	5.0	6.2
4	.01	1800	140	1.0	3.0	3.0	1.0	2.0
5	.04	1300	100	5.0	7.0	9.0	8.0	6.0
6	.04	1300	140	4.0	7.0	5.0	6.0	8.0
7	.04	1800	100	5.8	6.0	6.1	6.2	5.9
8	.04	1800	140	3.1	2.9	3.0	2.9	3.1

are eight treatments, with each treatment replicated five times. As the data are collected, the values are recorded as shown in Table 6.3. These data are also referred to as the *response values* since they show how the process or product responds to various treatments.

Note that the five values for a particular run are not all the same. This may be due to drift in the factor levels, variation in the measurement system, and the influence of noise factors. The variation observed in the readings for a particular run is referred to as *experimental error.* If the number of replications is decreased, the calculation of experimental error is less accurate, although the experiment has a lower total cost. If all the factors that impact the dependent variable are included in the experiment, and all measurements are exact, replication is not needed, and a very efficient experiment can be used. Thus, the accurate determination of experimental error and cost are competing design properties.

Planning and Organizing Experiments

When preparing to conduct an experiment, the first consideration is "What question are we seeking to answer?" In the previous example, the objective was to find the combination of process settings that minimizes the surface finish reading. Examples of other experimental objectives include:

- Find the inspection procedure that provides optimum precision.
- Find the combination of mail and media ads that produces the most sales.
- Find the cake recipe that produces the most consistent taste in the presence of oven temperature variation.
- Find the combination of valve dimensions that produces the most linear output.

Sometimes, the objective derives from a question. For example, the question "What's causing the excess variation in hardness at the rolling mill?" could generate the objective "Identify the factors responsible for the hardness variation and find the settings that minimize it." The objective must be measurable, so the next step is to establish an appropriate measurement system. If there is a tolerance on the objective quantity, the *rule of 10* measurement principle says that the finest resolution of the measurement system must be less than or equal to 1/10 of the tolerance. The measurement system must also be reasonably simple and easy to operate.

Once the objective and a measurement system have been determined, the factors and levels are selected. People with the most experience and knowledge about the product or process are asked to name the adjustments they'd make to achieve the objective. Their responses should include the things they would change (factors) and the various values (levels) they would recommend. From these recommendations, the list of factors and the levels for each factor are determined.

The next step is to choose the appropriate design. The selection of the design may be constrained by such things as affordability and time available. At this stage, some experimenters establish a budget of time and other resources that may be used to reach the objective. If production equipment and personnel must be used, how much time is available? How much product and other consumables are available? Typically, 20 to 40 percent of the available budget should be allocated to the first experiment because it seldom meets the objective and in fact often raises as many questions as it answers. Typical new questions are:

- What if an additional level had been used for factor A?
- What if an additional factor had been included instead of factor B?

Therefore, rather than designing a massive experiment involving many variables and levels, it is usually best to begin with more modest screening designs whose purpose is to determine the variables and levels that need further study.

The next few sections discuss various designs.

Randomization

Returning to the surface finish example, there are eight treatments with five replications per treatment. This produces 40 tests. The tests should be performed in random order. The purpose of randomization is to spread out the variation caused by noise variables. The 40 tests may be randomized in several ways. Here are two possibilities:

1. Number the tests from one to 40 and randomize those numbers to obtain the order in which tests are performed. This is referred to as a *completely randomized design*.
2. Randomize the run order, but once a run is set up, make all five replicates for that run.

Although it usually requires more time and effort, the first method is better. To see that this is true, suppose time of day is a noise factor such that products made before noon are different from those made after noon. With the completely randomized design, each run will likely have parts from both morning and afternoon.

Blocking

If it is not possible to run all 40 tests under the same conditions, the experimenting team may use a technique called *blocking*. For example, if the 40 tests must be spread over two shifts, the team would be concerned about the impact the shift difference could have on the results. In this experiment, the coolant temperature is probably the most difficult to adjust, so the team may be tempted to perform all the 100-degree runs during the first shift and the 140-degree runs on the second shift. The obvious problem here is that what appears to be the impact of the change in coolant temperature may in part reflect the impact of the change in shift. A better approach would be to randomly select the runs to be performed during each shift. This is called a *randomized block design*. For example, the random selection might put runs 1, 4, 5, and 8 in first shift, and 2, 3, 6, and 7 in second shift. Another method that might be used to nullify the impact of the shift change would be to do the first three replicates of each run during the first shift and the remaining two replicates of each run during the second shift.

Once the data are collected as shown in Table 6.3, the next step is to find the average of the five replication responses for each run. These averages are shown in Table 6.4.

Main Effects

The first step in calculating the main effects, sometimes called *average main effects*, is to average the results for each level of each factor. This is accomplished by averaging the results of the four runs for that level. For example, $F_{.01}$ (feed at the .01 in/min level) is calculated by averaging the results of the four runs in which feed was set at the .01 level. These were runs 1, 2, 3, and 4, so

$$F_{.01} = (10 + 4 + 6 + 2) \div 4 = 5.5$$

$$\text{Similarly, } F_{.04} = (7 + 6 + 6 + 3) \div 4 = 5.5.$$

Table 6.4 A 2^3 full-factorial data collection sheet with run averages.

Run #	Feed	Speed	C temp	Average surface finish reading
1	.01	1300	100	10
2	.01	1300	140	4
3	.01	1800	100	6
4	.01	1800	140	2
5	.04	1300	100	7
6	.04	1300	140	6
7	.04	1800	100	6
8	.04	1800	140	3

Runs numbered 1, 2, 5, and 6 had S at 1300 rev/min, so

$$S1300 = (10 + 4 + 7 + 6) \div 4 = 6.75$$

$$\text{and } S1800 = (6 + 2 + 6 + 3) \div 4 = 4.25$$

$$C100 = (10 + 6 + 7 + 6) \div 4 = 7.25$$

$$C140 = (4 + 2 + 6 + 3) \div 4 = 3.75.$$

The main effects may be graphed as shown in Figure 6.10.

Because the better surface finish has the lowest score, the team would choose the level of each factor that produces the lowest result. The team would suggest using a speed of 1800 rev/min and coolant temp of 140 °F. What feed rate should be recommended? Since both $F_{.01}$ and $F_{.04}$ are 5.5, the feed rate doesn't impact surface finish in this range. The team would recommend a feed rate of .04 since it will result in a faster operation.

Factors with the greater difference between the high and low results are the factors with the greatest impact on the quality characteristic of interest (surface finish in this case). Most authors refer to the main effect as *the high-level result minus the low-level result for the factor.* For example,

$$\text{Main effect of factor } F = F_{.04} - F_{.01} = 5.5 - 5.5 = 0$$

$$\text{Similarly, main effect of } S = S_{1800} - S_{1300} = 4.25 - 6.75 = -2.50$$

$$\text{and } C = C_{140} - C_{100} = 3.75 - 7.25 = -3.50$$

Using this definition of main effect, the larger the absolute value of the main effect, the more influence that factor has on the quality characteristic. It is possible that the perceived difference between high and low results is not statistically significant. This would occur if the experimental error is so large that it would be impossible to determine whether the difference between the high and low values is due to a real difference in the dependent variable or due to experimental error. This may be determined by using *analysis of variance* (ANOVA) procedures. For analysis of data from an experiment, the null hypothesis is that changing the factor level does not make a statistically significant difference in the dependent variable. The β-risk is the probability that the analysis will show that there is a significant difference when there is not. The β-risk is the probability that the analysis will show

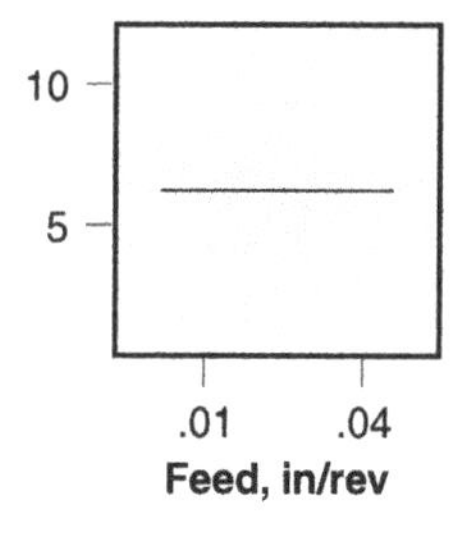

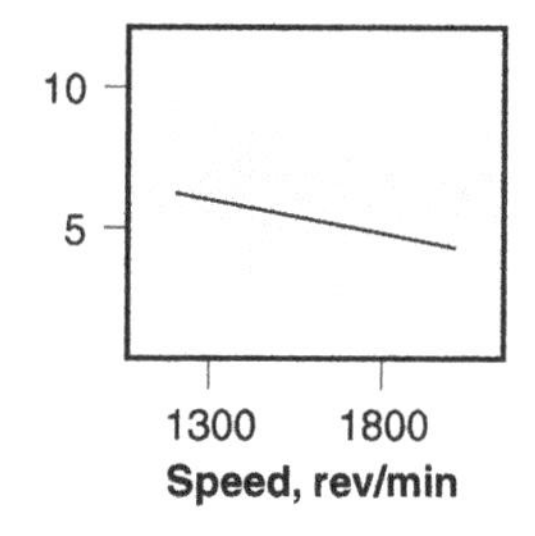

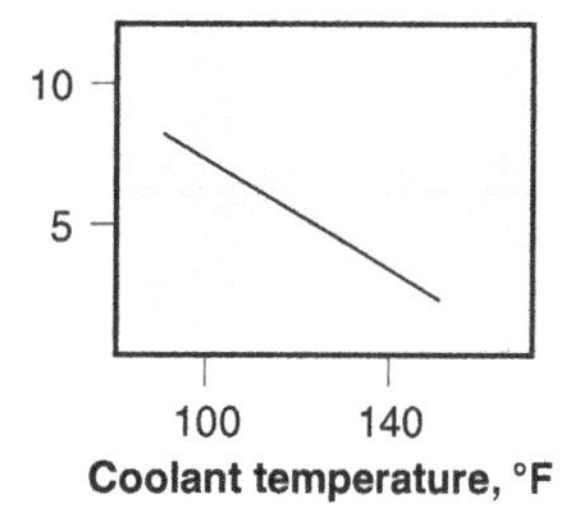

Figure 6.10 Plot of main effects.

that there is no significant difference when there is. The *power* of the experiment is defined as $1 - \beta$, so the higher the power of the experiment, the lower the β-risk. In general, a higher number of replications or a larger sample size provides a more precise estimate of experimental error, which in turn reduces the β-risk.

Interaction Effects

To assess the interaction effects, return to the original experimental design matrix, replacing each high level with "+" and each low level with "–" as shown in Table 6.5.

To find an entry in the column labeled "$F \times S$," multiply the entries in the F and S columns, using the multiplication rule "If the signs are the same, the result is positive; otherwise, the result is negative." Fill the other interaction columns the same way. To fill the $F \times S \times C$ column, multiply the $F \times S$ column by the C column (see Table 6.6).

Table 6.5 A 2^3 full-factorial design using + and – format.

Run	*F*	*S*	*C*	*F* × *S*	*F* × *C*	*S* × *C*	*F* × *S* × *C*
1	–	–	–				
2	–	–	+				
3	–	+	–				
4	–	+	+				
5	+	–	–				
6	+	–	+				
7	+	+	–				
8	+	+	+				

Table 6.6 A 2^3 full-factorial design showing interaction columns.

Run	*F*	*S*	*C*	*F* × *S*	*F* × *C*	*S* × *C*	*F* × *S* × *C*	Response
1	–	–	–	+	+	+	–	10
2	–	–	+	+	–	–	+	4
3	–	+	–	–	+	–	+	6
4	–	+	+	–	–	+	–	2
5	+	–	–	–	–	+	+	7
6	+	–	+	–	+	–	–	6
7	+	+	–	+	–	–	–	6
8	+	+	+	+	+	+	+	3

To calculate the effect of the interaction between factors F and S, first find $F \times S_+$ by averaging the results of the runs that have a "+" in the $F \times S$ column:

$$F \times S_+ = (10 + 4 + 6 + 3) \div 4 = 5.75$$

$$\text{Similarly, } F \times S_- = (6 + 2 + 7 + 6) \div 4 = 5.25$$

$$\text{The effect of the } F \times S \text{ interaction is } 5.75 - 5.25 = 0.50$$

$$\text{Similar calculations show that } F \times C = 1.50,\ S \times C = 0, \text{ and } F \times S \times C = -1$$

Interactions may be plotted in a manner similar to main effects, as indicated in Figure 6.11.

The presence of interactions indicates that the main effects aren't additive.

Now, suppose the experiment, as set up in the previous example, is considered too expensive, and the team must reduce costs. They can either reduce the number of replications for each run or reduce the number of runs by using what is called a *fractional factorial design*. It will be shown later that reducing the number of replications reduces the precision of the estimate of experimental error. So, the team decides to use a fractional factorial. They might choose the one illustrated in Table 6.7. This design uses only four of the eight possible runs; therefore, the experiment itself will consume only half the resources as the one shown in Table 6.6. It still has three factors at two levels each. It is traditional to call this a 2^{3-1} *design* because it has two levels and three factors but only $2^{3-1} = 2^2 = 4$ runs. It is also called a *half fraction of the full factorial* because it has half the number of runs as in the 2^3 full-factorial design.

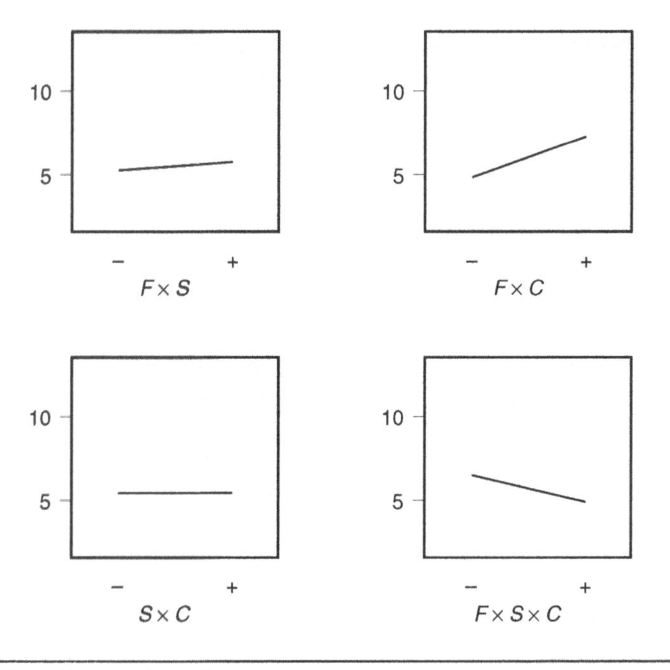

Figure 6.11 A plot of interaction effects.

Table 6.7 Half fraction of 2^3 (also called a 2^{3-1} design).

Run #	A	B	C
1	–	–	+
2	–	+	–
3	+	–	–
4	+	+	+

Table 6.8 Half fraction of 2^3 design with interaction columns to be filled in by the reader.

Run #	A	B	C	A × B	A × C	B × C	A × B × C
1	–	–	+				
2	–	+	–				
3	+	–	–				
4	+	+	+				

Balanced Designs

In Table 6.7, note that when factor A is at its low level in runs 1 and 2, factor B is tested once at its low level and once at its high level, and factor C is also tested once at each level. Furthermore, when factor A is at its high level in runs 3 and 4, factor B is tested once at its low level and once at its high level, and factor C is also tested once at each level. Likewise, when factor B is at its low level in runs 1 and 3, factor A is tested once at its low level and once at its high level, and factor C is also tested once at each level. And when factor C is at its low level in runs 2 and 3, factor B is tested once at its low level and once at its high level, and factor A is also tested once at each level. An experimental design is called *balanced* when each setting of each factor appears the same number of times with each setting of every other factor. The fractional factorial design in Table 6.7 is balanced.

The logical next question is, "Why use a full-factorial design when a fractional design uses a fraction of the resources?" To see the answer, add a column to the design for the A × B interaction, as shown in Table 6.8, and fill it using the multiplication rule.

Note that the A × B interaction column has the same configuration as the C column. Isn't that scary? This means that when the C main effect is calculated, it is not clear whether the effect is due to factor C or the interaction between A × B or, more likely, a combination of these two causes. Statisticians say that the main effect C is *confounded* with the interaction effect A × B. This confounding is the principal price the experimenter pays for the reduction in resource requirements of a fractional factorial. This is the source of much of the controversy about the fractional factorial methods advocated by Taguchi and others. It is interesting to calculate the A × C and B × C interactions. More fright! So when is it safe to use fractional factorial designs? Suppose the team has completed a number of full-factorial designs and determined that factors A, B, and C do not interact significantly in the ranges involved. Then there would be no significant confounding, and the fractional factorial would be an appropriate design.

Resolution

Table 6.9 shows a full-factorial 2^4 (two levels for four factors) design. Of course, the number of runs is $n = 2^4 = 16$.

Table 6.10 illustrates a half fraction of the 2^4 full-factorial design with interaction columns added. This half fraction was carefully selected to minimize the confounding of main effects with two-factor interactions.

Note that there are six two-factor interactions, four three-factor interactions, and one four-factor interaction. Also note that factor A is confounded with the BCD interaction because they have the same +/– pattern (although the A column has [–] signs where the BCD column has [+] signs, and vice versa).

Table 6.9 A 2^4 full-factorial design.

Run #	A	B	C	D
1	–	–	–	–
2	–	–	–	+
3	–	–	+	–
4	–	–	+	+
5	–	+	–	–
6	–	+	–	+
7	–	+	+	–
8	–	+	+	+
9	+	–	–	–
10	+	–	–	+
11	+	–	+	–
12	+	–	+	+
13	+	+	–	–
14	+	+	–	+
15	+	+	+	–
16	+	+	+	+

Table 6.10 A 2^{4-1} fractional factorial design with interactions.

Run #	A	B	C	D	AB	AC	AD	BC	BD	CD	ABC	BCD	ACD	ABD	ABCD
1	+	–	–	–	–	–	–	+	+	+	+	–	+	+	–
2	+	+	+	–	+	+	–	+	–	–	+	–	–	–	–
3	–	+	–	–	–	+	+	–	–	+	+	+	–	+	–
4	–	–	+	–	+	–	+	–	+	–	+	+	+	–	–
5	–	–	–	+	+	+	–	+	–	–	–	+	+	+	–
6	–	+	+	+	–	–	–	+	+	+	–	+	–	–	–
7	+	–	+	+	–	+	+	–	–	+	–	–	+	–	–
8	+	+	–	+	+	–	+	–	+	–	–	–	–	+	–

Similarly, factor B is confounded with the ACD interaction, factor C with the ABD interaction, and factor D with interaction ABC. The big advantage of this particular fractional factorial is that, although there is confounding, main effects are confounded with three-factor interactions only. Since three-factor interactions are often small, the confounding of main effects will usually be minor. Of course, if a three-factor interaction is significant, as it sometimes is, especially in chemical and metallurgical reactions, it will be missed with this design. Another downside of this design is that two-factor interactions are confounded with each other (AB with CD, and so on). This means that an accurate picture of two-factor interactions will not be possible. Fractional factorial designs fall into three categories:

1. *Resolution III* designs have main effects confounded with two-factor interactions.
2. *Resolution IV* designs have main effects confounded with three-factor interactions, and two-factor interactions confounded with each other. The example in Table 6.10 is a resolution IV design.
3. *Resolution V* designs have some two-factor interactions confounded with three-factor interactions, and some main effects confounded with four-factor interactions.

Recall that full-factorial designs have no confounding.

One-Factor Experiments

A process can be run at 180 °F, 200 °F, or 220 °F. Does the temperature significantly affect the product's moisture content? To answer the question, the experimenting team decided to produce four batches at each of the temperatures. The 12 tests could be completely randomized by numbering them from 1 to 12 and randomizing the twelve numbers to obtain the order in which the tests are to be run. Several experimental designs are possible. The least expensive design to execute would be to run the four 180-degree batches and then the four 200-degree batches, followed by the four 220-degree batches. This would reduce the wait time for oven cool-down that more-random designs have, but would tend to confound any time-of-day effects with temperature effects. A chart showing testing order would look like the following table, where test #1 is done first, and so on:

Temperature, °F		
180	200	220
#1	#5	#9
#2	#6	#10
#3	#7	#11
#4	#8	#12

A completely randomized design would have a chart like the following to show the testing order, where test #1 is done first, and so on:

Temperature, °F		
180	200	220
#3	#11	#8
#7	#5	#1
#12	#9	#2
#6	#4	#10

If the team decided to produce one batch at each temperature each day for four days, they would randomize the order of the temperatures each day, thus using a randomized block design. The test order chart could look like the following:

	Temperature, °F		
Day	180	200	220
1	#3	#1	#2
2	#1	#3	#2
3	#1	#2	#3
4	#2	#1	#3

The team might decide to block for two noise variables: the day the test was performed and the machine the test was performed on. In this case, a Latin square design could be used. However, these designs require that the number of levels of each of the noise factors is equal to the number of treatments. Since they have decided to test at three temperatures, they must use three days and three machines. This design is shown in Table 6.11.

Assume that the team decides on the completely randomized design and runs the 12 tests with the following results:

Temperature, °F		
180	200	220
10.8	11.4	14.3
10.4	11.9	12.6
11.2	11.6	13.0
9.9	12.0	14.2

The averages of the three columns are 10.6, 11.7, and 13.5, respectively.

A scatter plot of these data is shown in Figure 6.12.

The graph suggests that an increase in temperature does cause an increase in moisture content. The vertical spread of the dots on each temperature raises some concern. If the dots were spread vertically too much, the within-treatment noise would shed doubt on the conclusion. How much spread is too much? That question is best answered by using ANOVA procedures.

A 2^2 full-factorial experiment has two factors, with two levels for each factor. For instance, to help determine the effect that acidity and bromine have on

Table 6.11 Latin square design.

Day	Machine #1	Machine #2	Machine #3
1	180	200	220
2	200	220	180
3	220	180	200

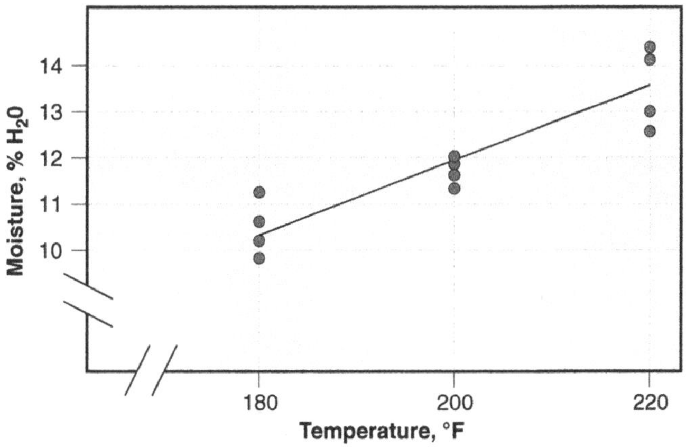

Figure 6.12 Dot plot of data. The heavy line connects the averages for each temperature.

nitrogen oxide (NOx) emissions, two levels of acidity and two levels of bromine are established. The two levels are denoted – and + in each case. One scheme for listing the $2^2 = 4$ combinations is

Run	A	B
1	–	–
2	–	+
3	+	–
4	+	+

That is, run #1 consists of measuring NOx emissions with acidity and bromine both at low levels. As stated earlier, it is important to repeat (or replicate) each run several times to get a handle on experimental error. If the measured results of the replications of a particular run are not consistent, a larger experimental error is indicated. The concept of experimental error is quantified by using ANOVA. Assume in this case that each run will be replicated three times. Once again, it is important to randomize the order in which the 12 tests are done. Following is an example of a completely randomized experimental design:

Run	A	B	Replicates		
1	–	–	11	8	5
2	–	+	2	6	4
3	+	–	7	10	9
4	+	+	1	3	12

The number 1 in the last row indicates that for the first test, acidity and bromine are both set at their high levels and the resultant NOx emission is measured.

Full-Factorial Experiments

A 2^2 full-factorial completely randomized experiment is conducted, with the results shown in Table 6.12.

The first step is to find the mean response for each run and calculate the interaction column as shown in Table 6.13.

The main effect of factor A is $(24.7 + 37.3) \div 2 - (28.4 + 33) \div 2 = 0.3$

The main effect of factor B is $(33.0 + 37.3) \div 2 - (28.4 + 24.7) \div 2 = 8.45$

The interaction effect $A \times B$ is $(28.4 + 37.3) \div 2 - (33.0 + 24.7) \div 2 = 4.0$

The next issue is whether these effects are statistically significant or merely the result of experimental error. The larger the effect, the more likely that it is significant. Intuitively, it appears that factor B may be significant and factor A probably

Table 6.12 A 2^2 full-factorial completely randomized experiment with results.

Run #	A	B	Response, y		
1	–	–	28.3	28.6	28.2
2	–	+	33.5	32.7	32.9
3	+	–	24.6	24.6	24.8
4	+	+	37.2	37.6	37.0

Table 6.13 A 2^2 full-factorial completely randomized experiment with results and mean results shown.

Run #	A	B	A × B	Response, y			$\bar{y}$
1	–	–	+	28.3	28.6	28.2	28.4
2	–	+	–	33.5	32.7	32.9	33.0
3	+	–	–	24.6	24.6	24.8	24.7
4	+	+	+	37.2	37.6	37.0	37.3

isn't. It's not too clear whether the interaction A × B is significant. The definitive answer to the question can be found by conducting a two-way ANOVA on the data. The calculations are quite cumbersome, so software packages are often employed. The following illustrates the use of MS Excel. Excel requires the data in a slightly different format, as shown in Table 6.14. The values obtained when factor A is at its low level are shown in the top two boxes. The values obtained when factor A is at its high level are shown in the bottom two boxes. The values obtained when factor B is at its low level are shown in the left two boxes. So, the values obtained when A is low and B is low are in the top left corner box.

These numbers and labels are put in the first rows and columns of the spreadsheet, and the two-way ANOVA function is invoked. The result is illustrated in Table 6.15.

The printout from the ANOVA function contains additional information, but the portion shown in Table 6.15 provides the statistical significance results. The P-value column contains information relating to the statistical significance of the factors. P is the probability that the source of variation is not significant. Using a hypothesis test model, the null hypothesis would be that the source of variation is not statistically significant. Assuming $\alpha = 0.05$, the null hypothesis should be

Table 6.14 Format for entering data into an Excel spreadsheet in preparation for two-way ANOVA.

		B –	B +
A	–	28.3 28.6 28.2	33.5 32.7 32.9
A	+	24.6 24.6 24.8	37.2 37.6 37.0

Table 6.15 ANOVA printout from Microsoft Excel. ANOVA is invoked through the Tools>Data analysis menus.

ANOVA

Source of variation	SS	df	MS	*F*	P-value	*F* criteria
Acidity	0.213333	1	0.213333	2.639175	0.142912	5.317645
Bromine	223.6033	1	223.6033	2766.227	1.89E-11	5.317645
Interaction	47.20333	1	47.20333	583.9588	9.17E-09	5.317645
Within	0.646667	8	0.080833			
Total	271.6667	11				

rejected if $P \leq 0.05$. In this case, the P-values for the sources of variation labeled "bromine" and "interaction" are small enough to reject the null hypothesis and declare these factors—B and the A × B interaction—to be statistically significant at the 0.05 significance level. The row labeled "within" in Table 6.15 calculates what is known as within-treatment variation, that is, the variation that occurs between replicates within the same run. The value of 0.080833 shown in the MS column of the "within" row is the estimate of the variance of these replicate values. If replication hadn't been used, this number, which indicates the size of the experimental error, would not have been available. What the ANOVA test really does is compare the between-treatment variation with the within-treatment variation. The *F*-statistic shown in the column labeled "*F*" in Table 6.15 is obtained by dividing the corresponding MS value by the within-MS value. The first *F*-ratio, about 2.6, shows that the variation due to changing the levels of factor A is about 2.6 times the within-treatment error.

Two-Level Fractional Factorial Experiments

The full-factorial experiments described in the previous section require a large number of runs, especially if several factors or several levels are involved. Recall that the formula for number of runs in a full-factorial experiment is

$$\text{Number of runs} = L^F$$

where

L = Number of levels

F = Number of factors

For example, an experiment with eight two-level factors has $2^8 = 256$ runs, and an experiment with five three-level factors has $3^5 = 243$ runs. If runs are replicated, the number of tests will be multiples of these values. If the experiment is testing the effect of various agricultural factors on crop production, a plot of ground is divided into the required number of subplots, and all runs and replicates can be conducted simultaneously. For instance, a 243-run experiment with four replications of each run would require 972 plots of ground. If a total of an acre is available for experimentation, each plot could be approximately 45 ft^2.

If, however, the experiment is testing the effect of various factors on product quality in a manufacturing process, the tests typically must be run sequentially rather than simultaneously. A full-factorial experiment with several factors and/or levels may require the piece of production equipment to be taken out of production for a considerable amount of time. Because of the extensive resource requirements of full-factorial experiments, fractional factorial experimental designs were developed.

Robustness Concepts

Robustness means resistance to the effect of variation of some factor. For example, if brand A chocolate bar is very soft at 100 °F and brittle at 40 °F, and brand B maintains the same level of hardness at these temperature extremes, it could be said that brand B is more robust to temperature changes in this range. If a painting process produces the same color on moist wood as on dry wood, the color is robust to variation in moisture content. The changes in temperature and humidity are referred to as *noise*. Products and processes that are robust to noise of various kinds are clearly desirable. The Japanese engineer Genichi Taguchi is credited with developing techniques for improving robustness of products and processes.

One approach to improving robustness is illustrated in Table 6.16.

As usual, the average value for each run is calculated and is labeled $\bar{y}$. In addition, the standard deviation of the values in the run is calculated and shown in the column labeled "S."

Now the experimenter can complete the usual main effects calculations to determine the levels of each of the factors that will optimize the response value y. In addition, the main effects calculations can be run using the values in the S column to find the levels of each of the factors that will minimize the S value. If these two combinations of levels do not agree, then a compromise between optimizing the response and minimizing the variation must be made. One way to approach the compromising process is through what Taguchi called the *signal-to-noise* (S/N) *ratio*. If it is desirable to maximize y, the signal-to-noise ratio may be calculated for each run, using

$$S/N = \frac{\bar{y}}{S}$$

The main effects may then be calculated using the S/N ratios to find the best levels for each factor. If, instead, it is desirable to make y as small as possible, the S/N ratio can be defined as

$$S/N = \frac{1}{\bar{y}S}$$

Table 6.16 Robustness example using signal-to-noise ratio.

A	B	C	Replications				$\bar{y}$	S
–	–	–	34	29	38	25	31.5	5.7
–	–	+	42	47	39	38	41.5	4.0
–	+	–	54	41	48	43	46.5	5.8
–	+	+	35	31	32	34	33.0	1.8
+	–	–	62	68	63	69	65.5	3.5
+	–	+	25	33	36	21	28.8	6.9
+	+	–	58	54	58	60	57.5	2.5
+	+	+	39	35	42	45	40.3	4.3

If it is desirable to make y as close to some nominal value N as possible, the S/N ratio can be defined as

$$S/N = \frac{1}{|\bar{y} - N|S}$$

Note that the S/N ratio is an attempt to find a useful compromise between two competing goals, optimizing y and minimizing S. It does not necessarily accomplish either of these goals, so it should be used with a bit of judgment.

Another technique Taguchi used for improving robustness is called the inner/outer array design. In this procedure, the uncontrolled factors—those factors that the experimenter either can not or chooses not to control—are placed in separate columns next to the controllable factors, as shown in Table 6.17.

In this example, hardness of the steel and the ambient temperature are the uncontrolled factors. These factors could conceivably be controlled by putting the machine in an environmental enclosure and putting a tighter specification on the steel, but the experimenter chooses not to do either of these. Instead, anticipated extremes of hardness and ambient temperature are used for the experiment to determine settings of the controllable variables that will minimize variation in the output quality characteristic.

When the first run of the design in Table 6.17 is executed, the feed, speed, and coolant temperature are all set at their low levels. One part is made with low-hardness steel and low ambient temperature, and the value of the quality characteristic is entered in the spot labeled "a." When all 32 values have been entered, the averages and standard deviations are calculated. In this inner/outer array approach, the design intentionally causes perturbations in the uncontrollable factors to find level combinations for the controllable factors that will minimize

Table 6.17 Illustration of inner and outer arrays.

Inner array			Outer array						
			Hardness	–	–	+	+		
Feed	Speed	Coolant temp.	Ambient temp.	–	+	–	+	$\bar{y}$	S
–	–	–		a					
–	–	+							
–	+	–							
–	+	+							
+	–	–							
+	–	+							
+	+	–							
+	+	+							

Part III.A.7

the variation in the quality characteristics under the anticipated hardness and ambient temperature variation. One might ask why hardness and ambient temperature are not merely added to the factor list, making five factors at two levels, which would require $2^5 = 32$ tests, exactly the number required in this example. That approach would establish the best settings for hardness and ambient temperature. But it would also require a tighter spec on hardness and ambient temperature. Instead, the inner/outer array design determines optimum levels for the control factors in the presence of variation in the outer array factors.

8. FAULT TOLERANCE

> Define and describe fault tolerance and the reliability methods used to maintain system functionality. (Understand)
>
> **Body of Knowledge III.A.8**

Fault tolerance refers to the ability of a system to perform according to specifications even when undesired changes occur in its internal or external environment. The reliability engineer's task here is to use knowledge of system design to prevent a fault from causing a failure. Single-fault tolerance can be effected by a system design that will permit a component (hardware or software) to be removed and replaced without causing system failure. Multiple-fault tolerant systems are designed to continue operating in the presence of more than one concurrent failure.

Various techniques have been employed to produce fault-tolerant systems.

Redundancy

Redundancy is a standard technique for mitigation of system failure. The theory is that if two or more parallel channels are available for the system, the likelihood of failure is reduced. Redundancy alone is often insufficient because the common cause variation may overwhelm all branches of the system.

Examples:

- The catastrophic failure that caused United flight 232 to attempt a landing at Sioux City in 1989. The DC-10 was designed with three independent hydraulic systems so that if one or two failed, the surviving system could manage flight control surfaces. Unfortunately all three lines were at one point in close proximity. Shrapnel from a damaged engine passed through that point and severed all three lines, causing loss of control.
- The use of parallel CPUs in computers performing critical tasks. Some systems are set up to process every command by two different

processors and compare the results. If the results do not match, an error flag is set, and the system is put in a safe mode and maintenance is performed. A problem here is that if the software is defective, the two results may agree but be erroneous. The common cause in this case is the software bug. It caused errors in both branches of the parallel design. A variant on this approach is to have simultaneous calculations performed by several processors. When the results are compared, a *voter* module compares the results and goes with the one produced by the majority of the processors.

- Improper maintenance of all branches of a parallel system.
- Use of a process material that corrodes all redundant branches.
- Computer backup systems using identical disk drives in an identical environment (which often results in identical failures).

Diversity and separation are two techniques that have been used to make redundancy more effective.

Diversity is employed to construct parallel channels different in ways that will make them immune to failure by the same cause. This implies an understanding of the common causes—an understanding that will only come from considerable study. In the case of the parallel CPUs mentioned above, a different piece of software could be employed to make the calculation for one of the CPUs.

Separation can apply to branches that have separate physical environments, electronic isolation, or other means of preventing a common cause from causing multiple-branch failure. In the case of the DC-10 mentioned earlier, the hydraulic lines were not adequately separated.

9. RELIABILITY OPTIMIZATION

> Use various approaches, including redundancy, derating, trade studies, etc., to optimize reliability within the constraints of cost, schedule, weight design requirements, etc. (Apply)
>
> **Body of Knowledge III.A.9**

The reliability engineer involved in the design process will frequently be doing some sort of balancing act. The standard ways to improve reliability, redundancy (as discussed in Section 8 of this chapter), derating (as discussed in Section 3 of Chapter 7), and environmental robustness (as discussed in Section 7 of this chapter) tend to increase cost and weight, and delay the design process. The prioritization matrix discussed in Example 6.8 can aid in this balancing act.

EXAMPLE 6.8

A design team is charged with optimization of the motor/drive train for a commercial air handler. The reliability engineer is considering several options, which are listed in the left column of the matrix in Figure 6.13. The other columns are self-explanatory, and additional columns may be necessary to cover additional constraints. Once the matrix has been constructed, the decisions become a little more quantitative. In this case the team might decide that the hardened motor can be sold as an alternative but not be included in the standard product. The additional motor might be eliminated from the product line.

Option	Advantage	Cost ($)	Weight (kg)	Schedule delay (days)
Derate motor	Increase motor life Handle excess load	900	25	0
Derate drive belt	Extend maintenance schedule	100	2	0
Add 2nd drive belt	Allow operation with failed belt	400	10	5
Harden motor	Allow use in caustic environment	2000	95	10
Add 2nd motor	Allow use after motor failure	2000	300	15

Figure 6.13 Prioritization matrix.

10. HUMAN FACTORS

Describe the relationship between human factors and reliability engineering. (Understand)

Body of Knowledge III.A.10

Reliability testing must consider variation introduced by differences between people. These differences can be categorized as usage factors, installation factors, and process management factors.

Use Factors

When conducting tests on a home laundry appliance, consideration must be given to the differences in the way people slam doors, choose washing detergent, maintain cleanliness, load the washer, and so on. Automotive components also experience a wide variety of operator habits. If the testing protocol calls for 100,000 door slams using a particular force, it has not accommodated all the variation that customers may employ.

Installation Factors

Products may be installed in a wide variety of environments. For building materials, automotive spare parts, stationary equipment, and many other products, the useful lifetime is impacted by the quality of the installation and the location parameters. For instance, if a window is installed slightly out of plumb, it may not be as reliable as a properly installed unit. This has implications for design and testing procedures because the window should be robust to minor out-of-plumbness. Knowledge in this area also will impact installation instructions.

Process Management Factors

If reliability testing is conducted on products that are produced by an ideal process, the variation introduced by people in the production process may not be taken into consideration. Operators or process managers can influence product characteristics by the way they hold the paint gun or welding stick, the promptness with which a tank is evacuated, the delay between molding cycles, and so on.

Recommendations

From the testing viewpoint, it would be easier if all this human variation could be removed by specifying usage, installation, and processing parameters, and in fact that is typically done. A balance must be struck, of course, between the narrowness of the specifications and the attendant loss in flexibility. If, for instance, an installation specification states that the ambient air must be between 65 °F and 75 °F, the testing will be simplified, but some customers will be lost, and others will ignore the specification and be disappointed with product performance. Therefore, the product design team should strive for robustness to the variation introduced by these human factors, keeping the specifications as broad as possible. This means that the reliability testing procedures must include tests throughout the specification spectrum.

11. DESIGN FOR X (DFX)

> Apply DFX techniques such as design for assembly, testability, maintainability, environment (recycling and disposal), etc., to enhance a product's producibility and serviceability. (Apply)
>
> **Body of Knowledge III.A.11**

Every design team has constraints within which they must function. The items listed in the Body of Knowledge excerpt above constitute design constraints. The relative importance of each of these characteristics has an impact on the resulting design:

- *Design for assembly* refers to the conscious thought given to the assembly process by the product design team. Decisions by the team affect the simplicity and ease of the assembly process by giving consideration to visual and mechanical access to fasteners, special tooling or fixturing that may be required, safeguards against incorrect assembly, and so forth.
- *Design for manufacturability* is similar to design for assembly but tends to put more emphasis on the fabrication or primary functions that precede assembly. Here, consideration is given to narrowness of tolerances and difficult configurations such as deep dead-end holes, thin walls, and so on.
- *Design for testability* looks at testing procedures and tries to develop a design for which important characteristics are easily and accurately measured.
- *Design for cost* puts emphasis on final cost of the product and makes this a strong constraint on product or process design.
- *Design for serviceability* considers the ease and simplicity of installing replacement parts, as well as standard servicing requirements.
- *Design for reliability* emphasizes the long-term usefulness of the product to the customer.
- *Design for environment* considers the activity beyond the useful life of the product. While considering the various trade-offs in the design process, it is becoming increasingly important to study issues of recycling, reclamation, and disposal. Issues to take into account include contractual obligations/warranties and regulatory and certification requirements (the 2015 version of ISO 9001 may have a section on this topic).

Many of these constraints are applied to typical designs, and often the extent of the constraint is a matter of emphasis.

12. RELIABILITY APPORTIONMENT (ALLOCATION) TECHNIQUES

> Use these techniques to specify subsystem and component reliability requirements. (Analyze)
>
> **Body of Knowledge III.A.12**

Reliability requirements are usually specified for the system level. The requirement might be a failure rate or an MTBF. *Reliability apportionment* is a technique used to allocate the system-level reliability requirement to the various subsystems. Each subsystem can then allocate reliability requirements to each of the various components that comprise the subsystem. If each component achieves its allocated reliability requirement, the subsystem will meet its requirement. And if each subsystem achieves its allocated requirement, the final system will meet the system-level requirement. Reliability apportionment should begin as soon as possible in the design process. Reliability apportionment can begin as soon as a preliminary design and engineering drawings are available.

The allocation program forces the design team to understand the relationship between component, subsystem, and system reliability requirements. If reliability is made a characteristic of the design, it will be given the same consideration as other characteristics such as power consumption, weight, or performance. The allocation process helps to ensure that an adequate effort is made to design reliability into the system.

An essential component of the apportionment/allocation process is a system reliability model or block diagram. The system reliability requirement is then allocated to each subsystem. The method used to allocate could depend on the stage of development of the design.

In order to involve reliability engineering in the early stage of development, it might be appropriate to assign to each subsystem an equal part of the system reliability requirement. This method is known as *equal allocation*. As the design becomes more mature and more information about the design is available, an allocation process that takes into account the complexity of each subsystem along with prior information on the various subsystems should be used. One method of weighted allocation known as the *ARINC allocation method* can be used. This method, developed by the ARINC Research Corporation, a subsidiary of Aeronautical Radio, Inc., allocates individual subsystem reliability requirements based on the predicted attained reliability of all the subsystems.

The method assumes a series reliability model and exponential times to failure for each subsystem. Using these assumptions, the system failure rate is the sum of the subsystem failure rates. Predicted failure rate values (λ) of each

EXAMPLE 6.9

A hydraulic flow pressure reducer has the reliability model shown in Figure 6.14.

The reliability requirement for the system is a failure rate equal to 20 failures per million hours:

$$\lambda^* = 20/10^6 \text{ hours}$$

Using historical information, information from similar designs, and information from the various subsystem vendors, the predicted failure rates of the subsystems are

$$\lambda_1 = 6.8/10^6 \text{ hours}$$

$$\lambda_2 = 5.4/10^6 \text{ hours}$$

$$\lambda_3 = 14.3/10^6 \text{ hours}$$

$$\lambda_4 = 3.5/10^6 \text{ hours.}$$

The attained failure rate of the system is

$$\lambda_{\text{System}} = \Sigma\lambda_i = (6.8 + 5.4 + 14.3 + 3.5)\ 10^{-6} \text{ hours}$$

$$= 30/10^6 \text{ hours.}$$

The attained system failure rate is greater than the required system failure rate. Reliability allocation is necessary.

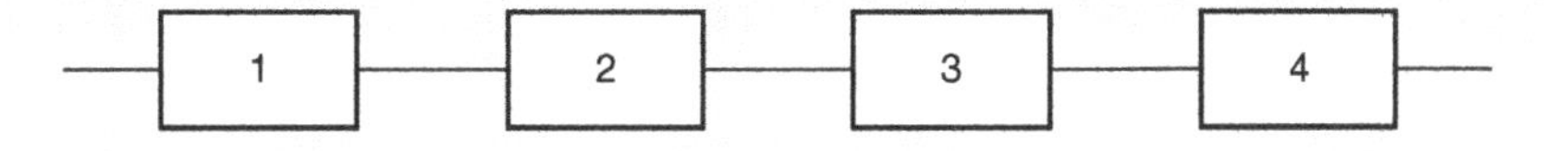

Figure 6.14 Series reliability model for allocation.

Part III.A.12

subsystem are used to predict the system failure rate. The predicted system failure rate is compared to the system requirement (λ^*). If the predicted value exceeds the requirement ($\Sigma\lambda > \lambda_i^*$), allocation becomes necessary.

For the ith subsystem a weighted allocation can be determined:

$$\lambda_i^* = \frac{\lambda_i}{\Sigma\lambda_i}(\lambda^*)$$

The following example illustrates the use of both equal allocation and the ARINC allocation method.

Equal Allocation

If equal allocation is used, the requirement for each subsystem will be

$$\lambda i = \lambda^*/4 = 5/10^6 \text{ hours.}$$

This allocation method does not consider any differences in the complexity of the subsystems, any differences in the stresses each subsystem will experience, or the design maturity of the various subsystems.

This method of allocation gives no incentive to improve the reliability of subsystem 4, and perhaps imposes an unattainable goal on subsystem 3.

ARINC Allocation

The requirement for each subsystem is allocated based on the weighted values

$$\lambda_1^* = [6.8/30]\ 20/10^6 = 4.53/10^6 \text{ hours}$$

$$\lambda_2^* = [5.4/30]\ 20/10^6 = 3.60/10^6 \text{ hours}$$

$$\lambda_3^* = [14.3/30]\ 20/10^6 = 9.53/10^6 \text{ hours}$$

$$\lambda_4^* = [3.5/30]\ 20/10^6 = 2.33/10^6 \text{ hours.}$$

This method is fair and gives each subsystem a goal for reliability improvement.

The allocation is not final. It is possible that the subsystem 3 failure rate could be substantially reduced. It is often possible to reduce a high failure rate easier than a low failure rate. As additional information becomes available, a reallocation should occur.

Chapter 7

B. Parts and Materials Management

1. SELECTION, STANDARDIZATION, AND REUSE

> Apply techniques for materials selection, parts standardization and reduction, parallel modeling, software reuse, including commercial off-the-shelf (COTS) software, etc. (Apply)
>
> Body of Knowledge III.B.1

Parts standardization refers to the use of the same components in several products. An example in the automotive industry is the use of the same chassis or engine for several automobile models. In some families of electronic products the same circuit board may be used, with different functions activated depending on the application. A *parts reduction* initiative makes an effort to reduce the number of parts required to perform a function. If, for instance, a two-bar linkage can be used to replace a three-bar linkage, there will be an associated simplification in parts, assembly, and maintenance. When product families are essentially the same except for size, they are sometimes referred to as *parallel models.* A valve producer, for instance, may offer essentially the same geometry in valves for pipes ranging from one inch to ten inches. During the fabrication process, if parts are maintained in a state where they may be used in several models and then customized as they are needed, the *raw as possible* (RAP) principle is being used. Example: A dome-shaped sheet metal part requires two, three, or four triangular holes depending on the product. Rather than producing an inventory of each type, it might be useful to use one die to cut and form the dome shape and put a small press on the assembly line to punch the triangular holes as needed.

If a module of software can be structured so that it is useable in more than one application, the costs of generating and testing the module are reduced.

There are obvious cost advantages associated with these techniques in terms of inventory, purchasing efficiencies, storage, accounting, and auditing. Reduced costs of reliability testing is an often overlooked advantage. If the function of three very similar items can be handled by a single item, the costs for reliability testing may be reduced by 67 percent. And if that item is one that has already been tested, further savings are realized.

These testing cost savings can be very substantial, so design teams are well advised to look to families of parts and consider potential standardization.

During product design it is important to consider the variation in materials. Probabilistic methods permit the design team to produce products that are robust to this variation. This is accomplished by constructing distribution models that describe the variability in materials.

EXAMPLE 7.1

A design team is selecting building materials for a set of livestock shelters and is considering the use of a clip that slips over two two-by-fours.

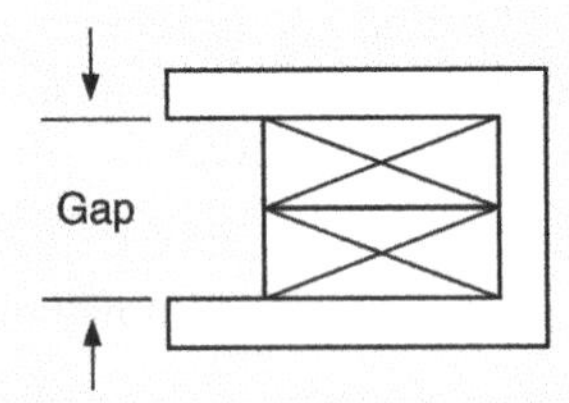

The failure mode of concern occurs if the gap is so narrow that the two-by-fours won't fit. The nominal thickness of the two-by-fours is 1.5 inches, and the gap is 3.010 ± .005. What percent of the clips will be too narrow?

Solution:

The team finds that the thickness of the two-by-fours to be used has $\mu_{2\times4} = 1.500$ and $\sigma_{2\times4} = 0.001$ and is normally distributed. The gap in the proposed clip has $\mu_{Gap} = 3.011$ and $\sigma_{Gap} = 0.002$ and is normally distributed. The pair of two-by-fours has a total thickness that is normally distributed, with $\mu_{Pair} = 3.000$ and

$$\sigma_{Pair} = \sqrt{\sigma_{2\times4}^2 + \sigma_{2\times4}^2} \approx .0014.$$

The clearance between the clip and the pair of two-by-fours is normally distributed, with

$$\mu_{Clear} = \mu_{Gap} - \mu_{Pair} = 3.011 - 3.000 = .011$$

$$\sigma_{Clear} = \sqrt{\sigma_{2\times4}^2 + \sigma_{2\times4}^2 + \sigma_{Gap}^2} \approx .0024.$$

Then,

$$z = \frac{\mu_{Clear}}{\sigma_{Clear}} \approx .46.$$

From a normal distribution table, about 32 percent of the clips will be too narrow. The design team returns to the drawing board.

In some situations the distributions of the materials are not known, and software packages may be used to fit sample data to distribution models.

Software Reuse

Software packages are generated from specifications. These specs are used to test and verify that the program performs correctly. Extreme caution must be exercised when attempting to use a program for an application not in the specifications used to produce the program. An effort to modify a program that has been written and tested runs the serious risk of undermining its capability to satisfactorily perform the originally specified tasks. Most IT professionals are familiar with examples such as that of the process control program modified to perform accounting functions, which was then unable to do either. The advice here is that software should be reused only if the new use is within its tested capability. Particular caution is advised with commercial off-the-shelf (COTS) software, which typically can't be modified. In some cases configuration is permitted, but software specifications are usually not provided.

2. DERATING METHODS AND PRINCIPLES

Use methods such as S-N diagram, stress-life relationship, etc., to determine the relationship between applied stress and rated value, and to improve design. (Analyze)

Body of Knowledge III.B.2

In some fields, especially electronics, standard ratings are available for components. These include stress limits, environmental conditions, and other characteristics. The rating is the maximum level for stresses such as temperature, voltage, pressure, and so on. Operating beyond the rated value risks part damage or failure. Some common standards are MIL-HDBK-217, Bellcore (SR-22), NSWC-98(LEI), China 299B, and RDF 2000. *Derating* is the practice of operating components at lower stress levels than those specified by the standards. It usually refers to limiting electrical, thermal, or mechanical stresses. In general, the more a part is derated, the lower its failure rate. The extent to which derating impacts failure rate is technology dependent. Paradoxically, derating is also used to permit operating at a higher stress level. For instance, rather than install a more powerful, noisier fan, an electronics designer may choose to use a component with higher thermal rating. To avoid excessive costs, derating should be done selectively rather than using a mass indiscriminate derating scheme.

Underclocking of synchronous electronic devices can be thought of as another form of derating. This refers to running the device, such as a CPU, at a slower than designed speed. This may decrease the rate of heat generation, therefore prolonging life and/or decreasing the ventilation or heat sinking requirements.

Stress-Life Relationships

The analysis of reliability data can be used to generate a life distribution. This distribution is typically defined for a given stress level. In situations where the stress may change, another dimension is added to the life distribution. This is depicted in Figure 7.1, which illustrates the fact that there may be a different life distribution for different stress levels.

3. PARTS OBSOLESCENCE MANAGEMENT

> Explain the implications of parts obsolescence and requirements for parts or system requalification. Develop risk mitigation plans such as lifetime buy, backwards compatibility, etc. (Apply)
>
> Body of Knowledge III.B.3

Product improvement often means redesigning components. The integration of new components requires an orderly procedure if genuine improvement is to occur. The relatively short product lifetimes in areas such as electronics and software has caused increased emphasis on obsolescence management. The following paragraphs identify ways to reduce the risks associated with obsolescence and redesigned components.

Obsolescence from the Producer's Viewpoint

When any component of a system is changed, the result is a new system, which must be tested to verify that it performs as specified. In many cases, qualifying

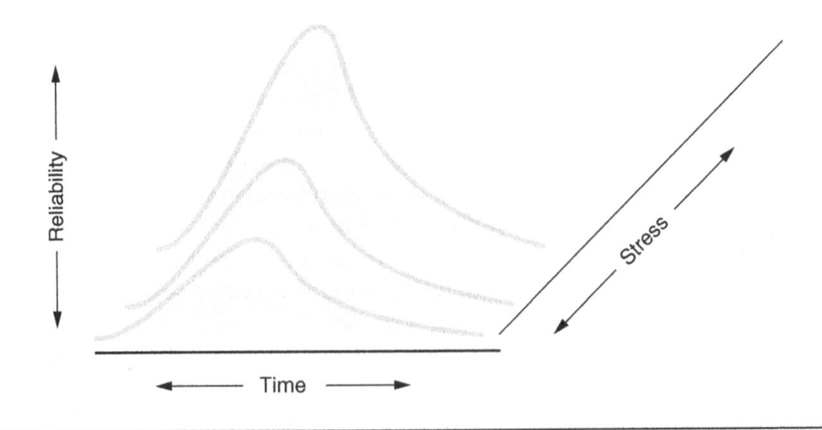

Figure 7.1 Stress-life distributions.

the component independently from the system will miss undesirable interactions with other components. Judgment is required, of course, in deciding how much of the new system must be requalified. Redesign of a pivot pin in a vehicle door latch linkage does not require requalification of an entire vehicle, for example. It is necessary, however, that the new latch system be tested not only on a lab bench but also on vehicles in their designed environment.

Once the decision has been made to make the change, the disposition of obsolete parts must be managed. In some cases the change occurs when the supply of obsolete parts has been exhausted. If not, the inventory of obsolete parts must be segregated to prevent its use in the production process.

The repair and maintenance of parts in the field is another consideration. Will the new design be installed, or are the obsolete parts to be used?

Obsolescence from the Customer's Viewpoint

What strategies can a customer use when an environment of continuous improvement means component changes? The warranty documents the responsibility that the supplier has to the customer. Some provisions that are useful include:

a. Commitment to support the product with replacement parts for a stated period of time.

b. *Lifetime buy*, which is the practice of providing the buyer with enough replacement parts to cover the needs for the lifetime of the system. This option requires a careful forecasting methodology.

c. Backwards compatibility, guaranteeing that new or redesigned components will function within the existing system.

4. ESTABLISHING SPECIFICATIONS

> Develop metrics for reliability, maintainability, and serviceability (e.g., MTBF, MTBR, MTBUMA, service interval) for product specifications. (Create)
>
> **Body of Knowledge III.B.4**

Some Common Reliability Metrics

Failure rate is defined as the number of failures per unit of time. The Greek letter lambda (λ) is the symbol used for failure rate. We will denote failure rate as λ or $\lambda(t)$. $\lambda(t)$ is also called the *hazard function*.

Mean time to failure (MTTF) is defined as the average time elapsed until the product is no longer performing its function. If the item is repairable, the *mean time between failures* (MTBF) is used. MTTF and MTBF are reciprocals of l:

$$\text{MTTF} = \frac{1}{\lambda} \text{ or } \text{MTBF} = \frac{1}{\lambda}$$

Example: If $\lambda = 0.00023$ failures per hour, MTBF $\approx$ 4348 hours.

Reliability R(t) is defined as

$$R(t) = \frac{\text{Number of units functioning at the end of the time period}}{\text{Number of units that were functioning at the start of the test}}$$

BX *life,* or B(X) *life,* is the amount of time that has elapsed when X percent of the population has failed. For example B(10) = 367 hours means R(367) = .90.

Mean time between repairs (MTBR) provides another measure of the reliability of a product. MTBR data should include information about the type of repair and resources required. These data are helpful in determining spare parts inventories, planning for resources, and scheduling preventive maintenance.

Mean time between unplanned maintenance action (MTBUMA) indicates the level of confidence that can be placed in the machine when it is placed in a vital role. If MTBUMA is relatively short, redundant equipment may be needed.

Service interval refers to the recommended time between routine checks and replacements. Familiar examples include lubrication and filter change schedules.

EXAMPLE 7.2

A set of 283 non-repairable units are tested, and the number of failures during each 100 block of time is recorded. The test produced the following data:

Time interval (hrs)	# failures
0–99	0
100–199	2
200–299	10
300–399	30

Calculate $\lambda(t)$, MTBF, and R(t) for each time block.

Solution:

Using the formulas given in the definitions:

Time interval (hrs)	# failures	# surviving	$\lambda(t)$	MTBF (hrs)	R(t)
0–99	0	283	0.0000	Undefined	1.000
100–199	2	281	0.0001	10,000	0.993
200–299	10	271	0.356	28.1	0.958
300–399	30	241	0.1107	9.0	0.852

EXAMPLE 7.3

An automotive company had specified black rubber door seals, giving content, hardness, profile dimensions, and other characteristics. Instead, they now specify reliability requirements such as resistance to UV exposure for specified amounts of time and passing rain tests for specified time periods. The customer has left other details up to the supplier and has found that their research requirements are reduced, the new seal does a better job, and the price is slightly reduced because the rubber supplier designed a product that is also easier for them to manufacture.

Relationship to Product Specifications

It is customary to specify dimensions, weights, carbon content, and so forth, for products. From the reliability engineering standpoint, it would often be more productive to specify one or more of the reliability metrics listed above. This has proven especially useful when specifying purchased parts. This is because suppliers often know more about the characteristics of their products than the customer. Although Example 7.3 is anecdotal and somewhat subjective, it illustrates this point.

Maintainability

Maintainability is a metric usually specified as a probability that a particular maintenance activity can be accomplished in a stated period of time. It is made up of two components: serviceability and repairability. *Serviceability* refers to the level of difficulty encountered when performing the maintenance activity, and *repairability* refers to the level of difficulty when returning a failed item to usefulness. The ultimate goal is to develop a *reliability-centered maintenance* (RCM) program. Steps in that direction include:

1. Develop a functional hierarchy of the machines and their components.
2. Make a complete list of failure modes for the machines.
3. Perform an evaluation of each *preventive maintenance* (PM) activity to determine whether it really aids in preventing failure modes.
4. Reorganize PM activities to align with the information from steps 1–3.

Part IV

Reliability Modeling and Predictions

Chapter 8

A. Reliability Modeling

1. SOURCES AND USES OF RELIABILITY DATA

> Describe sources of reliability data (prototype, development, test, field, warranty, published, etc.) their advantages and limitations, and how the data can be used to measure and enhance product reliability. (Apply)
>
> **Body of Knowledge IV.A.1**

Reliability data are generally available to a manufacturer from several sources, both external and internal to the company. A valuable external source of component reliability data is the supplier. Other users of the same component can also supply reliability data. A comprehensive external data source available to military contractors is the Government–Industry Data Exchange Program (GIDEP). This program generates and distributes reliability data on commercially available off-the-shelf units that are used by the various contractors. Failure data for units such as motors, pumps, relays, and so on, are exchanged through the GIDEP program. Information can be obtained from GIDEP Operations Center, Corona, California. Professional organizations such as the Institute of Electrical and Electronics Engineers (IEEE) develop and maintain reliability data on hardware of various types. The IEEE can be contacted at 3 Park Avenue 17th Floor, New York, NY 10016-5997.

Information on nonelectrical parts (NPRD) can be obtained from the Rome Air Development Center, National Technical Information Service, Springfield, VA.

Many times, internally generated data are given higher credibility; the conditions used to generate the data are known and can be controlled. Capabilities of the manufacturing process, the quality control methods employed, and the specified operation environment of the design, as well as other factors specific to the manufacturer, are reflected in internally generated data. One example of internally generated data is from analysis of prototypes. Early in the design process it is often useful to produce a model of the item and use it for reliability testing. Prototype testing can provide early warning of design flaws, material requirements, and other changes that can improve reliability. Throughout the development process, the reliability engineer can be subjecting proposed parts to preliminary stress studies and other tests. Testing of final products provides another opportunity to collect data for warranty calculations, preventive maintenance, and possible future designs.

Another valuable source of reliability data is the field service facility. Most manufacturers maintain data on the repair and maintenance services provided to their customers. This could be warranty data, data from dealers or distributors, or data from customers or users of the product. Each return or maintenance activity should result in a report. These data should be collected and distributed. Failure analyses performed on failed parts, if they are available, will provide data as to the root cause of the failure. It is a reliability engineering function to collect and distribute field failure data. The data need to be available to reliability engineering, quality engineering, product design, testing, marketing, service, and other appropriate engineering and support functions.

Care must be taken to use the data appropriately. The conditions under which data generated internally are developed will be known. The true test time, the test acceleration numbers if used, the environment of the test, the actual failure mode, and other conditions will be documented. Data that are generated externally might be suspect, as the actual test or use conditions may not be known.

The various sources are summarized in Table 8.1.

Table 8.1 Sources of reliability data.

Source	Comments	Advantages	Limitations	Potential contribution to product reliability
Prototype	Early in design process	Provides visible and testable item	Spending effort on preliminary design	Can head off early design flaws
Development	Throughout development process	Can track reliability changes with each design	Time/resources required	Can head off early design flaws
Final product	Test as soon as available from manufacturing	Early basis for warranty calculations	May not be able to simulate environmental conditions	More extensive testing may flag weaknesses
Field warranty	Need complete data	Customer viewpoint	Data often sketchy	Helps put dollar sign on reliability

2. RELIABILITY BLOCK DIAGRAMS AND MODELS

> Generate and analyze various types of block diagrams and models, including series, parallel, partial redundancy, time-dependent, etc. (Evaluation)
>
> Body of Knowledge IV.A.2

A system can be modeled for reliability analysis using block diagrams. A system consists of subsystems connected to perform given functions. Systems can become complex, making reliability analysis difficult. A math model reduces the system to a graphical representation of the interconnection of its subsystems. The system reliability can then be modeled using the reliability of the various subsystems. There are several advantages to modeling the system, including predicting system reliability using reliability predictions for the various subsystems. The math model can be used to assist in making changes to the system for reliability improvement. The model can be used to identify weak links in the system and to indicate where reliability improvement activities should be introduced. The model can be used to determine test and maintenance procedures. Modeling of the system should be initiated as soon as preliminary designs are completed, and the model should be updated as design changes are made to the system.

Static System Reliability Models

Series System. A system block diagram reduces the system to its subsystems, and provides a tool for understanding the effect on the system of a subsystem failure. The most basic of the static block diagrams is the *series model.* A reliability series model block diagram shows that the successful operation of the system depends on the successful operation of each of the subsystems. The failure of any subsystem will result in the failure of the system. This is a natural way to design and build systems. Consequently, most systems are series systems unless some effort is made to incorporate redundancy into the design.

In the series system block diagram, each block represents a subsystem, as shown in Figure 8.1. There is only one path for system success. If any subsystem fails, the system will fail. It is important not to overlook the subsystem interfaces as they may be a source of failure.

To analyze the system reliability using a series block diagram, it is necessary to assume that the probabilities of failure for the individual subsystems are indepen-

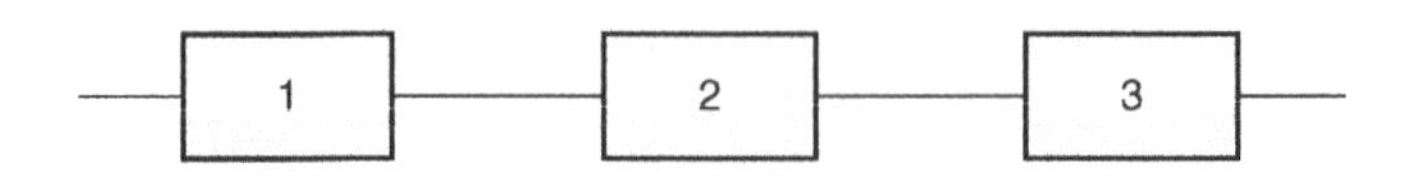

Figure 8.1 The series system.

dent. The assumption of independence is reasonable, and need only apply until the time of first failure. Any secondary failure, although it may be a safety consideration, does not affect the reliability analysis. The system has already failed. This is not true of all models. Other models require the user to assume that the probabilities of subsystem failure are completely independent for the entire mission.

It is necessary that all subsystems survive the mission if the system is to survive the mission. This is the joint occurrence of the success of the subsystems, and the product law for the joint occurrence of independent events is used to calculate system reliability.

The reliability of the system [$R_{System}(t)$] is the probability of success of the system for the mission time, t. The reliability of a subsystem i [$R_i(t)$] is the probability of the success of that subsystem for a mission of time t. If there are n subsystems, and the reliability of each subsystem is known, the system reliability can be found as

$$R_{System}(t) = [R_1(t)] \times [R_2(t)] \times \ldots \times [R_n(t)].$$

The reliability of each subsystem is less than 1. See Example 8.1. The reliability of the system will be less than the reliability of any subsystem:

$$R_{System}(t) < R_i(t)$$

System modeling will assist in identifying reliability problems and the implementation of a reliability improvement effort. If significant reliability improvement is to be made to a series system, the subsystem with the minimum reliability must be improved. Maximum improvement in system reliability will be achieved by increasing the reliability of the subsystem with the minimum reliability.

If the reliability of subsystem 1 in Example 8.1 is improved to 0.999, the improvement in system reliability is marginal.

$$R_{System}(t) = (0.999)(0.98)(0.94) = 0.92$$

Regardless of the amount of improvement in subsystems 1 and 2, the system reliability can not exceed 0.94. The focus for system reliability improvement should be on subsystem 3.

EXAMPLE 8.1

A system consists of three subsystems connected in series.

The reliability for each subsystem for a mission time of t is:

$$R_1(t) = 0.99$$

$$R_2(t) = 0.98$$

$$R_3(t) = 0.94$$

The system reliability for mission time t:

$$R_{System}(t) = (0.99)(0.98)(0.94) = 0.91$$

There are other block diagram models that could result in higher system reliability. These models will be considered. However, it should be noted that there are many highly reliable series systems in use. The series design has some advantages over other designs. A series system will require a minimum number of parts, consume minimum power and therefore dissipate less heat, take up less room, add less weight, and be cheaper to build than other system configurations.

Parallel System. Redundancy can be designed into a system to increase system reliability. A parallel system provides more than one path for system success, as shown in Figure 8.2. An active redundancy system has subsystems online that can individually perform the functions required for system success. If a subsystem fails, system success can be accomplished with the successful operation of a remaining subsystem. The system fails only when all the redundant subsystems fail.

To analyze a parallel system it is necessary to assume that the probabilities of failure for the various subsystems are totally independent for the entire mission time. This requires that the redundant system design be engineered to ensure that this assumption is valid. If a single event can cause more than one subsystem to fail, or if the failure of one subsystem can cause a secondary failure of a redundant subsystem, the desired improvement in system reliability due to the redundancy is lost. For example, a single power supply failure might cause the failure of redundant navigational systems. This situation is referred to as a *single-point failure*. Also, particular attention must be given to the interconnection points of the redundant subsystems. Many times, these are the source of single-point failures. Reliability engineering tools such as FMEA that can be used to identify single-point failures or failure modes and remove them from the design are covered in Chapter 6.

A parallel system will fail only if all subsystems fail. If a redundant system consists of n independent subsystems, and if the reliability of each subsystem is known, then system reliability can be calculated as

$$R_{System}(t) = 1 - [1 - R_1(t)] \times [1 - R_2(t)] \times \ldots \times [1 - R_n(t)]$$

Active redundancy is an important reliability tool available to the system designer. See Example 8.2. It should not, however, be used to improve the reliability of a poor design. It is much more efficient to engineer the design for the highest

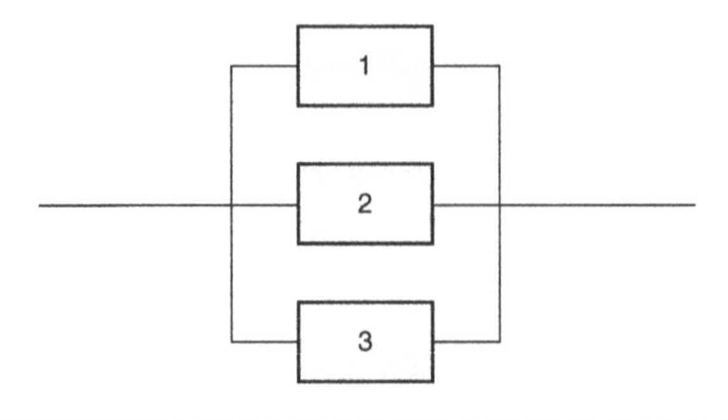

Figure 8.2 The parallel system.

EXAMPLE 8.2

An active parallel system has three independent subsystems.

The reliability for each subsystem for a mission time of t is:

$$R_1(t) = 0.99$$

$$R_2(t) = 0.98$$

$$R_3(t) = 0.94$$

The reliability of the system for mission time t is equal to

$$R_{System}(t) = 1 - (1 - .99)(1 - .98)(1 - .94) = .99998.$$

It should be noted that the reliability of the system is greater than the reliability of any of the redundant subsystems.

$$R_{System} > R_i \text{ for all } i = 1 \text{ to } n$$

possible reliability and then use active redundancy if the desired reliability is still not achieved.

Series–Parallel Model. A system may be modeled as a combination of series and parallel subsystems. For this model the same assumptions apply as for individual series or parallel systems. The combined system reliability can be found by converting the system to an equivalent series or equivalent parallel model.

EXAMPLE 8.3

Figure 8.3 shows a series–parallel system. The reliabilities for each subsystem are shown on the diagram. Find the reliability for the system.

The reliability for the two parallel subsystems is:

$$R_1 = 1 - (1 - .99)(1 - .99) = 1 - (1 - .99)^2$$

$$R_1 = 0.9999$$

The reliability for the three parallel subsystems is:

$$R_3 = 1 - (1 - .89)(1 - .92)(1 - .90)$$

$$R_3 = 0.99912$$

The system is now an equivalent series system with R_1 and R_3 and $R_2 = 0.999$.

The system reliability can be found as:

$$R_{System} = R_1 \times R_2 \times R_3 = (0.9999)(0.999)(0.99912)$$

$$R_{System} = 0.998$$

Continued

Continued

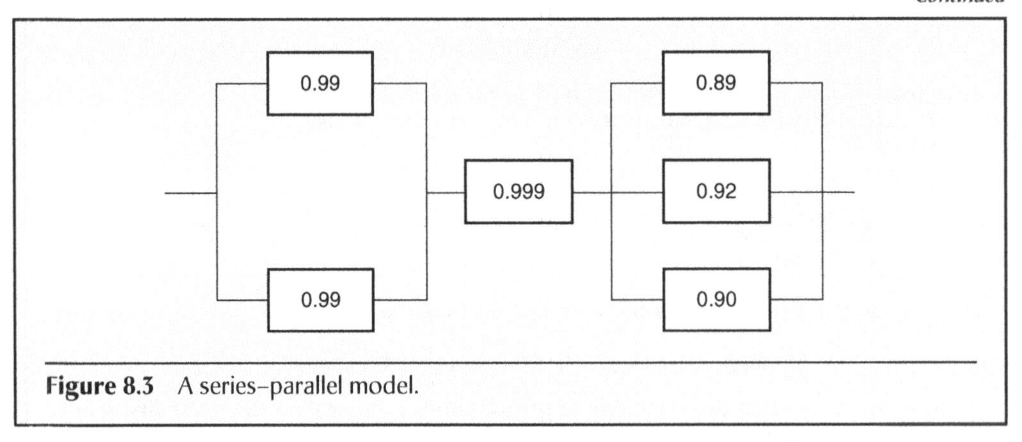

Figure 8.3 A series–parallel model.

Special Case Reliability Modeling

m *out of* n *System Model.* A special case of the parallel system is the m out of n system. This is a parallel system of n equivalent subsystems. System success requires that at least m ($m < n$) of the subsystems not fail. For this system, m can be any positive integer less than n; however, if $m = 1$, the system reduces to an active parallel system. The binomial distribution, along with the addition law for mutually exclusive events, is used to find the reliability of the system.

If R is the reliability for each of the n redundant subsystems, and m subsystems are required for system success,

$$R_{System}(t) = \Sigma C_i^n (R)^i (1 - R)^{n-i} \text{ for } i = m \text{ to } n$$

where

$$C_i^n = \frac{n!}{[i!(n-i)!]}$$ is the number of combinations of n items taken i at a time

$n!$ (defined only for nonnegative integers) is the product of all the positive integers up to and including n

$6! = 2 \times 3 \times 4 \times 5 \times 6 = 720$

$0! = 1$ by definition

Significance. The reliability engineer must always guard against presenting reliability results with more precision than the data warrant. All engineers need to know how to express results using correct significance. A legitimate concern might be raised about the significance of the results in the examples in this section. In order to illustrate the method of solving the problems, the results are many times carried to more significant digits than can be justified by the data used in the calculations. In the next examples, the results of the calculations are carried to the fourth decimal place and then rounded. This exceeds any significance that can be justified by the values used. This is done to show that the methods will give the same results.

EXAMPLE 8.4

Eight units each with a reliability of 0.85 are connected in a parallel configuration. At least six units are required for system success. What is the reliability of the system?

The reliability of the system is the probability that six of the units are successful and two of the units fail or that seven of the units are successful and that one of the units fails or that all eight of the units are successful. These probabilities are mutually exclusive and can be added to give the probability of system success (see Chapter 4).

$$R_{System}(t) = C_6^8(.85)^6(1-.85)^2 + C_7^8(.85)^7(1-.85) + C_8^8(.85)^8$$

$$= (28)(.85)^6(.15)^2 + (8)(.85)^7(.15) + (.85)^8$$

$$= .895$$

System Model Using Bayes's Theorem. Another special case is a coherent system, a connection of subsystems that can not be reduced to a series or parallel model. One method to solve for system reliability of such a system is known as *Bayesian analysis.*

Choose one subsystem. With that subsystem assumed to be first in a success state, then in a failed state, the remaining system should reduce into a series–parallel arrangement. If it does not, choose another subsystem. With the chosen subsystem assumed to be in a success state, find the reliability of the remaining system and multiply by the probability that the subsystem has not failed. With the chosen subsystem assumed to be in a failed state, find the reliability of the remaining system and multiply by the probability that the subsystem has failed. The sum of these two values is the system reliability.

EXAMPLE 8.5

The system model shown in Figure 8.4 can not be reduced to an equivalent series or equivalent parallel model. Find the reliability of the system using Bayes's theorem.

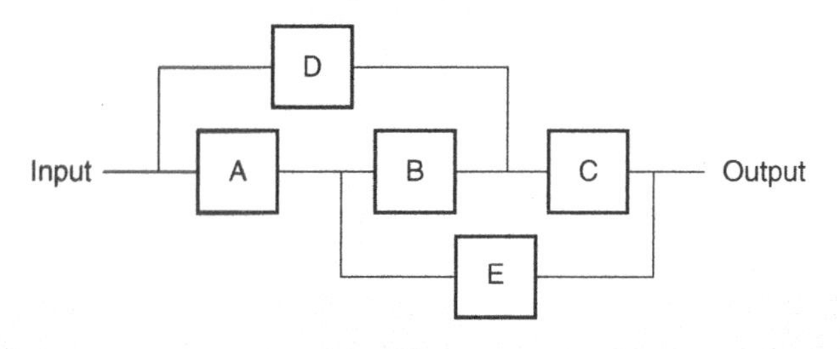

Figure 8.4 System model for Bayesian analysis.

Continued

Continued

The subsystem reliability values are:

$$R_A = R_B = R_C = 0.95$$

$$R_D = R_E = 0.99$$

Choose subsystem E.

Assume E is in a success state. The system model will reduce to

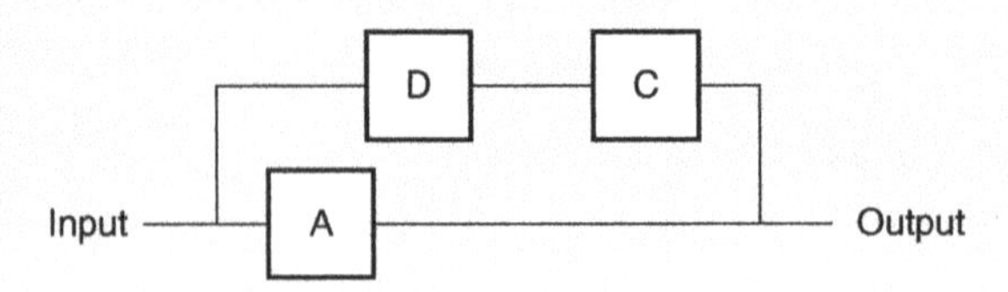

and the system reliability with E good is

$$R_{(E\ success)} = 1 - [1 - (.99)(.95)][1 - .95] = 0.9970.$$

Assume E is in a failed state. The system model will reduce to

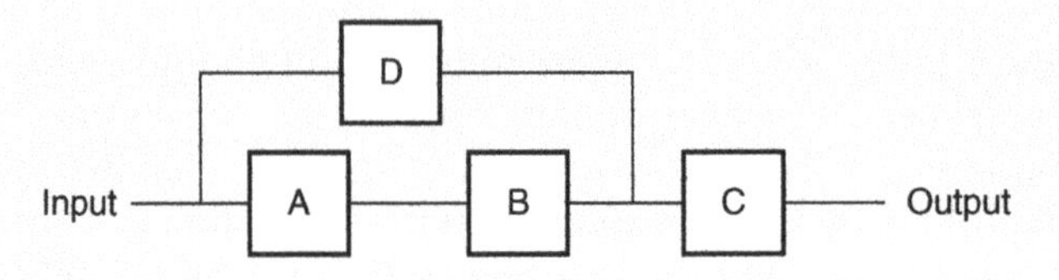

and the system reliability with E failed is

$$R_{(E\ failed)} = [1 - (1 - .99)(1 - (.95)(.95))]\ [.95] = 0.9491.$$

The system reliability is

$$R_{System} = R_{(E\ success)} \times P(E\ success) + R_{(E\ failed)} \times P(E\ failed)$$

$$R_{System} = (0.9970)(0.99) + (0.9491)(1 - 0.99)$$

$$R_{System} = 0.9965.$$

Truth Table Method. System reliability can be found from any block diagram—when the subsystem reliabilities are known—using a systematic method to identify all the events that will result in system success. The total events can be identified by finding all the combinations of subsystem success (S) and subsystem failure (F). It can then be determined if an event will result in system success or in system failure. The probability that each event will occur can be determined. The events are mutually exclusive. The probability of system success is the sum of the probabilities of the events that result in system success. The total number of events for a system modeled using k subsystems is 2^k.

EXAMPLE 8.6

Find the reliability of the system shown in Figure 8.5 using the truth table method.

Subsystem reliabilities are:

$$R_A = 0.85$$

$$R_B = 0.90$$

$$R_C = 0.92$$

The total number of events is $2^3 = 8$.

Construct a truth table using S for success and F for failure.

	Subsystem				
Event #	A	B	C	System	Event probability
1	S	S	S	S	(.85)(.90)(.92) = 0.7038
2	S	S	F	S	(.85)(.90)(.08) = 0.0612
3	S	F	S	S	(.85)(.10)(.92) = 0.0782
4	S	F	F	S	(.85)(.10)(.08) = 0.0068
5	F	S	S	S	(.15)(.90)(.92) = 0.1242
6	F	S	F	F	
7	F	F	S	F	
8	F	F	F	F	

The system reliability is the sum of the probabilities of the events that result in system success:

$$R_{System} = 0.7038 + 0.0612 + 0.0782 + 0.0068 + 0.1242 = 0.9742$$

This can be verified using the series–parallel model.

$$R_{System} = 1 - [1 - (.90)(.92)] \times [1 - .85] = 0.9742$$

For an exercise, apply this method to the problem in Example 8.5 and verify that the system reliability is 0.9965. There are $2^5 = 32$ total events. Fifteen events result in system success.

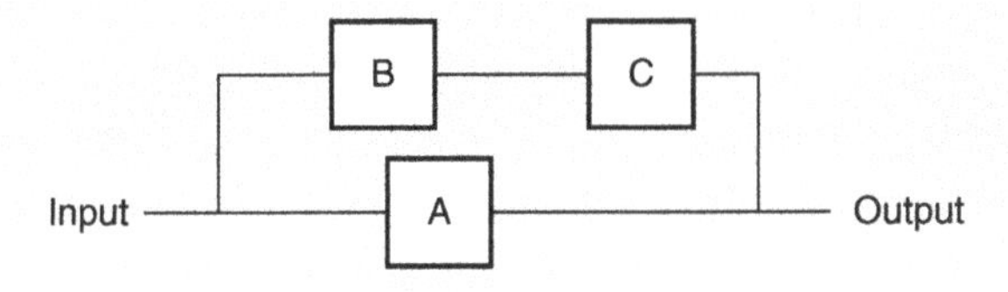

Figure 8.5 Series–parallel model for truth table method.

The method will be demonstrated on a simple system that can be easily verified. The method can be used for complex system diagrams but can become cumbersome as the number of subsystems increases. A computer model can be used for complex systems. See Example 8.6.

Part IV.A.2

It should be noted that the solutions for system reliability for all the above static math models are distribution free. This means that no assumption is made about the distributions that describe the subsystem failures. Each subsystem failure could be described by a different distribution. The subsystem reliabilities can be determined independently, and the system reliability can be calculated using the laws of probability.

It might require a complex analysis to determine the distribution describing the system failures. The exception to this is the series model with subsystem failures described by the exponential distribution (constant failure rate). In this case the system has a constant failure rate, and the system failure rate is the sum of the subsystem failure rates.

Series Model (Constant Failure Rate). A series system is shown in Figure 8.6. Assume that the failure distribution for each of the subsystems is exponential. Each subsystem i has a constant failure rate λ_i.

The system failure rate is also constant and is equal to the sum of the subsystem failure rates:

$$\lambda_{\text{System}} = \sum_{n} \lambda_i$$

The system reliability is

$$R_{\text{System}} = e^{-(\lambda_{\text{System}})t}.$$

The MTTF/MTBF of the system is

$$\theta = \frac{1}{\lambda_{\text{System}}}.$$

See Example 8.7.

Dynamic System Models

Dynamic math models are time dependent. To analyze a dynamic model, it is necessary to assume a failure distribution for the subsystems. For the following models the exponential distribution is assumed to describe the failures of the subsystems. Each subsystem is assumed to have a constant failure rate.

Load Sharing Model. The subsystems of an active parallel system are connected such that each shares equally in the total load. The subsystems are derated so that each is operating at less than its maximum load capacity. The failure rate of each subsystem is lowered and reliability is improved because the units are operating at a lower stress level. If a failure occurs, the remaining subsystems have enough

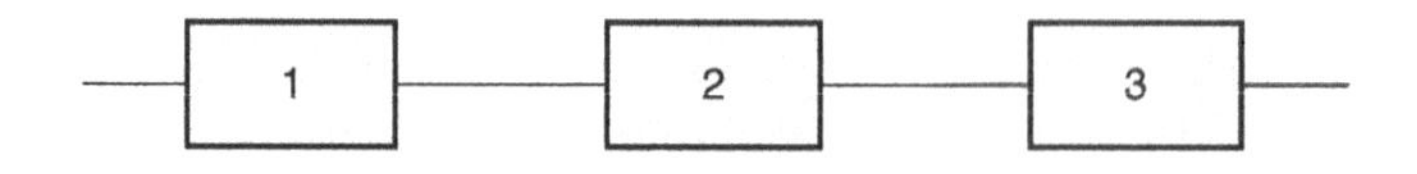

Figure 8.6 Series system.

EXAMPLE 8.7

For the system in Figure 8.6:

$$\lambda_1 = 100 \times 10^{-6} \text{ failures/hour}$$

$$\lambda_2 = 80 \times 10^{-6} \text{ failures/hour}$$

$$\lambda_3 = 20 \times 10^{-6} \text{ failures/hour}$$

What is the system reliability for $t = 100$ hours?

$$\lambda_{\text{System}} = \lambda_1 + \lambda_2 + \lambda_3$$

$$\lambda_{\text{System}} = (100 + 80 + 20) \times 10^{-6} = 200 \times 10^{-6} \text{ failures/hour} = 200\text{E}{-6} \text{ failures/hour}$$

$$\text{Reliability: } R_{(t=100)} = e^{-(200\text{E}-6 \times 100)} = 0.98$$

$$\text{MTBF: } \theta = 1/(200 \times 10^{-6}) = 5000 \text{ hours}$$

An alternate method would be to first find the reliability of each subsystem and then find the system reliability using the series model:

$$R_1 = e^{-(100\text{E}-6 \times 100)} = 0.99$$

$$R_2 = e^{-(80\text{E}-6 \times 100)} = 0.992$$

$$R_3 = e^{-(20\text{E}-6 \times 100)} = 0.998$$

$$R_{\text{System}} = (0.99)(0.992)(0.998) = 0.98$$

capacity to carry the system load, but they will be operating at a higher stress level and therefore at a higher failure rate.

For the case of two units in a load-sharing configuration, each subsystem operates with a constant failure rate of λ_1 as long as both are operating. If one subsystem fails, the remaining subsystem will continue to operate but at an increased failure rate $\lambda_2 > \lambda_1$.

System success for a mission time of t would be described as both subsystems operating successfully for the entire time t at failure rates equal to λ_1, or a failure occurring at time $t_1 < t$, and the remaining subsystem operating for a time of $t - t_1$ at a failure rate of λ_2:

$$R_{\text{System}(t)} = e^{-2\lambda_1 t} + \frac{2\lambda_1\left(e^{-\lambda_2 t} - e^{-2\lambda_1 t}\right)}{2\lambda_1 - \lambda_2} \quad \text{if } 2\lambda_1 - \lambda_2 \neq 0$$

If $2\lambda_1 = \lambda_2$, the denominator of the above equation becomes zero.

The reliability equation reduces to

$$R_{\text{System}(t)} = e^{-2\lambda_1 t} + 2\lambda_1 t e^{-\lambda_2 t}.$$

The reliability equation can be expanded to include systems with more than two units.

Standby Redundant Systems. A system that has parallel units that are utilized only in the event of a failure is a *standby redundant system.*

Figure 8.7 shows a primary unit performing a function. If the primary unit were to fail, a sensor could detect the failure, and a standby unit could be switched in to allow the system to continue to perform the function. The standby unit must be capable of performing the function but it might not be identical to the primary unit. The sensing and switching system may be an automatic part of the system or may require some manual interface. An example of automatic sensing and switching and nonidentical redundant units is the backup battery in an alarm clock. An example requiring manual switching is the replacement of a failed tire on an automobile. These are examples of standby redundant systems because the secondary units become a functioning part of the system only after the primary unit fails.

A failure distribution for the subsystems must be assumed to analyze the system for reliability. Other factors to be considered are the reliability of the sensing and switching system and the probability that the secondary subsystem could fail before it is needed.

The simplest system is one with both the primary and secondary units identical, perfect sensing and switching, zero probability of failure of the secondary unit in its quiescent mode, and a constant failure rate to describe the probability of failure for the units.

For such a system, the failure rate of the primary and secondary units is λ, and the mission time is t. The reliability of the system, $R_{System}(t)$, is the probability that the primary unit will operate successfully for time t, or the primary unit will fail at $t_1 < t$, and the secondary unit will operate successfully for the time $t - t_1$.

The reliability equation is

$$R_{System}(t) = e^{-\lambda t}(1 + \lambda t).$$

If the sensing and switching are not perfect ($R_{s/s} < 1$), the equation becomes

$$R_{System}(t) = e^{-\lambda t}(1 + R_{s/s}\lambda t).$$

The equation can be expanded to include multiple standby units.

The reliability for a system with two standby units and perfect switching is

$$R_{System}(t) = e^{-\lambda t}\left(1 + \lambda t + \frac{(\lambda t)^2}{2}\right).$$

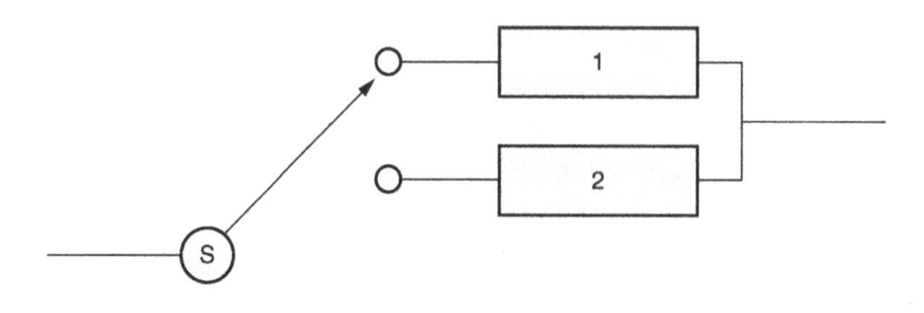

Figure 8.7 Standby redundant model.

3. PHYSICS OF FAILURE MODELS

> Identify various failure mechanisms (e.g., fracture, corrosion, memory corruption) and select appropriate theoretical models (e.g., Arrhenius, S-N curve) to assess their impact. (Apply)
>
> **Body of Knowledge IV.A.3**

The physics-of-failure approach to reliability uses data regarding the actual environmental conditions and loading during the life of the unit to make predictions about the unit's current "health" and its expected reliability. This requires a data collection mechanism, usually via various sensors, some sort of data mining algorithm, and an analysis of the failure modes. The goals of the research in this field are to reduce life cycle costs by:

- Generating alarms for impending failures
- Reducing unscheduled maintenance
- Extending preventive maintenance times
- Reducing inspection costs and replacement part inventory

One technique for sensing the accumulated stress on a unit is to add an additional device that will provide an alarm that a critical level has been reached. For example, to help detect the accumulated effect of oxidation on the undercarriage of a marine vessel, a separate metal tab could act as an early warning such that when oxidation of the tab reached a certain point, an alarm would be set. Other devices may be used to detect the effect of vibration, chemistry, temperature, and so on, on the unit's reliability.

There are two more or less competing approaches to the physics of failure: model based and data driven.

Model-Based Approaches

These systems attempt to derive a mathematical model to describe unit performance and to generate reliability predictions. Much of the work in this field has been done in the area of electronics and circuit board reliability. Pecht and Gu (2009) identify the circuit board failure mechanisms and their associated theoretical models, as shown in Table 8.2

Data-Driven Approaches

A disadvantage of the model-based approach is that detailed knowledge about the geometry and composition of the unit are needed but not always available. The data-driven approach skirts this issue by relying on a software system that learns

Table 8.2 Failure modes and models for circuit boards.

Failure sites	Failure modes	Failure models
Fracture	Die attach, wirebond, solder leads, bond pads, traces, interfaces	Nonlinear power law
Corrosion	Metalizations	Eyring
Electromigration	Metalizations	Eyring
Conductive filament migration	Between metalizations	Power law
Stress driven diffusion voiding	Metal traces	Eyring
Time-dependent dielectric breakdown	Dielectric layers	Arrhenius

the data patterns that lead to failure. A disadvantage of this method is that historical data must be available to establish correlations and detect patterns.

Physics of failure is associated in the literature with *prognostics and health management* (PHM), where the word "health" refers to a methodology for continuous assessment of the remaining useful life of the unit. A good source in this field is the PHM Society (http://www.phmsociety.org).

Memory corruption is somewhat independent of the more physical causes of failure. This failure refers to software activity that modifies a section of computer memory. The corruption, when unintentional, is due to human error by the programmer. In some early programming systems, variables were global, meaning that programmers needed to check with each other when they wanted to use a particular variable to make sure no one else was using it. Failure to make that check produced errors that were sporadic and extremely difficult to debug. More recent examples of memory corruption are far more sophisticated but produce similarly sporadic results and, sometimes, system crashes, which, again, are difficult to detect and fix. In some cases malware exploits memory corruption in commercial software to cause it to execute damaging code.

4. SIMULATION TECHNIQUES

> Describe the advantages and limitations of the Monte Carlo and Markov models. (Apply)
>
> **Body of Knowledge IV.A.4**

Modeling a dynamic system to predict its performance can become complex. The dynamic models discussed previously assumed a constant failure rate for each block of the model, and the only parameter generated was reliability for a given mission time.

EXAMPLE 8.8

The strength of a unit is normally distributed with a mean (μ_S) of 2600 psi and a standard deviation (σ_S) of 300 psi. The unit is exposed to a stress that is normally distributed with a mean (μ_s) of 2000 psi and a standard deviation (σ_s) of 200 psi. What is the reliability of the unit?

A Monte Carlo simulation can be performed on the unit by randomly selecting values from the strength and stress distributions and finding their difference. Failure of the unit will occur when the difference between strength and stress is negative (the stress is greater than the strength). Table 8.1 shows a printout of the first several steps of an Excel computer run using these values. The printout shows a failure at replication number 9. These steps would be replicated thousands of times. N is the number of replications. The reliability of the unit could then be estimated:

$$\hat{R} = \frac{\#\ \text{failures}}{N}$$

Table 8.3 Computer-generated values for a Monte Carlo simulation.

Replication number	Stress s $\mu = 2000$ $\sigma = 200$	Strength S $\mu = 2600$ $\sigma = 300$	Difference $x(S) - x(s)$ d	Failure = 1 $d < 0 = 1$
1	1942	2278	335	0
2	2091	2566	475	0
3	1599	2196	597	0
4	2013	2572	559	0
5	1934	2842	908	0
6	2177	2416	239	0
7	1936	2788	853	0
8	1952	2110	157	0
9	2042	2029	–13	1
10	1950	2167	218	0
11	1789	2743	954	0
12	2141	2306	165	0
13	2010	2619	610	0
14	2030	2362	332	0
15	1786	2593	806	0
16	1805	2961	1156	0

System reliability may be dependent on failure distributions other than the exponential (constant failure rate), or the desired system performance parameter may include availability (see Chapter 13). Availability is a function of the probability that the system remains in a usable state (reliability) and the probability of restoring the system to a usable state in a given period of time if a failure were to occur (maintainability). To analyze these systems, a simulation technique can be used. To use simulation, the defining parameters of each distribution must be known.

In a *Monte Carlo* simulation, repeated calculations of system performance are made using randomly selected values based on the probability distributions that describe each element of the model. The large number of values of system performance generated can be used to develop a probability distribution of system performance. Monte Carlo simulation does not involve complex mathematics; however, it requires an extensive use of computer time as each possible event for each unit of the model must be repeatedly sampled over the desired mission time.

Markov Analysis

A complex system can be analyzed as a Markov process. The Markov process can be used to find the probability of being in a given state at some time in the future if the probability of moving from one state to another state is known and the probability remains constant. The system can exist in only one state at a time, and except for the immediately preceding state, all future states are independent of past states.

One use of a Markov analysis is to determine the future probability of a repairable system being in a success state when the failure probability and probability of restoring the system are known. A simple system consisting of one unit can be considered to be in a success state (not failed) or in a failed state. If it is in the failed state, it is waiting to be returned to the success state. The transition probability from one state to the other is determined by the failure rate and the rate at which the unit can be restored to the success state after a failure (see Figure 8.8).

P_{S-F} is the probability of transition from state S (success) to state F (failure) in a given time interval.

P_{F-S} is the probability of transition from state F to state S in the same time interval.

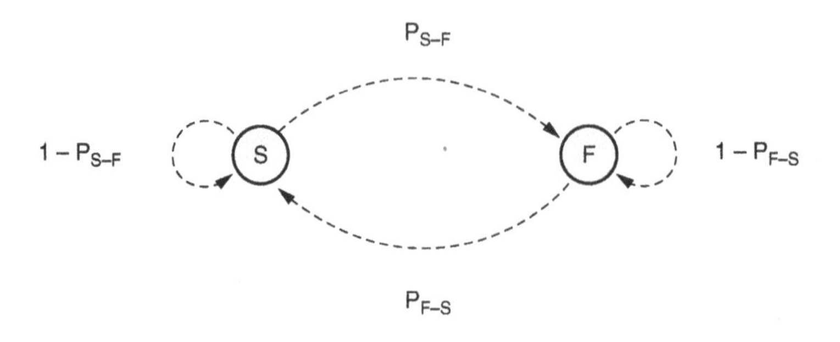

Figure 8.8 A transition diagram for a system with two states.

$1 - P_{S-F}$ is the probability of remaining in state S if the system is in state S.

$1 - P_{F-S}$ is the probability of remaining in state F if the system is in state F.

A Markov analysis is not limited to a system with only two states. There can be more than one success state. All the possible states of the system are identified. All of the transitions from one state to another state are identified, assuming one action at a time. The transition rates from one state to another state are determined. The probability of the system being in a success state can be determined by considering all possible states and the rates of transition from state to state. A Markov analysis requires the use of differential equations, Laplace transforms, and the solution of a series of linear equations using matrix algebra. Even though the analysis is exact, the assumptions necessary can affect the credibility of the results. A tree diagram can be used to model a simple system.

Each probability path represents a mutually exclusive event. The probability of the machine being in the success state (state S) at the end of two time intervals can be calculated as the summation of the probability paths to state S:

$$P(\text{State } S) = (0.7)(0.1) + (0.9)(0.9) = 0.88$$

EXAMPLE 8.9

Refer to the system in Figure 8.9.

A machine in state S (success) is operating successfully. There is a probability of 0.9 that the machine will remain in state S for a given time interval. There is a probability of 1 – 0.9 = 0.1 that the machine will fail and transition to state F (failure). If the machine is in state F, there is a probability of 0.7 that it will be restored in the given time interval and transition back to state S. There is a probability of 1 – 0.7 = 0.3 that it will not be restored in the given time interval and will remain in state F.

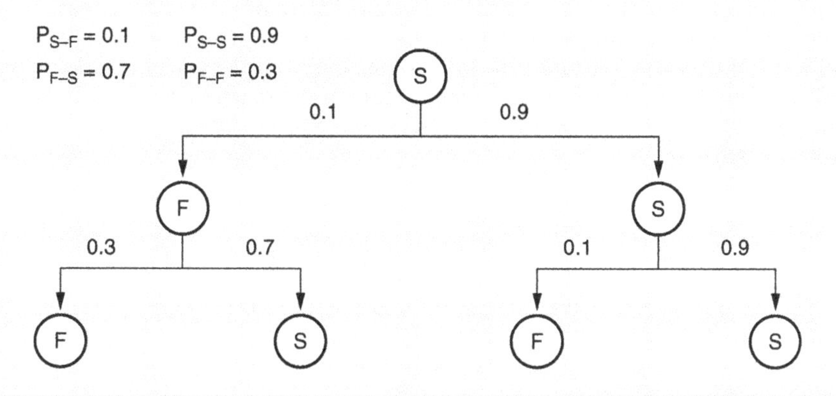

Figure 8.9 Tree diagram for Markov analysis for two time intervals.

Part IV.A.4

As an exercise, expand the tree diagram to cover three time intervals. Verify that the probability of being in the success state (state S) at the end of three time intervals is

$$P(\text{State S}) = (0.7)(0.3)(0.1) + (2)(0.9)(0.7)(0.1) + (0.9)(0.9)(0.9) = 0.876.$$

The tree diagram and the manual calculations become complex if the system has several success states. A Markov analysis will result in a transitional probability matrix that can be evaluated to find the state probabilities over many periods of time.

$$T = \begin{bmatrix} x_{1,1} & x_{1,2} \\ x_{2,1} & x_{2,2} \end{bmatrix}$$

EXAMPLE 8.10

Using the values in Example 8.9, the transitional matrix is

$$T = \begin{bmatrix} 1-P_{S-F} & P_{S-F} \\ P_{F-S} & 1-P_{F-S} \end{bmatrix} = \begin{bmatrix} 0.9 & 0.1 \\ 0.7 & 0.3 \end{bmatrix}.$$

T^K will give the state probabilities after the kth time interval:

$$T^2 = \begin{bmatrix} 0.9 & 0.1 \\ 0.7 & 0.3 \end{bmatrix}^2 = \begin{bmatrix} 0.88 & 0.12 \\ 0.84 & 0.16 \end{bmatrix}$$

and

$$T^3 = \begin{bmatrix} 0.9 & 0.1 \\ 0.7 & 0.3 \end{bmatrix}^3 = \begin{bmatrix} 0.876 & 0.124 \\ 0.868 & 0.132 \end{bmatrix}$$

The value in the first row and the first column, $x_{1,1} = 0.88$ when $k = 2$ and $x_{1,1} = 0.876$ when $k = 3$, is the probability of the machine being in the success state (state S) at the end of the second and third time intervals if the original state of the machine was success. The value in the second row and the first column, $x_{2,1} = 0.84$ when $k = 2$ and $x_{2,1} = 0.868$ when $k = 3$, is the probability of the machine being in the success state at the end of the second and third time intervals if the original state of the machine was failure (state F). As k increases, these values will converge, becoming the steady state value of the availability of the machine:

$$T^6 = \begin{bmatrix} 0.875008 & 0.124992 \\ 0.874994 & 0.125056 \end{bmatrix}$$

The values $x_{1,1}$ and $x_{2,1}$ are converging to 0.875, the steady state value of the availability of the machine.

There are some handheld calculators and many software programs that can be used to do the matrix algebra.

5. DYNAMIC RELIABILITY

> Describe dynamic reliability as it relates to failure criteria that change over time or under different conditions. (Understand)
>
> Body of Knowledge IV.A.5

The standard statement of reliability lists four components:

- Probability
- Required function
- Stated conditions
- Specified period of time

The usual examples list a *fixed* function and a *fixed* set of conditions, and therefore the probability of survival varies with the elapsed time. The standard metrics such as MTTF are predicated on this situation. Dynamic reliability considers the cases in which function or the conditions (or both) are variable. The challenge then becomes, given some data about the required function or the stated conditions, what predictions can be made about the system reliability.

Stated Conditions As a Variable

Consider the example of a wind generator rated at 1.5 megawatts. It can be determined that with given wind velocity, ambient temperature, humidity, and so on, the system will generate at least 1.4 MW for five years with 0.95 reliability. The environmental conditions may even be stated as a range, for example, wind velocity between 5 mph and 40 mph. The effective management of a wind generation farm requires preventive replacement of various components of the system. If a five-year period goes by, and the wind velocity remained in the 5 to 10 mph range, is the 0.95 figure useful? Or, more generally, if data are available on the environmental conditions, can a better preventive replacement schedule be built?

Required Function As a Variable

Continuing the wind generation farm example, suppose some of the units were required to generate only 0.5 MW throughout the five-year period. Is the 0.95 figure useful? Or, more generally, if data are available on the required function for a particular unit, can a better preventive replacement schedule be built?

The field of dynamic reliability builds theory and techniques that generate reliability calculations based on information about the nature of the changing conditions and function for the unit. This is a rich field for research.

Chapter 9

B. Reliability Predictions

1. PART COUNT PREDICTIONS AND PART STRESS ANALYSIS

> Use parts failure rate data to estimate system- and subsystem-level reliability. (Apply)
>
> **Body of Knowledge IV.B.1**

A reliability prediction is a design tool to be used early in the design and development stage. To be effective, the prediction is started before the design is completed, and completed before production tooling is set and hardware is procured for production. Many times, the prediction is done without any actual reliability data on the particular system, although there may be data on some of the components. The reliability prediction can be used to determine the feasibility of the design to meet the system reliability goals, to focus attention to "weak links" in the design, to assess the impact of design changes on the system reliability, to compare the reliability of competing designs early in the development stage, and to assist in establishing maintenance procedures. The reliability prediction is not a reliability estimate, which requires data, and should not be used as a measure of the actual achieved system reliability.

The best known and most widely used source of predictive data for electrical/ electronic systems is MIL-HDBK-217. The handbook assumes electronics can be modeled using a constant failure rate, and contains failure rate data for electrical and electronic components. The handbook contains data for all passive devices: resistors, inductors, capacitors, transformers, and so on. It also contains data for active elements: transistors, diodes, field-effect transistors (FETs), and so on, as well as digital and analog intergraded circuits. There are methods to adjust the base failure rate of the components depending on the number of leads, the number of gates in an IC, the use environment the component will experience, the quality control requirements the customer can impose on the supplier, and other factors that could affect the component failure rate. The multipliers used to adjust the base failure rate are referred to as π (pi) factors.

EXAMPLE 9.1
THE USE OF MIL-HDBK-217

The failure rate model for a resistor is $\lambda_P = \lambda_b \times \pi_E \times \pi_R \times \pi_Q$ failures/million hours.

Tables in MIL-HDBK-217 contain λ_b, the base failure rate, based on the standard derating used by the designer.

Tables also contain the pi factors:

π_E is the environmental factor reflecting how the resistor is used.

π_R is a factor based on the value of the resistor.

π_Q is a quality factor and is dependent on the amount of control the customer has in the production of the component.

These factors and the base failure rate are combined to give the predicted failure rate of the resistor.

EXAMPLE 9.2

A 10,000-ohm carbon resistor is used in a communications receiver located in the crew compartment of a commercial aircraft. Assume a standard design derating of 40 percent at 60 °C.

From MIL-HDBK-217:

$\lambda_b = 0.0012/10^6$	Base failure rate
$\pi_E = 3$	Environmental stress factor
$\pi_Q = 1$	Quality factor
$\pi_R = 1$	Resistance factor
$\lambda_p = (3)(1)(1)(0.0012/10^6) = 0.0036/10^6$	Predicted failure rate

If the same resistor were used in a radar detector located on the deck of a naval ship:

$\pi_E = 12$	Environmental stress factor
$\lambda_p = (12)(1)(1)(0.0012/10^6) = 0.0144/10^6$	Predicted failure rate

The handbook was last revised in 1994 and was issued as MIL-HDBK-217F. It is common knowledge that predictions made using MIL-HDBK-217 will be pessimistic, and the product will have a failure rate less than the predicted value.

A circuit board (subsystem) failure rate could be predicted by adding the predicted failure rates of all the components on the board. This assumes a series model for the circuit board and will give a worst-case prediction. This is referred to as the *part count method*. A system failure rate could be predicted by adding the predicted failure rates of all the subsystems. This also assumes a series model.

Using the assumption of a series model, the system will have a constant failure rate if all the components have a constant failure rate.

In an attempt to overcome some of the deficiencies of MIL-HDBK-217, the PRISM reliability prediction method was developed by the Reliability Analysis Center (RAC).

Instead of using just multipliers to model component failures, PRISM uses combinations of both additive and multiplicative factors. The factors are called π factors, and the model could be of the form

$$\lambda_S = \lambda_i (\pi_1 \pi_2 + \pi_3 \pi_4 + \pi_5)$$

where λ_i is the initial failure rate and the π values are based on the part. PRISM also allows for consideration of early-life failures as well as failures due to a constant failure rate.

The 217(plus) method is an outgrowth of PRISM. It contains new part type failure rate models including connectors, switches, and relays. Software programs for utilizing PRISM and 217(plus) are available from the Reliability Analysis Center (RAC) (http://theRAC.org).

For predicting the reliability of mechanical systems, the *Handbook for Reliability Prediction for Mechanical Systems* (NSWC-07) is available. Information can be obtained from the Naval Surface Warfare Center, Carderock Division, Bethesda, Maryland.

The handbook follows the format of MIL-HDBK-217 and contains base failure rates for mechanical components, which can be adjusted using multipliers or "c" factors depending on the type of material, physical configuration, heat treatment, the use environment, and other factors that would affect the probability of failure. Examples of specific mechanical devices included are springs, bearings, brakes, and clutches, seals, pumps, and other nonelectrical devices. This handbook assumes the constant failure rate model.

Non-electronic Parts Reliability Data (NPRD-2011) contains failure rates for mechanical and electromechanical parts. This method allows the engineer to consider failures due to wear-out.

Telecordia SR-332 (once known as Bellcore) has prediction data in tables and graphs for communication equipment with commercial applications. These Telecordia models were developed by AT&T Bell Labs, and the method is based on MIL-HDBK-217.

Several international standards are also available. The methodology used in these standards is similar to that in MIL-HDBK-217.

Reliability predictions can also be based on the engineer's experience with components used in previous designs of similar systems. In some cases, the component vendor may supply typical failure rate data.

The Government-Industry Data Exchange Program (GIDEP) is a source available to military contractors that contains failure reports on commercially available subsystems such as motors, compressors, pumps, and so on. System reliability values obtained using any of these sources should be treated only as prediction values.

2. RELIABILITY PREDICTION METHODS

> Use various reliability prediction methods for both repairable and non-repairable components and systems, incorporating test and field reliability data when available. (Apply)
>
> **Body of Knowledge IV.B.2**

The reliability of a product needs to be continually evaluated using the available information during all stages of design, development, and production. This is necessary to assure that the product will meet the system reliability requirements. During design, information on components and parts, along with the system model, can be used to predict the reliability. As the product moves through the development and production stages, information from tests can be used to estimate product reliability.

Early reliability predictions using the system model can be very useful to the reliability engineer. Predictions can be used to determine weak spots in the design and initiate a redesign effort. Predictions can be used to choose between alternate designs. Predictions can be used to evaluate the effect on reliability of a design change. Predictions can be used to evaluate the feasibility of meeting the final system-level reliability requirement with the present design.

Predictions have the limitation that they are based on the simplifying assumption that each component has an inherent constant failure rate. Predictions can consider environmental stress levels, the complexity of the component, the manufacturing capability of the component manufacturer, and other factors that might affect failure. However, predictions can not take into account the human factors that most likely cause failure. Factors such as the skill of the operator using the product, the ability of the design group to anticipate how the product will be used, the motivation of the designer to design a reliable product, the manufacturing capability of the system manufacturer, and the skill of maintenance personnel may be much more significant to product reliability.

A prediction, unlike an estimate, does not have relevant experimental data for support. A prediction has no statistical confidence. Reliability prediction is, at best, an exercise in uncertainty. Databases containing typical reliability values such as failure rates are available. However, the reliability of a specific product is not a characteristic that can be inherently predicted with high precision. It is necessary that the reliability engineer realize that even though complex models and mathematical relations exist for making predictions, the usefulness of a prediction is limited.

If testing results are available, reliability predictions can be made based on the distribution of times to failure and the estimated parameters. If the exponential distribution models the times to failure, the estimated parameter is the mean. This is called the *mean time to failure* (θ) or MTTF. If n units are tested, and during the test r units fail, the estimate of MTTF is

$$\hat{\theta} = \frac{T}{r}$$

where

T is the total time accumulated on the units, including the units that failed and the units that did not fail

r is the number of failures

The estimated failure rate

$$\hat{\lambda} = \frac{r}{T}$$

is the reciprocal of the mean.

The estimated reliability for a mission time of t is

$$\hat{R} = e^{-(t/\theta)} = e^{-\lambda t}$$

A lower confidence limit ($\theta_{(\alpha)}$) can be calculated for the estimate of the mean:

$$\alpha = 1 - \text{Confidence (See Chapter 5)}$$

The estimate of reliability for a mission time t at a confidence level of $1 - \alpha$ is

$$R_{(\alpha)} = e^{-(t/\theta(\alpha))}.$$

For a repairable system, if the exponential distribution model is appropriate, the above discussion is valid up to the first failure. Predicting reliability for a repairable system becomes complicated because the system can be restored to use after a failure. A system that has undergone a series of restore (repair) actions comprises subsystems and components that have acquired different operating times. Even though a repair might return a subsystem to a new state, the system is not in a new state. This means that the system can not be modeled using a constant failure rate. Rather than making predictions of the reliability of a system that has experienced several repair cycles, predictions should be made as to the number of spare parts needed in order for the system to meet specified availability requirements.

Part V

Reliability Testing

Part V

Chapter 10

A. Reliability Test Planning

1. RELIABILITY TEST STRATEGIES

> Create and apply the appropriate test strategies (e.g., truncate, test-to-failure, degradation) for various product development phases. (Create)
>
> **Body of Knowledge V.A.1**

The *reliability* of a unit is a characteristic related to time or some other measure of product use. The results of some reliability tests are used to estimate reliability, to set confidence limits on the estimate, or to show conformance to some specified reliability value. These tests must be performed over an extended period of time. An exception to this is the testing of "one-shot" items, covered in Chapter 12 (see Attribute Testing). The results of other reliability tests are used to improve the reliability of a unit and may not require testing over long periods of time. If the primary strategy of a testing program is to use the results for any purpose related to reliability, the test is classified as a reliability test. It is also possible to obtain reliability information from other types of tests such as feasibility tests, functional tests, or quality assurance tests.

The strategies for various reliability tests are primarily determined by the stage of product development during which the testing is performed. During early development of a new product, no units are available, therefore no reliability testing can be performed.

As the preliminary design is available, and prototypes or engineering models are built, a *highly accelerated life testing* (HALT) program can begin. The strategy for this reliability testing program is to stress the units beyond the design limits in order to cause failures. A reliability improvement program to eliminate the weak components is then initiated based on analysis of these failures. The resulting design will become more robust, and failures due to marginal components will be eliminated, thus improving product reliability. Results from HALT testing can not be used to estimate reliability values since the units are stressed until they fail, and these stresses are beyond the design limits.

Reliability tests with a strategy to estimate the failure rate or mean time between failures (MTBF) of the product can begin once production units become available. These tests, referred to as *life tests,* accumulate test time on several units while recording failures. These tests are required to run for a period of time. In many cases the test time could be prohibitive, making some type of acceleration necessary.

Before the product is delivered to the customer, a *highly accelerated stress screening* (HASS) test should be performed. The strategy is to stress the entire production run to eliminate defective product from the population and to detect any shift in the production process. Units that pass this test are shipped to the customer. HASS is usually considered to be a quality engineering function. The test, however, has a positive effect on reliability by limiting early-life failures from reaching the customer.

If a compliance test is required, it must be performed before the product can be delivered to the customer. The strategy of this type of test is to verify that the product conforms to some minimum reliability value. This is done by accumulating time on several units, requiring this test to be performed over a span of time. Compliance testing can not be performed until the final design is set, production tooling is in place, and units have been produced. The strategy of compliance testing is *not* product improvement.

Planning for an integrated reliability test program must start at the beginning of the project. Planning must be thorough and timely if all the necessary elements are to be in place when they are needed. Planning must be done to assure that the number of units to be tested, the test facilities—including any special equipment—and the necessary time are available when the tests are to be performed. Scheduling of the various reliability tests is necessary so that the results can be used in a timely manner based on the strategy of the test. It is much more efficient to make a design change before the product is released to production.

A test plan needs to include the following: objectives of the testing program; provision of resources for facilities, test equipment and equipment calibration information, the time and necessary personnel to conduct the testing; test requirements and schedule, including the number of units to be tested and the test environments; procedures to make changes in the testing program as necessary; and documentation of the test results. If the strategy of the test is to record failures over time, a precise definition of product failure must be available to the test engineer as the output of some products can deteriorate over time.

Many sources list the necessary steps in developing a reliability test plan. The number of steps and the words describing the steps differ, but all essentially contain the same information. Dimitri Kececioglu (1993) lists nine steps:

1. Determine the test requirements and objectives.

2. Review existing data to determine if any requirements can be met without testing.

3. Review the list of tests to determine whether combining any tests would be economically feasible.

4. Determine the necessary tests.

5. Allocate the resources necessary to perform the tests.
6. Develop test specifications, and data handling and storage procedures. Review acceptance and qualification criteria. Establish procedures for making future changes in the test specifications.
7. Assign the responsibility for conducting the tests, analyzing the results, and providing the overall integration of the testing program.
8. Develop forms and procedures for reporting the test results.
9. Develop procedures for maintaining the test status information throughout the entire testing program.

Reliability tests can be placed in several different classifications. One way to classify tests is by the phase of the development/production cycle in which they are conducted. Four major test categories classified in this manner are:

- Product development tests
- Reliability performance tests
- Reliability acceptance tests
- Reliability verification tests

These tests have different objectives and are conducted at different times during the process of product development and production.

Product Development Tests

These tests are performed with the intent of taking action to improve the design in the event of failures, and to evaluate the system design, including the compatibility of subsystems. Results from product development testing may be used to demonstrate the functional capabilities of the product but not to determine reliability parameters such as failure rate or MTBF. The objective of product development testing is to cause failures. Only by analyzing failures and taking appropriate action to change the design can improvement be made to the product. Testing units for long periods of time and observing no failure is costly and does not result in product improvement.

Reliability Performance Tests

Sometimes referred to as *reliability qualification tests,* these tests are performed after the design is completed. They will demonstrate that the system can meet the specified requirements under the specified conditions of operation (including environmental). Generally, these tests do not provide the data to determine reliability parameters, but only to give assurance that performance under stated conditions can be met.

Reliability Acceptance Tests

Conducted during the production phase, these tests will demonstrate that reliability parameters of the design have not been compromised by the production process. These tests might be part of the overall quality program, but with an emphasis on the reliability parameters.

Reliability Verification Tests

These tests are performed to show compliance to stated reliability parameters such as MTBF or failure rate. It may be required to demonstrate compliance at a given confidence level. This type of testing is formal, statistical in nature, and requires standard procedures. Many units or long periods of test time could be required for these tests, making some form of acceleration necessary. These tests are based on acceptable and rejectable reliability values such as MTBF or failure rate, and decision risks (see Chapter 12).

Reliability tests could be classified according to the way the tests are conducted and the type of results that are recorded. The type of data to be taken and the way in which they are reported must be a part of the overall reliability test plan. Reliability tests can be continuous or they can be pass/fail. Results of continuous tests are recorded as variables data, and the results of pass/fail tests are recorded as attribute data.

Attribute data are recorded as success or failure for a given test. The units being tested might not have active operating times. These units are referred to as *one-shot* items. They are units that perform successfully or fail when they are called on to operate, such as sensors or fuses. Estimates of the reliability or the probability of success of one-shot items can be made from the attribute data. Confidence limits on the reliability can also be calculated using the data.

Attribute data might also be recorded even if the units being tested have active operating times. Units being cycled in an environmental chamber may pass or fail the test. Unless each unit is metered separately, it will not be known until the test is complete which units were successes and which units failed. The data that are recorded are the number of units that passed and the number of units that failed the test; the actual time to failure of the units that failed is not known. Confidence limits and estimates of the reliability of these units for a mission time equal to the equivalent time of the test can be calculated using the data.

The results of continuous tests are recorded as variables. The actual times to failure of each failed unit and the total time accumulated on the non-failed units is known. Life tests are conducted to determine reliability parameters such as MTBF or failure rate. The data from these tests should be recorded as variables. Estimates of the parameters can be made and confidence limits can be calculated from the variables data. Continuous testing may be conducted as replacement tests or as non-replacement tests. The advantage of a replacement test is that the number of units on test is constant, therefore generating data at a faster rate. In a non-replacement test, units that fail are not replaced, and the test population becomes smaller.

Compliance tests are performed to show that units have achieved some reliability value. Results of these tests could be recorded as variables data or as attribute data. Compliance to an MTBF or failure rate can be shown using variables data from a fixed-time test or from sequential testing. If required, the compliance could be shown at a given confidence level. Sequential testing will require, on the average, about one-half the total unit test time as a fixed-length test. The consequence of this is fewer units required or a decrease in the actual test time. The

EXAMPLE 10.1

Fifty integrated circuits (ICs) are cycled in an environmental test chamber for a test that is equivalent to 1000 hours of operation. At the completion of the test, it is found that two of the units failed during the test.

The binomial distribution is used for the estimate of reliability, and the *F*-distribution is used to calculate the confidence limit (see Chapter 5).

The estimate of the reliability of the IC for a mission time of 1000 hours is

$$\hat{R}_{(t=1000)} = \frac{n-r}{n} = \frac{50-2}{50} = 0.96.$$

From the *F*-distribution: $F_{(.10)\,6,96} = 1.82$.

The lower 90 percent confidence limit on the reliability of the IC for $t = 1000$ hours is

$$R_{L(\alpha=.10)} = \frac{48}{48+3(1.82)} = 0.90.$$

disadvantage of sequential testing is that the actual time required for the test is not known at the beginning of the test. Attribute data from a pass/fail test can be used to demonstrate compliance to a reliability value or a probability of success value for one-shot items at a given confidence level. The test requiring the fewest number of units is a no-failure test. This is sometimes referred to as *success testing.*

Reliability tests can be classified as to the strategy and the types and levels of accelerations that are used. The strategy for a HALT program is product reliability improvement. The design is in the early development stage and is not finalized. Units are tested at stresses exceeding the design limits. The stresses usually are environmental, such as temperature or vibration, but could be other loading stresses such as voltage. Stress is increased until some component in the product fails, resulting in product failure. A design change is made to eliminate the weak component. The process is then repeated on the product incorporating the new design. Reliability improvement is achieved as the marginal design is improved by eliminating the weak components. In some cases these types of tests are also used to extend the beginning of wear-out, known as the end of useful life. The HALT program, along with other early product development reliability functions, will insure that a robust and reliable design is released to production. The test failures are not typical, however, of the normal use of the product. Therefore, the results from the HALT program can not be used to estimate the reliability of the product.

The amount of time and the number of units required for a life test can become prohibitive. Unlike Example 10.1, few projects will have 50 units available for reliability testing, and the 1000 hours of testing in Example 10.2 is more than 40 days. If the product has high inherent reliability, it would not be unusual to need several thousand hours of total unit test time to get meaningful results. To perform these tests it becomes necessary to use some type of acceleration (see Accelerated Life Testing in Chapter 11).

EXAMPLE 10.2

A non-replacement life test that is of 1000 hours duration is conducted on 10 units in order to estimate the MTBF and set a lower 90 percent confidence limit.

One unit failed at 450 hours and a second unit failed at 800 hours. Eight units did not fail during the test. The test was time-censored at 1000 hours.

The exponential distribution is used to estimate the MTBF and the χ^2 distribution is used to set the confidence limit (see Chapter 5).

The total unit test time:

$$T = 450 + 800 + (8) \times (1000) = 9250 \text{ hours}$$

The estimate of MTBF:

$$\hat{\theta} = \frac{T}{r} = \frac{9250}{2} = 4625$$

From the χ^2 distribution:

$$\chi^2_{(0.10)(6)} = 10.645$$

The lower 90 percent confidence limit:

$$\theta_{L(\alpha=.10)} = \frac{2(9250)}{10.645} = 1740 \text{ hours}$$

HASS is performed on 100 percent of the production units before shipment to the customer. To perform HASS testing, the product is subjected to elevated stresses. Some of these stress levels may exceed the limits set by the customer requirements and stress the product beyond the levels expected in normal use. It is assumed that good units are unaffected by the test. All the units that pass the test are released for shipment. The product must be robust to the stresses; therefore, HASS testing can not be performed unless a HALT program has been part of the product development. HASS testing will remove early-life-failure units from the population, but the primary strategy of the test program is to detect any shift in the production process. Before a HASS program can be successful, it is necessary that the production process has already proven to be capable and in statistical control. The stresses, usually environmental, must be determined. The test equipment, including any special test chambers, must be available and in sufficient quantity, and test time allotted as part of the process in order not to slow production.

Life testing requires that the test be run on several units over a period of time to accumulate a total unit test time. Using the assumption of a constant failure rate, 100 units tested for 100 hours, 50 units tested for 200 hours, and 10 units tested for 1000 hours all result in 10,000 hours total unit test time. The data from these tests can be used to make estimates of product reliability measures, such as failure rate or the MTBF, and to determine the confidence limits of the estimate. It is important that the units tested be typical of the production units. Any changes in the design will affect the testing program, probably requiring more time or test units.

Before beginning a test, it is necessary to decide how the test will be terminated. Several possibilities are in use by reliability engineers. *Truncation* refers to terminating the test at some specified amount of time or other stress metric. For example, a part or sample of parts may be cycled a million times, or be cryogenically exposed for a certain amount of time and then analyzed. *Test-to-failure* refers to studies that record data about failure type and time. Some such tests continue until all units have failed. A *truncated test-to-failure* would test a sample of items until a certain number had failed (the truncation point). A *degradation test* terminates when a product reaches a predetermined reduced level of performance. The analysis of data from these different protocols requires special techniques. See Section 1 of Chapter 15 for a discussion of censored data. Statistical software packages typically are used.

2. TEST ENVIRONMENT

> Evaluate the environment in terms of system location and operational conditions to determine the most appropriate reliability test. (Evaluate)
>
> **Body of Knowledge V.A.2**

The strategy of the reliability test and the product itself will determine the environment of the test. If the desired result of the test is to produce failures typical of product use, the test must be conducted in an environment that reflects the environment of use. Sometimes, increased environmental stress can be used to accelerate these tests. Not all products react the same to a given environmental stress. Solid-state electronics are very sensitive to increased temperature. Mechanical units might be affected by vibration or increased contaminants such as salt spray or dust. Magnetic units might be affected by radiation.

Electronic circuit boards could be exposed to vibration to detect poor soldering, inadequate strength, or other mechanical defects. Radiation will also affect solid-state electronics and could be used to verify that shielding is adequate. Humidity, rapid temperature ramps, and shock are other environmental stresses used during reliability testing.

If the strategy of the reliability testing is to eliminate design weaknesses from the product, environmental stresses in excess of normal use may be used. The test strategy is to produce failures so that design changes can be introduced to improve reliability. The unit will fail in ways not typical of normal use. The elimination of the weak components will increase product reliability. The level of these stresses can greatly exceed the design limits of the product. No reliability information can be obtained by testing units at stress levels lower than the normal use environment.

Test chambers capable of the appropriate test environments are required. Test equipment is now available that will allow the simultaneous application of multiple environments. *Combined environmental reliability testing* (CERT) is an application of multiple environmental stresses simultaneously. The simultaneous application of temperature cycling and vibration is an example of CERT.

Harry W. McLean (2002) gives detail as to the maximum level of environmental stresses that are appropriate. The maximum levels are specific to the product, but typical operation ranges are:

Temperature:	–70 °C to + 100 °C
Temperature rate:	60 °C per minute
Vibration:	Up to 30 Grms

The most advanced vibration systems utilize six degrees of freedom at frequencies from 2 to 5000 Hz. Low-frequency energy is used to excite components with high mass; higher frequencies excite low-mass components. Solid-state components can be burned-in at 150 °C. Integrated circuits should withstand temperature ramp rates of 90 °C per minute.

More-typical test environments perform vibration using three degrees of freedom and a temperature ramp rate of 20 °C per minute. If the environmental testing is sequential, the order needs to be determined and might be important. Combined thermal and vibration testing eliminates this requirement. The results of the testing should be both accurate and precise. *Accuracy* implies that the data are on target, and *precision* relates to the tolerance limits of the equipment. This requires that the calibration of the test chambers, the measuring equipment, and the data recording devices be traceable to a standard. Periodic calibration is necessary and should be stated as part of the overall reliability test plan.

Chapter 11

B. Testing During Development

> Describe the purpose, advantages, and limitations of each of the following types of tests, and use common models to develop test plans, evaluate risks, and interpret test results. (Evaluate)
>
> **Body of Knowledge V.B**

1. ACCELERATED LIFE TESTS (E.G., SINGLE-STRESS, MULTIPLE-STRESS, SEQUENTIAL STRESS, STEP-STRESS)

The length of time required for a reliability life test can be prohibitive. To reduce the actual time of the test, accelerated life testing can be employed. The strategy is to increase the rate at which failures occur, not cause new types of failures. The assumption of accelerated life testing is that the failure modes are unchanged by the increased stress. Failure analysis can be used to determine whether new modes of failure result from the increased stress. This would be a reason to reduce the level or change the stress used to effect the acceleration.

Single-stress tests. Some devices can be accelerated in time by simply using them at a rate that exceeds the normal use rate. Automatic test facilities can be built that will continuously operate the units. The time of use and the times of failure can be recorded. A fixture operating at 120 strokes per minute could exercise keys on a computer keyboard one million times in less than six days. A home appliance such as a dishwasher that is cycled every two hours will accumulate 1200 cycles (average for eight years) in 100 days (less than three and one-half months).

Increased stress is often used to accelerate time. Suppose it is known that under normal use five percent of the population will fail in a time $t_{(n)}$. If the units are operated under an increased environmental stress, the same five percent will fail in a time $t_{(s)}$ ($t_{(s)} < t_{(n)}$). The ratio $t_{(n)}/t_{(s)}$ is defined as the *acceleration factor* (A_F). The actual time of the test (t) is referred to as the *test time* or *clock time* of the test. The equivalent time of the test is $t_{(Eq)}$ and is equal to the product of the acceleration factor and the test time:

$$t_{(Eq)} = (A_F) \times t$$

The acceleration factor is assumed to remain constant over the time of the test.

The models used to accelerate time will depend on several factors. Electrical and mechanical failure modes will need different models. Increasing the constant failure rate or reducing the time to the onset of wear will involve different models. Two commonly used models are the *Arrhenius model*, when temperature is used to accelerate time, and the *power law* (sometimes referred to as the *inverse power law*) *model* when stresses other than temperature are used. The acceleration factors are usually derived from the models using physics-of-failure analysis for the various failure modes, testing to determine the value of the acceleration factor, and then verifying the acceleration factor. Many times, the verification may be delayed until results from the field can be obtained and analyzed. The determination of acceleration factors can be a lengthy and costly process.

The Arrhenius Model

The Arrhenius acceleration model is based on the Arrhenius equation (Savanti Arrhenius, 1858–1927). The Arrhenius equation states that the rate of a chemical reaction increases as the temperature increases:

$$R = Ae^{\frac{-E_A}{KT}}$$

where

R is the reaction rate of the chemical reaction

A is a scaling factor and will divide out of the final result

K is Boltzmann's constant (8.617×10^{-5} electron volts/degree Kelvin)

T is the temperature in degrees Kelvin (°C + 273 degrees)

E_A is the activation energy in electron volts

Reaction rate can be thought of as being synonymous with failure rate. Increasing the test temperature will increase the constant failure rate of the units, which has the effect of accelerating time.

The acceleration factor is the ratio of the reaction rate at the increased temperature (T_S) and the reaction rate at the use temperature (T_U):

$$A_F = \frac{R_S}{R_U}$$

This reduces to

$$A_F = \exp\left[(E_A/K)(1/T_U - 1/T_S)\right].$$

The key to using the Arrhenius model is determining the appropriate activation energy. An approximation that can be used in the absence of knowing the activation energy is that every 10 °C increase in temperature doubles the failure rate. Activation energies usually fall in the range 0.5 eV to 2.0 eV. The activation energy is assumed to be constant for a given failure mode.

EXAMPLE 11.1
ARRHENIUS MODEL

An electronic device that normally operates at a temperature of 50 °C is subjected to a stress temperature of 100 °C. The activation energy for the failure mode is 0.8 electron volts (eV).

What is the acceleration factor using the Arrhenius equation?

$$AF = \exp\,[E_A/K\,(1/T_u - 1/T_s)]$$

$$AF = \exp\,[(0.8/8.617 \times 10^{-5})(1/323 - 1/373)] = 47$$

Assuming that the increased temperature did not cause new failure modes, two days of testing at 100 °C is equivalent to about three months of use.

To verify the value, it may be necessary to test units at various stress (temperature) levels. For example, units may be tested at a high temperature, an intermediate temperature, and a low temperature. The low temperature should represent the normal use value, but be high enough to result in a few failures during the test. The high temperature should be near the upper limit over which the failure modes remain unchanged. The intermediate temperature should be significantly different from the two other values. More units need to be assigned to the lower temperatures to ensure that failures will occur during the test. If a total of 100 units were available for test, a 4-2-1 allocation would result in 57 units tested at the low temperature, 29 units tested at the intermediate temperature, and 14 units tested at the high temperature. It may require more than one test to find these values. Failures should occur at a faster rate at the higher temperatures. If an analysis of the results shows that the same failure modes occurred at each temperature, then an acceleration factor could be determined by comparing the time for a given percentage of failures to occur. If five percent of failures occur at t_S for the high temperature and five percent of failures occur at t_L for the low temperature, the acceleration factor is t_L/t_S. This analysis could be done graphically assuming the Weibull distribution (see Meeker and Hahn 1985).

The Power Law Model

The power law model applies to units subjected to accelerating stresses that are not thermal. The law states that the life of the product is inversely proportional to the increased stress:

$$\text{(life at rated stress)/(life at accelerated stress)} = [\text{(accelerated stress)/(rated stress)}]^b$$

Testing at two stress levels will give data to solve for b. It is then assumed that b will remain constant over the range of applied stress.

Multiple-stress tests. Other models can be used with combined stresses. Temperature and humidity or temperature and vibration are examples of combined

EXAMPLE 11.2
POWER LAW MODEL

A device that normally operates with five volts applied is tested at two increased stress levels $V_1 = 15$ volts and $V_2 = 30$ volts.

Analysis of the data from the tests indicates that five percent of the population failed at 150 hours when the high stress voltage of 30 volts was used. When the stress voltage of 15 volts was applied, five percent of the population failed at 750 hours.

At what time t would five percent of the population be expected to fail under the normal stress of five volts?

$$[750/150] = [30/15]^b$$

$$\log [750/150] = b \log [30/15]$$

$$b = 2.3$$

$$[t/150] = [30/5]^{2.3}$$

$$t = 9240 \text{ hours.}$$

This means that an hour of testing at 30 volts is equivalent to 62 hours of use at five volts. This results in an acceleration factor of 62.

stresses that can be applied to accelerate failures and reduce the actual test time. The Eyring model can also be used when temperature is the acceleration stress. The Eyring model, the Arrhenius model, and the power law model can be combined to be used with multiple stresses. For an excellent discussion of combined stress models see Kececioglu (2001).

Step-stress testing is sometimes used to further accelerate a reliability test. This technique involves increasing the stress level at fixed points and recording the failures or degradation levels. For example, a test might apply five amps of current for the first 100 hours, then 10 amps for the next 100 hours, and so on, recording product performance at each step. The statistical analysis of the data is somewhat involved and is covered in Sheng-Tsaing (2000).

2. DISCOVERY TESTING (E.G., HALT, MARGIN TESTS, SAMPLE SIZE OF 1)

Highly accelerated life testing (HALT) is a reliability test program used in the early stages of product development. HALT is not a new concept. HALT testing grew out of the step-stress or overstress testing programs and uses the same strategy. The procedure is to continue to increase stress on a unit until failure occurs. The intent is to identify the weak link in the design in order to make a change to improve the design.

The unit is not expected to fail when exposed to stresses that are within the design limits. The stresses that are applied exceed the specification limits of the design, forcing the unit to fail. Failure analysis and design improvement follow each failure. The process ensures that a robust design is created during the design and development phase, and eliminates the need for design changes during the production phase. HALT testing many times uses combined environments utilizing temperature cycling, humidity, and vibration.

The purpose of HALT testing is to improve the reliability of the product. It must be employed early in product development. It is much more efficient and cost-effective to make changes before the design is completed, parts are ordered, and the processing tooling is set.

Product failures are analyzed for product improvement purposes. These failures may not be typical of the product when it is operated within its design specifications. The intent is to cause failures to occur quickly, using a small number of units for the test.

Times to failure are not recorded, and all the products tested might fail. Therefore, reliability measures such as *mean time between failures* (MTBF, or MTTF) or failure rate can not be calculated using the results of the test. The amount of reliability improvement due to the design change can not be quantified using the results of HALT. The product design will be robust, with a strength that not only exceeds the normal stresses it is expected to experience during use, but also in excess of the stresses in the tails of the stress distributions. It is these stresses that cause failure in otherwise well-designed products.

Highly accelerated life testing should be a standard component of the design and development phase if there is an expectation that the delivered product will perform with zero or very few failures. The details of the HALT program will differ with the product. The stresses to be used, the limits on the stresses, whether the testing is cycled or static, and other details are product dependent. The HALT stresses used during the design of a new type of stepper motor and radar antenna servo system will be different from those used during the design of an autopilot for a new military aircraft. Testing as an entire unit might be appropriate for some products, while the testing of subsystems might result in quicker reliability improvement for other products. The test plan and the test procedures need to be developed with input from Reliability Engineering, Test Engineering, Product Engineering, and Design to ensure the most efficient use of test time for the HALT process.

McLean (2002) describes three distinct phases of the HALT process.

Phase 1: Pre-HALT Phase

During this phase, testing procedures are documented. Test equipment is procured and made ready. Procedures for recording of data are in place. The availability of necessary resources is assured.

Phase 2: HALT Phase

The testing is performed during this phase. The test procedures developed in the pre-HALT phase are followed. The operating units are subjected to elevated stresses. The tests are monitored, the results are analyzed, and the data recorded as previously determined.

Phase 3: Post-HALT Phase

Each issue uncovered during the HALT phase is subjected to root cause analysis and corrective action. A person is assigned responsibility for each action. Each action is open-ended until it is closed by Reliability Engineering.

Phase 2 and Phase 3 can then be repeated as necessary.

Margin tests apply increasing stress levels in an effort to develop a margin above specification for a particular failure mode. When a failure occurs, the cause is remedied and the test continues until a design results that has a predetermined margin above design specification. The larger the margin, the lower the probability that that particular failure mode will occur under design stress.

3. RELIABILITY GROWTH TESTING (E.G., TEST, ANALYZE, AND FIX (TAAF), DUANE)

Reliability growth is the improvement in product reliability over a period of time. This improvement in reliability is due to changes in the product design. In the early stage of new product development, problems exist in the design that negatively affect reliability. Early-development reliability activities such as FMEA, design reviews, reliability prediction, and early testing on prototype and engineering models will begin to identify these problems. Changes to the product design will begin to eliminate these problems. Tests will then be run on product of the new design, and other problems will be identified and eliminated. This activity when repeated is referred to as *test, analyze, and fix* (TAAF) and will result in reliability growth. MIL-HDBK-189 provides an understanding of the concepts and principles of reliability growth testing.

The most frequently used models to track reliability growth are the *Duane model,* developed by James Duane in 1962, and the *Army Materiel Systems Analysis Activity* (AMSAA) *model* developed by Larry Crowe in 1978. Both use the observed linear relationship between the logarithm of cumulative MTBF and cumulative test time. The Duane model is based on observation and uses a deterministic method to track reliability growth. The AMSAA model allows for statistical procedures to be used to model reliability growth.

The data are cumulated as the TAAF process progresses. A quantity called the *cumulative MTBF* (θ_m) is calculated:

$$\theta_m = \frac{T}{r}.$$

where

T is the total unit test time including all testing

r is the total number of failures including all testing

In the Duane model,

$$\theta_m = K(T)^b.$$

As the TAAF process continues and T increases, θ_m will increase, reflecting reliability growth. This growth can be tracked:

$$\log \theta_m = \log K + b \log T$$

is a linear equation in logarithms, and will plot as a straight line on log-log graph paper. Measuring b from the graph will give the reliability growth rate.

The value of b can also be calculated. θ_0 is an initial value for the cumulative MTBF at cumulative unit test time of T_0. θ_1 is a cumulative MTBF value after a total unit test time of T_1 ($T_1 > T_0$):

$$b = [\log(\theta_1/\theta_0)/\log(T_1/T_0)]$$

In the Duane model, the true MTBF will grow at the same rate as the cumulative MTBF value. At any time during the testing, the true MTBF (θ) can be found as

$$\theta = \theta_m/(1 - b).$$

The growth rate b can be used to compare the growth of a given project with the growth of other similar projects. A higher than average growth indicates that resources are being used aggressively for that project. A lower than average growth is an indication that the use of reliability improvement resources is limited for the project.

A growth rate of about 0.25 to 0.4 is average for most projects. A higher growth rate shows that the effort to eliminate design weaknesses has been given top priority. A lower growth rate indicates that reliability improvement actions are taken to eliminate only the most obvious design flaws.

EXAMPLE 11.3
DUANE RELIABILITY GROWTH

A new speed sensor and control module for an ABS braking system is in development. Units are subjected to three cycles of growth development testing. After each set of tests the results are analyzed and corrective action is implemented. After each design change new units are built for the subsequent test. Each test is conducted as a replacement test utilizing 20 units. The equivalent test time for each test is 1000 hours, resulting in a total test time of 20,000 hours.

The following results were obtained:

Test Cycle 1

Equivalent test time $T_{(1)}$= 20,000 hours. The cumulative test time T = 20,000 hours.

Number of failures $r_{(1)}$ = 20. The cumulative number of failures r = 20.

$$\theta_{C(1)} = 20{,}000/20 = 1000 \text{ hours}$$

Test Cycle 2

Equivalent test time $T_{(2)}$= 20,000 hours. The cumulative test time T = 40,000 hours.

Number of failures $r_{(2)}$ = 12. The cumulative number of failures r = 32.

$$\theta_{C(2)} = 40{,}000/32 = 1250 \text{ hours}$$

The calculation of $\theta_{(2)}$ uses the data from test cycle 1 and test cycle 2.

Continued

Continued

Test Cycle 3

Equivalent test time $T_{(3)} = 20{,}000$ hours. The cumulative test time $T = 60{,}000$ hours.

Number of failures $r_{(3)} = 8$. The cumulative number of failures $r = 40$.

$$\theta_{C(3)} = 60{,}000/40 = 1500 \text{ hours}$$

The calculation of $\theta_{(3)}$ uses the data from test cycle 1, test cycle 2, and test cycle 3.

A plot of the data on log–log graph paper is shown in Figure 11.1. Cumulative MTBF (θ_C) is plotted on the *y*-axis versus cumulative time (T) on the *x*-axis. A linear measure of the slope of the line shows a growth rate of about .37.

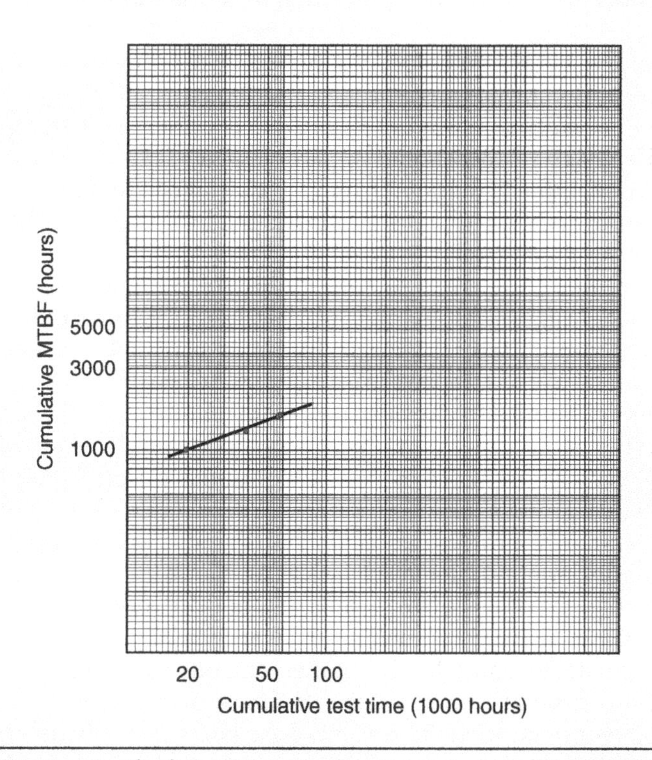

Figure 11.1 Duane growth plot.

4. SOFTWARE TESTING (E.G., WHITE-BOX, BLACK-BOX, OPERATIONAL PROFILE, AND FAULT INJECTION)

Various software testing strategies are used. To be the most effective at removing faults or "bugs" from the software, testing should be a function of each development phase. *White-box testing,* also referred to as *structural testing,* implies that the tests are designed and implemented with complete knowledge of the inner structure of the system being tested. It is testing the internal functions that must collectively result in the specified external output. The intent is to cover all paths and to test all branches within the system. White-box testing focuses on how the software works.

It is desired to predict the amount of testing required to remove the majority of faults from a new software package. The fault rate decreases as faults are found and eliminated from the software. To make a prediction as to the number of faults remaining at any given time in the testing program, it is necessary to have a prediction of the number of faults at the start of the testing program. One method, referred to as *fault injection,* is to seed the program with known faults and measure the test time to uncover a given percentage of these faults. Compare that to the number of true program faults that have been found in a given test time and, using the fault rate, make a prediction as to the number of original faults in the program.

Many of the faults contained in large software programs are due to poorly stated or misunderstood specifications. The focus of software reliability should be on defect prevention instead of defect detection and removal. A specification review and a design review should be part of every software development project.

A testing program is required to ensure that software is released with a minimum number of embedded faults. Testing is the most efficient method of finding and eliminating faults from software. O'Connor (2002) states that more than 50 percent of programming errors are due to the lack of understanding of the specifications. Good design controls and well-stated specifications reduce the probability of a programming error. Not introducing an error into the program is much preferred over trying to remove the error from the program.

As faults are found and eliminated, the probability of finding additional faults decreases. This is the early-failure part of the software bathtub curve. This testing activity is a part of the development phase, and differs from the early-failure part of the hardware bathtub curve. The early-failure part of the hardware bathtub curve occurs after the production phase. It is desirable to test the software package for every possible input under every condition. Error-free software would then be delivered to the customer if every fault that is uncovered is eliminated without adding additional faults. For complex software packages, all possible conditions may not be anticipated, and the elimination of faults could result in the creation of new faults. It is generally accepted that the most extensive testing programs will uncover and eliminate about 95 percent of the faults in a software program.

Testing software programs with respect to their external specifications is referred to as *black-box testing* or *functional testing.* Representative inputs are inserted, and the resulting outputs are compared to the requirements. The testing is to determine whether the software responds as intended. The intent is to test the completed system to determine whether the original requirements are met. Black-box testing is not concerned with internal functions. Black-box testing should be performed in the operating environment using the actual interface units.

Software testing should begin as the programs are written. Testing smaller programs is efficient and less complicated. Testing should continue as the programs are integrated into a system. White-box testing is concerned with how the system works. The individual programmers may be best suited for this type of testing as they understand their own codes. The program can be immediately corrected as errors are found.

Errors that are found by the customer after the software is released may not be immediately corrected. The customer may require use of the software in ways not anticipated by the test engineer. If the customer could test the software before release, these faults might be found and eliminated. For some applications, it is possible to involve the customer in testing the software before it is released. This is referred to as *beta testing*. For a flight control software package, this testing might be done on a flight simulator instead of an actual aircraft.

Operational profile testing exercises the software package under all the anticipated conditions of the users. Faults that remain in the software are obscure and difficult to detect. These faults are experienced in a random manner and will not immediately be corrected. This is the justification for considering the fault rate constant and for using the exponential distribution to model the reliability of software after it is released.

Chapter 12

C. Product Testing

> Describe the purpose, advantages, and limitations of each of the following types of tests, and use common models to develop product test plans, evaluate risks, and interpret test results. (Evaluate)
>
> **Body of Knowledge V.C**

1. QUALIFICATION/DEMONSTRATION TESTING (E.G., SEQUENTIAL TESTS, FIXED-LENGTH TESTS)

Qualification/demonstration reliability testing is commonly referred to as *compliance testing*. Compliance tests are used to demonstrate that a product parameter conforms to a given requirement. In reliability compliance testing, the parameter could be mean time between failures (MTBF) or mean time to failure (MTTF), a failure rate (failure intensity), or a reliability value. IEC standard 61124-2006 *Reliability Testing—Compliance Tests for Constant Failure Rate and Constant Failure Intensity* contains procedures for testing values of failure rate or MTBF/MTTF, and IEC standard 61123-1997 *Reliability Testing—Compliance Plans for Success Ratio* contains procedures for testing reliability values for success/fail items. IEC standard 61124 contains fixed-time test plans, and IEC 61123 contains fixed-trial test plans. Both standards contain truncated sequential test plans. The best-known military standard for reliability demonstration or compliance testing is MIL-HDBK-781 *Reliability Test Methods, Plans, and Environments for Engineering Development, Qualification, and Production*. The assumption for all compliance test plans that use time as the continuous variable is that the failure model is the exponential. This implies that the failure rate is constant and that MTBF/MTTF is equal to the reciprocal of the failure rate. The assumed failure model for the fixed-trial/failure test plans is the binomial.

All compliance test plans are described by an operating characteristic (OC) curve. An *operating characteristic curve* is a graph showing the probability of demonstrating compliance given the true value of the product parameter. For a test with time as the continuous variable, the OC curve will have the true value of the required parameter (failure rate [λ] or MTBF [m]) on the x-axis and the probability

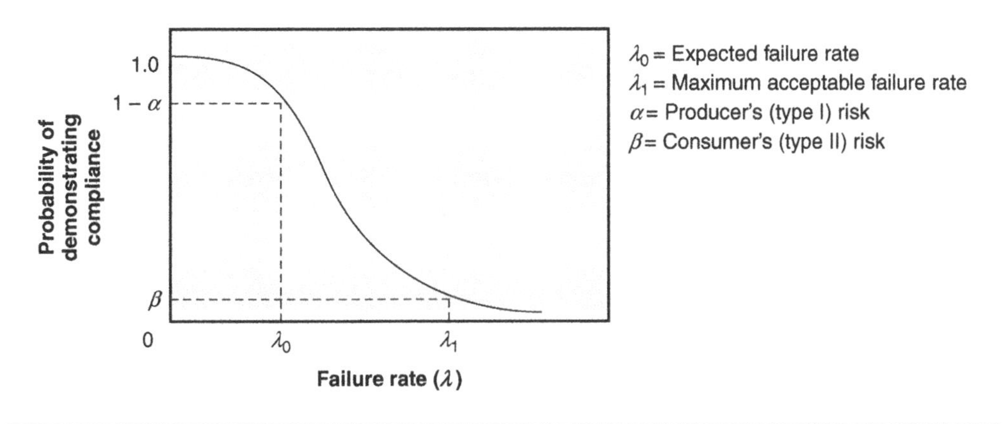

Figure 12.1 Operating characteristic curve for sampling plan.

of demonstrating compliance (P[A]) on the y-axis. The unit reliability (R), or the probability of success for each trial, is on the x-axis, and the probability of demonstrating compliance is on the y-axis for fixed-trial test plans.

If failure rate is used as the compliance requirement, the probability of demonstrating compliance will be high if the true failure rate is equal to or less than the requirement. The probability of demonstrating compliance will decrease as the true value of the failure rate increases. If MTBF is the compliance requirement, the probability of demonstrating compliance will be high if the true MTBF is equal to or greater than the requirement. The probability of demonstrating compliance will decrease as the true MTBF decreases. IEC standard 61124 uses MTBF as the compliance value.

A typical OC curve for failure rate is shown in Figure 12.1.

The OC curve shown uses failure rate (λ) as the compliance value. The test is defined by two points on the OC curve: an accept point defined by λ_0 and $1 - \alpha$, and a reject point defined by λ_1 and β. The product is in compliance if the true failure rate is equal to or better (less) than λ_0, and should pass the test. There is a risk of failing the test even though the product is in compliance with the requirement. This risk is α and is known as the *probability of a type I error.* The product is not in compliance if the true failure rate is worse (greater) than λ_1, and should fail the test. There is also a risk of passing the test even though the product is not in compliance with the requirement. This risk is β and is known as the *probability of a type II error.* The ratio of the two λ values (λ_1/λ_0) is referred to as the *discrimination ratio.* As the discrimination ratio approaches 1, the required amount of testing will increase. This could result in an increase of both the test time and the number of units on test.

MTBF/MTTF could be used as the compliance value. The accept point for this OC curve is defined by m_0 and $1 - \alpha$, and the reject point is defined by m_1 and β. The ratio m_0/m_1 is the discrimination ratio.

The Basics

Fixed-Time Test Plans. A fixed-time compliance test consists of placing units on test to accumulate test time (T) while recording failures. The criterion for acceptance (compliance) is based on accumulating a given amount of test time and having an acceptable number of failures. The maximum number of allowable failures for compliance is usually specified as *c*. The test is continued until either the required amount of test time has been accumulated (accept), or the allowable number of failures has been exceeded (reject).

Both the required amount of test time and the allowable number of failures are determined by the accept and reject points on the OC curve. Once the two points are defined, the amount of cumulative test time *T* required and the allowable number of failures *c* is set. The cumulative test time *T* is the total of the operating time of all the units during the test, including the units that fail as well as the units that do not fail.

Either a replacement test or a non-replacement test can be performed to acquire the required cumulative test time. In a replacement test, the units that fail are replaced with good units. This essentially keeps the same number of units on test at all times. If a non-replacement test is used, the units that fail are not replaced and the test population becomes smaller with each failure. The cumulative test

EXAMPLE 12.1

Ten units are on test. The units are not replaced when they fail (non-replacement).

One unit fails at $t_1 = 685$ hours, and a second unit fails at $t_2 = 1690$ hours.

The test is ended at $t = 2500$ hours with no additional failures.

What is the total accumulated test time?

$$T = 685 + 1690 + (8)(2500) = 22{,}375 \text{ hours}$$

EXAMPLE 12.2

An example using IEC standard 61124 will be shown.

It is desired to show compliance of a product MTBF.

The acceptable value $m_0 = 3000$ hours, and the rejectable value $m_1 = 1000$ hours.

The discrimination ratio $m_0/m_1 = (3000)/(1000) = 3$.

The α and β risks are both set to 0.10.

From Table 3 of IEC standard 61124 (Table 12.1) test plan B.7 can be used.

The OC curve for this plan is shown in Figure B.6 of the standard (Figure 12.2).

The test requires a cumulative test time of 9300 hours, and the allowable number of failures is five.

$$T = (3.1) \times (m_0) = (3.10) \times (3000) = 9300 \text{ hour}$$

$$c = 5$$

Continued

Continued

Table 12.1 Table of fixed-time test plans from IEC standard 61124.

Test plan no.	Characteristics of the plan: Nominal risks		Characteristics of the plan: Discrimination ratio	Test time for termination	Acceptable number of failures	True risks for $m = m_0$	True risks for $m = m_1$
	α%	β%	D	T_1/m_0	c	α%	β%
B.1	5	5	1.5	54.10	66	4.96	4.84
B.2	5	5	2	15.71	22	4.97	4.99
B.3	5	5	3	4.76	8	5.35	5.40
B.4	5	5	5	1.88	4	4.25	4.29
B.5	10	10	1.5	32.14	39	10.00	10.20
B.6	10	10	2	9.47	13	10.00	10.07
B.7	10	10	3	3.10	5	9.40	9.90
B.8	10	10	5	1.08	2	9.96	9.48

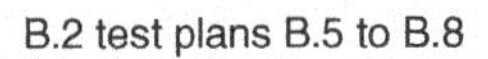

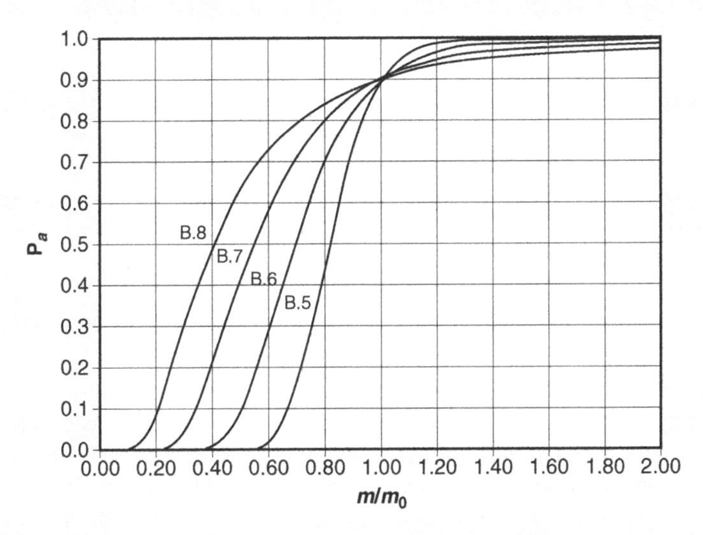

Figure 12.2 OC curves for fixed-time test plans from IEC standard 61124.

The accept and the reject points on the OC curve can be verified. Using the constant failure rate concept the expected number of failures is $\lambda \times T$. If the true MTBF is 3000 hours, $\lambda \times T = (1/3000) \times (9300) = 3.1$. This is the accept point. Using the cumulative Poisson tables, which can be justified because of the constant failure rate, the probability of five or fewer failures is 0.90. At the reject point, $\lambda \times T = (1/1000) \times (9300) = 9.3$. Again from the cumulative Poisson tables, the probability of five or fewer failures is 0.10.

EXAMPLE 12.3

If the rejectable MTBF is changed to 1500 hours (m_1 = 1500 hours), the discrimination ratio is now 2, and test plan B.6 can be used.

The required cumulative test time is 28,410 hours [(9.47) × (3000)] and the allowable number of failures is 13.

It is important to note the increase in required cumulative test hours as the discrimination ratio approaches 1. This is in spite of the fact that the MTBF value to be demonstrated remains the same.

time is the total of all the operating times of all the units during the test, no matter which type of test is used.

The time required to perform a demonstration test can become excessive. If accelerated testing can be used (see Chapter 11), the actual time required can be reduced. For Example 12.3, the required cumulative test time is 28,410 hours. This could be accomplished by placing 20 units on test and running the test for about 1420 hours (approximately two months). If a non-replacement test is performed, the test time would be extended with each failure. If an acceleration factor of 60 can be applied to the test, the actual test time (t) could be reduced to 23.7 hours to get the same cumulative test time (approximately one day using 20 units on test):

$$T = A_F(t)(n) = 60(23.7)(20) \approx 28410 \text{ hours.}$$

Sequential Test Plans. Sequential test or *probability ratio sequential test* (PRST) plans are different from the fixed-time test plans. For a fixed-time test, the decision to accept can not be made until the test is completed and the required cumulative test time is reached. The results of sequential tests are continuously assessed to arrive at one of three alternative decisions. Sequential testing results in a decision to accept compliance, reject compliance, or to continue testing at every value of cumulative test time. The intention is to make a decision to accept or reject compliance in the minimum amount of test time. A graph of a sequential test plan is shown in Figure 12.3.

The alternatives, described by the three regions on the graph, are *accept compliance, reject compliance,* or *continue testing.* The three regions are defined by parallel straight lines constructed using the values m_0 (acceptable MTBF) and m_1 (rejectable MTBF) along with α and β from a predetermined OC curve if MTBF is to be demonstrated. Acceptable failure rate (λ_0) and rejectable failure rate (λ_1) are used if the compliance value is a failure rate.

In order to ensure that the test does not continue for an indefinite time, there are rules for ending the test. The test could be terminated at a given time or on a given number of failures if a decision to accept or reject compliance has not been reached. This is referred to as a *truncated sequential test.*

A graph of a sequential test plan from IEC standard 61124 is shown in Figure 12.3. This plan has α and β risks equal to 0.10 and a discrimination ratio of 3. The OC curve for this test is shown in Figure 12.2 (test B.7). The equations for constructing the straight lines that identify the accept, reject, and continue testing regions, as well as the rules for termination, are included in the standard. On the

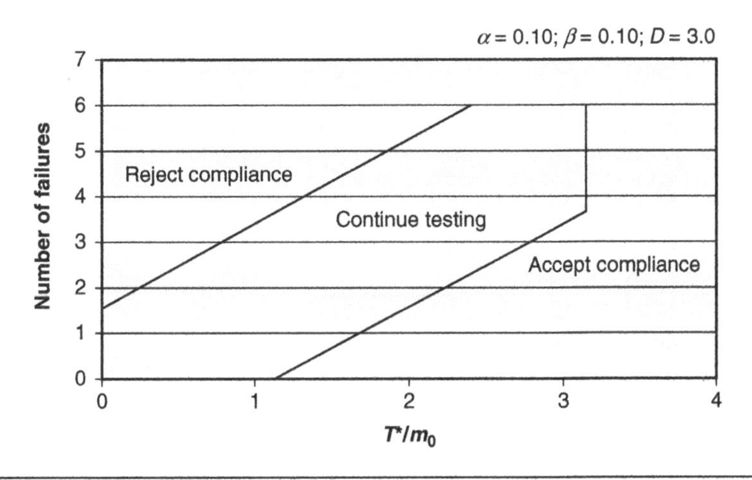

Figure 12.3 Sequential test plan.

average, an accept or reject decision will be reached with less cumulative test time than with fixed-time test plans.

2. PRODUCT RELIABILITY ACCEPTANCE TESTING (PRAT)

Product reliability acceptance testing (PRAT) will be done on reliability-critical elements during the production phase. This testing is to ensure that the inherent design reliability of the product is not compromised during production. This testing is performed on production units and is referred to as *production reliability testing.* If this testing is done to accept or reject production lots based on specified reliability requirements, it is referred to as PRAT. This testing is used to detect shifts in the process that would affect product reliability.

3. ONGOING RELIABILITY TESTING (E.G., SEQUENTIAL PROBABILITY RATIO TEST [SPRT])

Once the desired level of reliability has been achieved, the next challenge is to monitor production to determine whether that level is being maintained. The goal of the *sequential probability ratio test* (SPRT) is to get a cost-effective answer to this question. The procedure is to test a series of items one at a time. After each test, the cumulative number of failures c will be used to calculate a test statistic R. The R-value will be compared to two numbers L and U:

If $R \leq L$, the sequence of tests is terminated, and the product is considered satisfactory.

If $R \geq U$, the sequence of tests is terminated, and the product is considered unsatisfactory.

If $L < R < U$, the sequence of tests is continued.

Of course, the problem is to find the test statistic R and the set of L and U values. An example will help explain the method.

EXAMPLE 12.4

Assume that an MTTF of an item must exceed 100 hours. The expected MTTF is at least 200 hours. If the useful life is exponentially distributed, then the probability of survival is given by

$$f(x) = e^{-x/\mu}$$

where x = time and μ = MTTF

The items are tested one at a time for 20 hours. If the item is still working at the end of the test period it will be considered a success.

The probability of obtaining a success assuming $\mu = 200$ is

$$P_{200}(\text{success}) = e^{-20/200} \approx 0.905$$

The probability of obtaining a success assuming $\mu = 100$ is

$$P_{100}(\text{success}) = e^{-20/100} \approx 0.819$$

Assume n sequential parts have been tested. The binomial formula says that the probability of obtaining y successes is

$$P(y \text{ successes in } n \text{ trials}) = \binom{n}{y} p^y (1-p)^{n-y}$$

$$\text{Where } \binom{n}{y} = \text{the binomial coefficient } \frac{n!}{y!(n-y)!}$$

$$\text{So } P\,(y \text{ successes in } n \text{ trials assuming } \mu = 100) = \binom{n}{y} 0.819^y (0.181)^{n-y}$$

$$\text{and } P\,(y \text{ successes in } n \text{ trials assuming } \mu = 200) = \binom{n}{y} 0.905^y (0.095)^{n-y}$$

Define R as the ratio of these two probabilities:

$$R = \frac{P(y \text{ successes in } n \text{ trials assuming } \mu = 100)}{P(y \text{ successes in } n \text{ trials assuming } \mu = 200)} = \frac{0.819^y (0.181)^{n-y}}{0.905^y (0.095)^{n-y}}$$

For instance, suppose that after testing 10 items, six of the items have survived. Then,

$$R = \frac{0.819^6 (0.181)^4}{0.905^6 (0.095)^4} \approx 7.24$$

This R-value would then be compared to the L and U values for n tests.

The next problem is to define the L and U values. As one would expect, these values depend on the values for α and β, the probability of type I and type II error, respectively. It can be shown that the values of L and U are approximately

$$L \approx \frac{\beta}{1-\alpha} \text{ and } U \approx \frac{1-\beta}{\alpha}$$

Continued

Continued

For this example, use $\alpha = 0.05$ and $\beta = 0.10$ so L ≈ 0.1053 and U ≈ 18.

In the example with six items surviving after 10 items are tested, since $R = 7.24$, the sequential tests would be continued since 7.24 is between 0.1053 and 18.

The next step is to simplify the calculations to make them more practical to apply.

$$R = \frac{0.819^y (0.181)^{n-y}}{0.905^y (0.095)^{n-y}}$$

$$R = \left(\frac{0.819}{0.905}\right)^y \left(\frac{0.181}{0.985}\right)^{n-y}$$

$$R = 0.905^y 1.9053^{n-y}$$

Since this has exponential terms, take the ln of both sides

$$\ln(R) = y \ln(0.905) + (n - y) \ln(1.9053)$$

$$\ln(R) = -0.7444y + 0.6446n$$

One boundary for R is $L = 0.1053$, so the corresponding boundary for ln (R) is $\ln(L)$.

So, the formula for the boundary of $\ln(R)$ is

$$\ln(L) = -0.7444y + 0.6446n$$

and in our example $L = 0.1053$, so

$$\ln(L) = -2.251$$

$$-2.251 = -0.7444y + 0.6446n$$

$$\text{and } y = 0.87n + 3.02.$$

Similarly, the other boundary for R is

$$y = 0.87n - 3.882$$

Plotting these two lines provides a graphical tool as shown in Figure 12.4.

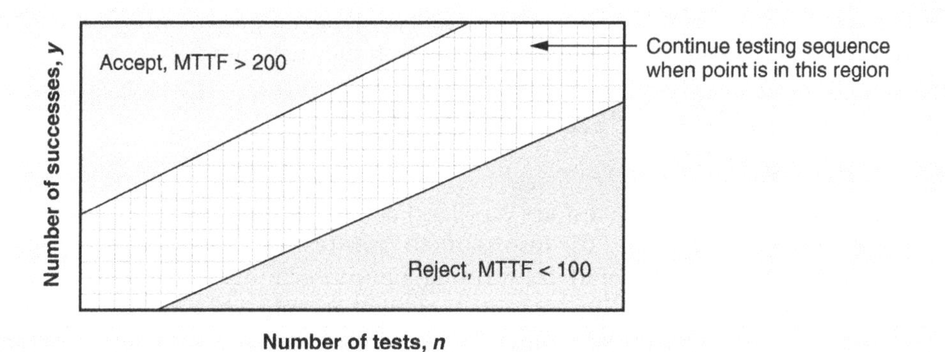

Figure 12.4 SPRT graph.

Part V.C.3

4. STRESS SCREENING (E.G., ESS, HASS, BURN-IN TESTS)

Stress screening is a final test performed on production units prior to release to the customer. *Burn-in testing* is allowing the product to accumulate use time before it is delivered. A burn-in or screening test program will reduce the occurrence of early-life failures after the product is delivered. During the burn-in, manufacturing-related nonconformities that cause early failure will be detected. To accelerate the rate at which the weak units fail (early-life failures), some stress (usually environmental) is applied during the test. This is referred to as *environmental stress screening* (ESS). This stress should not damage an acceptable unit. Units that do not fail will be delivered to customers.

Highly accelerated stress screening (HASS) (see HASS testing, Chapter 10) is a stress screening program to ensure that the final product will exceed its environmental requirements. The design has been proven to be robust using a *highly accelerated life test* (HALT) program. The environmental stresses of a HASS program may exceed the product design specifications. A HASS program can only be used if a HALT program was used during product development (see HALT testing Chapter 10). The HASS program will detect problems in the process that negatively affect product reliability. The HASS program is not intended to detect ongoing process problems. Its purpose is detecting a shift in a process that has previously been shown to be capable. HASS testing is not intended to compensate for a production process that is less than capable. Before HASS can be used successfully, accepted process quality procedures must be applied to the process. Processes must have capability measures, such as C_{pk}, at acceptable levels. Suitable SPC procedures must be in place. Modern test equipment allows the testing to be done using combined environments.

Once the process has demonstrated statistical control, it may be possible to replace the HASS test program, which is performed on 100 percent of the production units, with a stress audit program. This would reduce costs as it would require less test equipment and fewer personnel. It could also reduce time to market for high-volume production. A statistical approach to the audit that would involve stress screening a sample of the production units is called *highly accelerated stress audit* (HASA). A HASA test program will detect a process shift but has the risk of allowing some early-life failures to reach the customer.

5. ATTRIBUTE TESTING (E.G., BINOMIAL, HYPERGEOMETRIC)

The results of an attribute test are classified as one of two possible states. Examples of the classifications of the results of attribute testing include success/failure, acceptable/unacceptable, and conforming/nonconforming. Attribute testing is used to evaluate the reliability of units that must operate when called on to operate but have no active mission time. These units are referred to as *one-shot* items. These could be units such as sensors or igniters. For example, in the testing of a force-detecting sensor, the result of each test is success (the sensor detected the force) or failure (the sensor did not detect the force).

Results from a test in which a continuous variable, such as time, could be measured may sometimes be recorded as attributes. For example, several operating units are placed inside an environmental test chamber. At the end of the test, each unit is checked to see if it is still operating. The test result recorded for each unit is success or failure. If the unit failed, the exact time of failure is not known.

Binomial Distribution

The binomial distribution is usually used to estimate the reliability of units subjected to attribute testing. A necessary condition for using the binomial as a model is that the probability of success remains the same from trial to trial. This implies that each unit on test has the same probability of success.

A reliability estimate can be obtained from an attribute test. The test could be described as testing n units and recording r failures. The estimate of the reliability of the unit for the conditions of test is

$$\hat{R} = \frac{n-r}{n} \text{ for } r \geq 1.$$

A confidence limit could be placed on this estimate of reliability from an attribute test. The confidence limit will involve the use of the F-distribution. The confidence value is C. The risk or significance of the test is $\alpha = 1 - C$ (see Binomial Distribution, Chapter 5).

$$R_L = \frac{(n-r)}{(n-r)+(r+1)F_{[\alpha,2(r+1),2(n-r)]}}$$

The F value is the ratio of two χ^2 values and has two sets of degrees of freedom. The first set is for the numerator and is $2(r + 1)$. The second set is for the denominator and is $2(n - r)$.

EXAMPLE 12.5

Twenty sensors used in the airbag system of an automobile are tested at a force that should be detected. One of the sensors failed to detect the force and therefore was classified as a failure. What is the estimate of the reliability of the sensor for detecting that force?

$$n = 20$$

$$r = 1$$

$$\hat{R} = \frac{(20-1)}{20} = 0.95$$

EXAMPLE 12.6

Find the lower limit on reliability with confidence of 0.90 for the test results of Example 12.5.

From the F-distribution table, Appendix F:

$$F_{(0.10),\,4,38} = 2.1$$

$$R_L = \frac{19}{19+2(2.1)} \approx 0.82$$

The true reliability exceeds 0.82 with a confidence of 0.90.

Zero-Failure Test. A point estimate of reliability can not be made if the test results in zero failures. A lower confidence limit can be found for a zero-failure test. The lower confidence limit on the reliability value R_L for a test of n units with zero failures at a confidence value of C is

$$R_L = (1 - C)^{1/n}.$$

In the above equation (1 – C) is the risk and is referred to as the *significance* of the test. It may be replaced by α.

The equation becomes

$$R_L = \alpha^{1/n}.$$

This equation can be solved for the number of no-failure tests necessary to show a given reliability at a given confidence level. The test must be run without failures.

$$R_L = \alpha^{1/n}$$

$$\log R_L = (1/n) \log \alpha$$

$$n = \frac{\log \alpha}{\log R_L}$$

EXAMPLE 12.7

One hundred and fifty sensors used in the airbag system of an automobile are tested at a force that should be detected. None of the sensors failed to detect the force; therefore, no point estimate of the reliability of the sensor is possible. What is the lower 90 percent confidence limit on the reliability of the sensor for detecting the force?

$$R_L = (0.10)^{1/150} = 0.985$$

The true reliability exceeds 0.985 with a confidence of 0.90.

EXAMPLE 12.8

The compliance requirement is to demonstrate that the airbag sensor has a minimum reliability of 0.98 with a confidence of 0.90. How many sensors need to be tested if the test results in zero failures?

$$n = [\log(0.10)/\log(0.98)] = 114$$

Hypergeometric Distribution

If the probability of success does not remain the same from trial to trial, the binomial model is not applicable. The hypergeometric distribution can be used to model this case. An example would be sampling without replacement. The size of a finite population will decrease as sample units are removed. If the population contains a fixed number of defective items, the probability of finding a defective item changes as each individual unit is taken from the population.

The hypergeometric distribution gives the probability $P(d)$ of finding exactly d defective units in a sample of n units if the sample is taken from a population of N units that contained D defectives.

d = Number of defective units in the sample

D = Number of defective units in the population

n = Number of units in the sample

N = Number of units in the population

$$P(d) = \frac{C_d^D C_{n-d}^{N-D}}{C_n^N}$$

EXAMPLE 12.9

A production lot of 200 units contains 10 defective units. A sample of 30 is taken from the lot and inspected. What is the probability the sample will contain one or fewer defective units?

The probability that d is one or less is the probability that d is equal to 1 or that d is equal to zero. The probabilities are mutually exclusive, so they can be added.

$$P(d \le 1) = P(d = 0) + P(d = 1)$$

$$P(d \le 1) = \frac{C_0^{10} C_{30}^{190}}{C_{30}^{200}} + \frac{C_1^{10} C_{29}^{190}}{C_{30}^{200}}$$

$$P(d \le 1) = 0.189 + 0.352 = 0.541$$

Most scientific calculators will perform this calculation using the combination key.

C_d^D is the number of possible combinations that can be formed from D units if they are selected d at a time:

$$C_d^D = \frac{D!}{d!(n-d)!}$$

6. DEGRADATION (WEAR-TO-FAILURE) TESTING

It may be necessary to test units to determine the end of useful life. This is the beginning of wear-out. Some products may be expected to have extremely long lives. It is not uncommon to expect 150,000 miles of operation for some automotive systems, or 10 years for some electronic systems. To test for these times, some form of acceleration to increase the rate of wear is necessary. Rapid cycling to increase the rate of fatigue, salt spray to increase the rate of corrosion, temperature to increase the rate of degradation of solid-state electronics are a few of the acceleration stresses.

The failure rate of solid-state electronics will increase as the temperature increases. Good design practice for solid-state electronics dictates that heat be kept to a minimum and removed from the system. If the objective is to have units fail at a faster rate during testing, the addition of heat causes faster degradation, which can be equated to increasing the amount of test time. The Arrhenius model is often used to determine the equivalent time the unit is on test (see Accelerated Life Testing, Chapter 11).

It is sometimes possible to extrapolate to a failure time by observing the amount of degradation a unit has experienced in a given time. This would reduce the amount of test time necessary to determine the end of life. Information gained from measuring the amount of wear on a bearing, or the intensity of crack propagation due to fatigue at a given time could be used to predict the time to failure if the model for the rate of degradation is known or assumed. Models that are often used are the linear, exponential, and power law models. The model could be derived using physics of failure. Each product could be unique. For example, suppose the specification for the fabric covering of an office chair seat is 100,000 cycles of a 250-pound weight moving across then off the fabric. The failure mode is the visible color change when the backing surface shows. Previous testing has shown that the relationship of number of cycles versus fabric thickness lost is approximately linear. The problem is to find the fabric thickness that will meet the specification. To obtain the first estimate for the answer, a fabric is tested for 10,000 cycles and the thickness loss is measured. This would indicate that the first estimate for the design fabric thickness is 10 times the measured wear loss.

Part VI

Maintainability and Availability

Chapter 13

A. Management Strategies

1. PLANNING

> Develop plans for maintainability and availability that support reliability goals and objectives. (Create)
>
> **Body of Knowledge VI.A.1**

Repairable systems are systems that can be restored to service following a failure. The actions to restore a system to service following a failure are *corrective maintenance* actions. Most repairable systems can also have prescribed maintenance actions to keep the system in operating condition as well as reduce the probability of a failure due to wear-out. These actions are *preventive maintenance* actions.

Preventive maintenance (PM) includes all the actions performed to keep the system in an operating state by preventing wear-out failures. Preventive maintenance does not reduce the constant failure rate that is inherent to the system but tends to maintain the system at that level of failure probability. Preventive maintenance actions can be planned and, if possible, be performed when there is no demand to use the system.

Corrective maintenance (CM) includes all the actions required to return the system to an operating state once failure has occurred. Corrective maintenance actions can not be planned, and must be performed when the system fails. Corrective maintenance actions may be deferred in some cases; however, the system is not operable until the corrective maintenance is completed.

Availability is a measure of the likelihood that a system will be ready to operate (system is up) when it is called on to operate. Reasons the system would not be ready to operate (system is down) include the possibility that a failure has occurred and the corrective action has not been completed. Also, a system may not be operable because preventive maintenance actions are necessary. There could also be other logistic reasons that the system is not operable. Availability is a function of the number of corrective maintenance actions necessary and the time required to complete these actions, as well as the amount of maintenance required and the time necessary to restore the unit to service.

Availability is a function of reliability and maintainability. Reliability deals with reducing the frequency of unscheduled downtime. *Maintainability* is the ability of an item to be retained in or restored to specified conditions when maintenance actions are necessary. Maintainability deals with reducing the duration of downtime. This downtime includes both scheduled and unscheduled actions. If the preventive maintenance actions can be planned and performed when there will be no demand to use the system, the time to perform these actions will not affect the availability. Availability can be perceived as uptime (the time the system is operable) divided by total time (the time during which there could be a demand to use the system).

The mean time required to bring a system back to an operable state after a failure is expressed as the *mean time to repair* (MTTR). This value is the mean of all corrective maintenance actions and includes the probability that the action is necessary and the time required to complete the action:

$$\text{MTTR} = \frac{\Sigma(\lambda_i t_i)}{\Sigma\lambda_i}$$

λ_i is the failure rate (probability of occurrence) for the ith failure mode, and t_i is the time to repair the system after the failure has occurred. The distribution usually used to model times to repair is the lognormal. Most repairs can be made in a time less than the mean.

A mean for preventive maintenance actions can be found by replacing the failure rate for each failure mode with the frequency of occurrence of each preventive maintenance action and the time to perform that maintenance action.

The steady state or inherent availability of a system with a constant failure rate is

$$\text{A} = \frac{\text{MTBF}}{\text{MTBF} + \text{MTTR}}.$$

In order to increase the availability of a system, the system reliability could be increased (increase *mean time between failures* [MTBF]), or the time to restore the system to service could be reduced (decrease MTTR).

The MTTR is the mean time to repair the system to service after a failure. In other forms of the availability relationship, MTBF is replaced with a term that expresses the mean time between system downtimes for any reason, and MTTR is replaced with a term that includes corrective maintenance, preventive maintenance, logistic, and administrative downtimes.

2. MAINTENANCE STRATEGIES

> Identify the advantages and limitations of various maintenance strategies (e.g., reliability-centered maintenance (RCM), predictive maintenance, repair or replace decision making), and determine which strategy to use in specific situations. (Apply)
>
> **Body of Knowledge VI.A.2**

Maintenance strategies should be chosen to ensure a high level of availability while controlling costs. Preventive maintenance does not improve the inherent reliability of the system. Preventive maintenance will maintain the reliability level of useful life, keeping the failure rate low. It will also delay the onset of wear, thus increasing the length of useful life. Preventive maintenance actions can also introduce a fault into the system, causing the system to fail and the need for corrective maintenance. A study should be done to determine the most cost-effective scheduling of maintenance actions necessary to keep the system operational. Maintenance actions that are unnecessary should be eliminated.

A single unit, such as a pump, could be considered to be a system. Or, the pump could be considered to be a component of a larger system. In either case, system reliability is at the highest level if all units are in their useful life phase. Preventive maintenance strategies such as *reliability-centered maintenance* and *predictive maintenance* are proactive and require the replacement or refurbishment of the components of the system that are nearing the end of their useful life or entering wear-out. Replacement is made before failure occurs. *Condition-based maintenance* is reactive and is a corrective maintenance strategy. Replacement is made after failure occurs.

The appropriate strategy for replacement needs to be chosen. The information needed in order to choose the most appropriate replacement strategy includes:

- The failure distribution of the unit
- The cost associated with the failure of the unit
- Any safety issues associated with the failure
- The cost of the replacement unit
- The cost associated with scheduled replacements
- The cost of inspection or test

Predictive and Reliability-Centered Maintenance

Predictive maintenance assumes that the operator can detect the imminent failure of a unit. This detection can be by observation, analysis, or using test equipment. The analysis of an oil sample or the measurement of increased vibration might indicate wear and an increasing failure probability. If the increasing failure probability can not be detected by the operator, a *reliability-centered maintenance* strategy could be adopted. This strategy uses the predicted failure distributions to determine the optimum replacement time for units about to enter the wear-out phase. Replacement of units that are entering the wear-out portion of the bathtub curve will maintain the system reliability at the useful life level.

It is possible that any maintenance action can negatively impact reliability by inducing failure modes due to the maintenance action. These maintenance-induced failures are similar to the early-life failures due to the manufacturing or installation of a system. An effort should be made during design to ensure that standard preventive maintenance actions can be performed quickly and with a low risk of introducing problems due to the maintenance actions.

Maintainability/Apportionment/Allocation

Reliability deals with reducing failures of the system and thus reducing the frequency of unscheduled maintenance actions. Maintainability deals with reducing the duration of the downtime that is a result of both scheduled and unscheduled maintenance actions. Maintainability is often thought of as a technique for making repairs easy. It is, however, the engineering involved in minimizing the total downtime.

MIL-HDBK-472 gives the following definition of maintainability:

> *Maintainability is the ability of an item to be retained in or restored to specified conditions when maintenance action is performed by personnel having specified skill levels and using prescribed procedures and resources at each prescribed level of maintenance and repair.*

The maintenance actions are to be performed by skilled personnel, using proper procedures, with the proper tools, and having access to standard replacement parts. The objective is best accomplished if an effort is made to eliminate unscheduled downtime and reduce the duration of scheduled downtime.

A system-level maintainability requirement may need to be allocated to lower levels of the system. Maintainability allocation is a continuing process of apportioning requirements at the system level to subsystem levels. This will give values the designer can work toward. This is similar to a system reliability requirement being allocated to the various subsystems. To allocate the maintainability requirements, the failure rate and the MTTR are used.

MTTR Allocation

- A series system has three subsystems.
- The failure rate for subsystem i is λ_i, and the MTTR for subsystem i is t_i.
- The system failure rate is $\lambda = \lambda_1 + \lambda_2 + \lambda_3$.
- The system-level MTTR requirement is t^*.

$$\text{MTTR}_{(\text{System})} = t = \frac{(\lambda_1 t_1 + \lambda_2 t_2 + \lambda_3 t_3)}{(\lambda_1 + \lambda_2 + \lambda_3)} = \frac{\lambda_1 t_1}{\lambda} + \frac{\lambda_2 t_2}{\lambda} + \frac{\lambda_3 t_3}{\lambda}$$

EXAMPLE 13.1

A series system has three subsystems.

The failure rates of the subsystems are:

$\lambda_1 = 10 \times 10^{-6}$ failures/hr

$\lambda_2 = 30 \times 10^{-6}$ failures/hr

$\lambda_3 = 60 \times 10^{-6}$ failures/hr

The subsystem MTTR values are:

$t_1 = 150$ min

$t_2 = 100$ min

$t_3 = 50$ min

The system MTTR requirement is one hour ($t^* = 60$ min).

$$\text{MTTR}_{\text{(System)}} = (10/100)(150) + (30/100)(100) + (60/100)(50) = 75 \text{ min}$$

The attained value of the system MTTR is greater than the requirement.

In order to meet the system requirement, the maintainability allocation to each subsystem will be:

$t_1^* = (150/75) \times 60$ min $= 120$ min

$t_2^* = (100/75) \times 60$ min $= 80$ min

$t_3^* = (50/75) \times 60$ min $= 40$ min

In order to meet the maintainability goal, an MTTR allocation to each subsystem must be made if the $\text{MTTR}_{\text{(System)}} > t^*$. The MTTR allocation to subsystem i is

$$t_i^* = [t_i/t] \times t^*$$

The designer can work to subsystem MTTR requirements better than to system-level MTTR requirements.

3. AVAILABILITY TRADEOFFS

Describe various types of availability (e.g., inherent, operational), and the tradeoffs in reliability and maintainability that might be required to achieve availability goals. (Apply)

Body of Knowledge VI.A.3

System availability is determined by reliability (the probability of the system not failing) and maintainability (the ability to restore the system to service).

Availability is the measure of the time the system is in an operating state compared to the total time. There are several commonly used measures of availability. The various availabilities depend on the downtimes that are included in the total time. In the following equations the failure rate of the system is assumed to be constant.

Inherent availability excludes preventive maintenance and any logistic downtime. Included in the total time is downtime due to corrective maintenance.

$$A = \frac{MTBF}{MTBF + MTTR}$$

MTTR is the mean time to repair and includes the time for corrective maintenance actions along with the probability of the occurrence of the failure.

Achieved availability includes the downtime due to corrective and preventive maintenance actions. Only the active time is included. Not included is any delay time in acquiring supplies and administrative downtime. The *mean time between maintenance actions* (MTBMA) includes both scheduled and unscheduled maintenance actions.

MTBMA is a function of the failure rate (λ) and the preventive maintenance rate (μ). If the preventive maintenance rate is constant,

$$MTBMA = \frac{1}{\lambda + \mu}.$$

The *mean active maintenance time* (MAMT) includes the average (mean) corrective maintenance time and the average time to perform preventive maintenance. Included in the calculation is the frequency at which the actions will occur. The achieved availability can be calculated as

$$A = \frac{MTBMA}{MTBMA + MAMT}.$$

Operational availability includes all downtime. The *mean downtime* (MDT) includes logistic time, time waiting for replacement parts, and administrative downtime. Operational availability is calculated as

$$A = \frac{MTBMA}{MTBMA + MDT}.$$

The effort during design should be to ensure a high probability that the equipment is ready to be used when the customer demands its use. This requires that both reliability and the ability to maintain the equipment be specified characteristics of the design.

Chapter 14

B. Maintenance and Testing Analysis

1. PREVENTIVE MAINTENANCE (PM) ANALYSIS

> Define and use PM tasks, optimum PM intervals, and other elements of this analysis, and identify situations in which PM analysis is not appropriate. (Apply)
>
> **Body of Knowledge VI.B.1**

The goal of preventive maintenance is to optimize system reliability. Procedures must be established for preventive maintenance taking into consideration the failure rate pattern of the system or its various subsystems. Cost savings can be realized if preventive maintenance (replacement) is performed on systems with increasing failure rates. A preventive maintenance action can also be advantageous for a unit with a constant failure rate if that failure rate will begin to increase if the maintenance is not performed. For example, if a unit in its useful life phase is not lubricated periodically, the failure rate could increase due to excessive wear. If the preventive maintenance action is to replace a working unit, this would only be advantageous if the unit is nearing the end of its useful life or entering wear-out. It is possible that the maintenance action could induce a failure, causing a negative impact on the reliability of the system.

An optimum maintenance interval can be determined by considering the cost of performing the maintenance, the cost of a failure if the maintenance is not performed, and the cost associated with system downtime. If the interval is too short, the maintenance costs increase. If the interval is too long, the costs due to failure increase.

The cost due to system downtime can be minimized if preventive maintenance can be scheduled when there will be no demand to use the system. Downtime can also be reduced if scheduling will allow multiple maintenance actions to be performed once the system is taken off-line.

It is possible that the preventive maintenance action will consist of replacing a unit before failure can occur. O'Connor (2002) lists the following as necessary to determining the optimum replacement time:

1. The time-to-failure distribution and the parameters of that distribution.
2. The effect on the system of the failure.
3. The cost of the failure. This includes the cost of downtime because of the failure and any cost incurred due to safety considerations because of the failure.
4. The cost of the scheduled maintenance, including the cost of the replaced unit.
5. The effect scheduled maintenance has on reliability. Will the maintenance activity introduce failures into the system?
6. Is the potential failure detectable by an operator? Can the operator take corrective action before the failure propagates throughout the system, causing other failures?
7. The cost of inspection and testing.

See Section 2 in Chapter 13.

2. CORRECTIVE MAINTENANCE ANALYSIS

> Define the elements of corrective maintenance analysis (e.g., fault-isolation time, repair/replace time, skill level, crew hours) and apply them in specific situations. (Apply)
>
> **Body of Knowledge VI.B.2**

Corrective maintenance is performed in the event of a failure or malfunction of the system. The corrective maintenance actions can not be planned, as the time the system will fail is not known. The goal is to return the system to an operable state while incurring a minimum amount of downtime. The total downtime includes both active maintenance time and inactive or delay time. The active corrective maintenance time can be analyzed as seven steps (MIL-HDBK-472). This active corrective maintenance time can be quantified as time to repair. A value referred to as the *mean time to repair* (MTTR) can be determined using the probability that the repair will be necessary and the average time required to perform that repair.

The seven steps are:

1. *Localization.* Determining the system fault without using test equipment.
2. *Isolation.* Verification of the system fault using test equipment.
3. *Disassembly.* Accessing the fault.
4. *Interchange.* Replacing or repairing the fault.

5. *Reassembly.* Rebuilding the item.
6. *Alignment.* Adjusting the item per specifications.
7. *Checkout.* Performing standard QA checks to verify correct operation.

In order to minimize the repair time, it is necessary to minimize the time required for each step. The time to work on reducing maintenance time is during the design phase of product development. The MTTR for each failure could be found.

A designer would not intentionally design an automotive system in such a way that it was necessary to remove the engine in order to replace a broken fan belt. However, unless there is active maintainability engineering input at the design stage, such absurdities can be sent to the customer. Good design practice would dictate that repairs with long MTTR values will rarely be necessary.

The task of isolating faults in complex systems, especially systems that are computer controlled, can be difficult and time-consuming. Computer-controlled automotive engines, flight control systems, electronic-controlled weapons systems, and many other complex systems can be designed to include *built-in testing* (BIT). The design could include test ports to connect diagnostic testing equipment to the system, or have indicators to display the probable fault. Built-in testing can increase the availability of the system by reducing the time to locate and isolate the fault so the process of restoring the system to service can begin earlier.

Failure of the built-in test system can negatively impact reliability. The BIT system in many cases consists of additional hardware. The probability of failure can increase as additional complexity is designed and built into the system. Improvement in reliability can be achieved if it is possible to use software to monitor the system and report faults instead of using mechanical or electrical sensors.

In order to minimize the total downtime, it is also necessary to minimize the inactive maintenance time. Inactive maintenance time includes the time waiting to obtain spare parts, tools, or test equipment, any delay in delivering the failed system to the repair facility, and administrative delay time.

3. NON-DESTRUCTIVE EVALUATION

> Describe the types and uses of these tools (e.g., fatigue, delamination, vibration signature analysis) to look for potential defects. (Understand)
>
> **Body of Knowledge VI.B.3**

Non-destructive evaluation encompasses a number of individual techniques. They provide information about a product beyond what can be seen with the naked eye. These tests are especially useful in evaluating systems designed for extreme environments or where failure has unusually dire consequences. These techniques are used to detect faults in newly produced items as well as products that have been in use or stressed in some way.

Eddy current testing involves the application of an alternating current (AC) passing through a coil that is placed near the unit being evaluated. The signature of the induced current can be used to detect cracks, seams, pits, and other surface flaws. Eddy current testing can also be used to measure the results of heat treatment, including hardness, case depth, and pattern. Eddy current testing is limited to conducting materials, but it can be used to measure thin layers of nonconductive coatings on metal parts. The eddy current device may vary from a fairly large-diameter coil for covering large surfaces to a pencil-like coil for inspecting hole and tubing inside diameters.

Ultrasonic testing can be used to detect surface defects. It consists of aiming high-frequency sound at an item and studying the deflected wave. Sonic devices may also be used to measure plastic wall thickness nondestructively as well as to detect delamination in some metal products.

X-ray testing is used to detect internal features and defects. This technique is used heavily to evaluate weld processes. *X-ray fluorescence analyzers* (XRFs), often in the form of an XRF gun, are used to determine the alloy makeup of an item.

Liquid penetrant may be painted or sprayed on a part to display cracks, corrosion, and other surface irregularities, as shown in Figure 14.1

Magnetic particle inspection can be used to examine items made from ferromagnetic materials for surface defects. The item to be inspected is first magnetized, then iron filings are applied to the surface either in dry form or as a wet slurry. The magnetic field lines typically form below the surface of the metal, but when cracks occur, magnetic flux "leaks," attracting the iron filings as shown in Figure 14.2.

Vibration signature analysis typically uses an acoustic detector linked to signal processing software. It may be used to inspect rotating equipment for malfunction, misalignment, fatigue, and wear. It may be used to test newly manufactured products or to do continuous or intermittent analysis of operating equipment. The software typically compares the detected spectrum with a set of alarm levels. An example is shown in Figure 14.3.

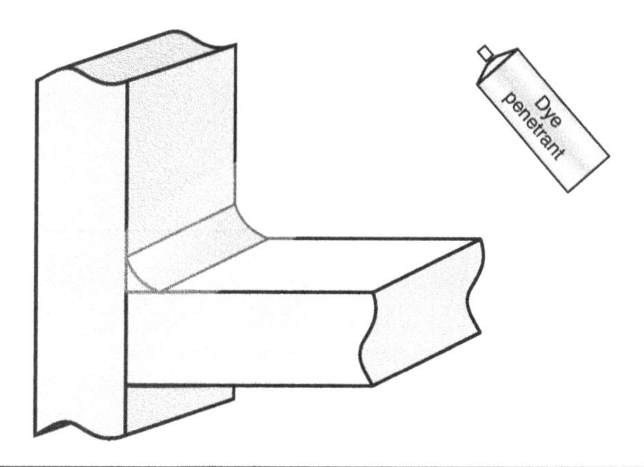

Figure 14.1 Liquid penetrant being sprayed on a weld to detect cracks and voids.

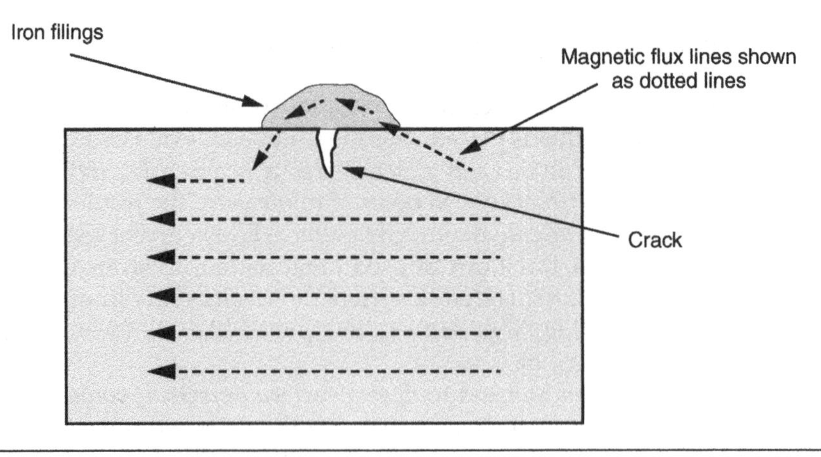

Figure 14.2 Magnetic particle inspection.

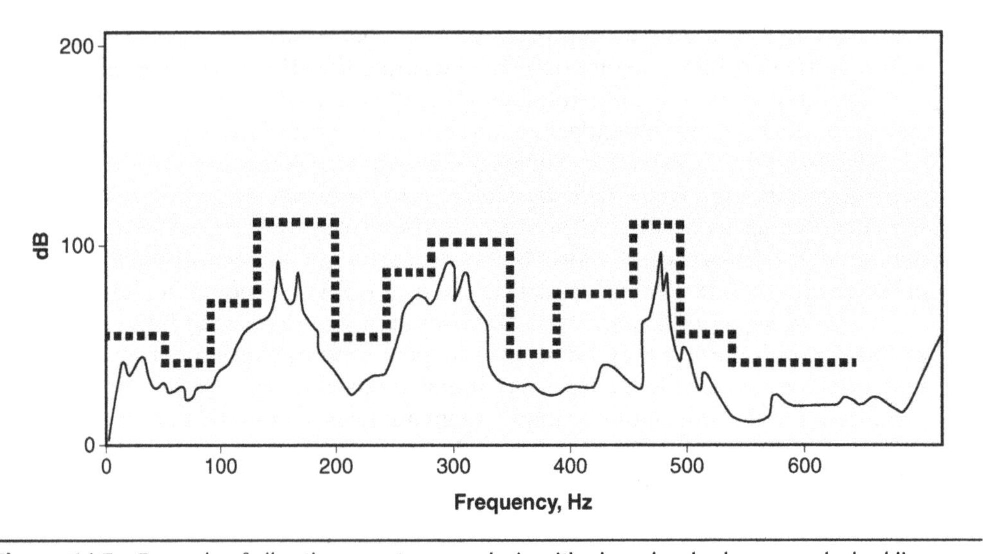

Figure 14.3 Example of vibration spectrum analysis with alarm levels shown as dashed lines.

4. TESTABILITY

> Use various testability requirements and methods (e.g., built in tests (BITs), false-alarm rates, diagnostics, error codes, fault tolerance) to achieve reliability goals. (Apply)
>
> **Body of Knowledge VI.B.4**

The maintainability of a system is influenced by the ability to detect and repair a system fault. Therefore, the testing requirements for a system should provide for a method to detect a system fault and isolate the failed component. In some cases, such as for a complex system, the requirement could be met with a built-in test system. This could add additional complexity to the system, and care must be taken to assure the reliability of the test system.

The testability requirements of systems can also be met by providing easily accessible test points for critical measurements taken using external test equipment. It is important that a maintainability engineering effort is included in the design stage to ensure testability of the system. The testability of a system must be developed as part of the system design.

At the system or subsystem level, the testability requirements should reflect the need to be able to detect failure quickly and isolate the failure so that replacement or repair can be accomplished in an acceptable amount of time.

When designing for testability it is important to consider the various conditions when testing is required, such as during new product development, conducting safety analysis or fatigue analysis, and other times, depending on the product.

Fault detection capability is a measure of the faults detected by the fault detection system (usually built-in test) compared to the total number of system faults.

Fault isolation capability is a measure of the percentage of time the failure can be isolated to a given number of replaceable (repairable) components. Fault isolation can be accomplished by a diagnostic analysis, built-in test, or using external test equipment.

False-alarm rate is a measure of the rate at which the system declares the detection of a failure when no failure has occurred.

A *fault recording and analysis system* that identifies specific components will aid in improving reliability. Particular attention should be paid to the possibility of dependence among fault types. The discovery of a cascading pattern among faults permits work on the root cause rather than on symptoms.

5. SPARE PARTS ANALYSIS

> Describe the relationship between spare parts requirements and reliability, maintainability, and availability requirements. Forecast spare parts requirements using field data, production lead time data, inventory, and other prediction tools, etc. (Analyze)
>
> **Body of Knowledge VI.B.5**

The required number of spares necessary for a given period of time is a function of the expected number of failures and the number of spares required for preventive maintenance. The number of failures is a function of the unit operating time and the failure rate. If any spares are required for preventive maintenance, the number is a function of operating time and the preventive maintenance cycle.

The MTTR calculation assumes that spare parts are readily available. This means that spares must be in inventory when they are needed. If there is waiting time to obtain a spare, the unit downtime will increase, and availability will suffer. If the unit has a constant failure rate, the probability of requiring no more than r replacement units can be found using the cumulative Poisson distribution.

The probability of having r or fewer failures in operating time t if the unit is operating with a constant failure rate of λ is

$$P(r)=\sum_{n=0}^{r}\frac{(\lambda t)^n e^{-\lambda t}}{n!}.$$

Tables are available to evaluate the cumulative Poisson distribution.

EXAMPLE 14.1

A unit that operates an average of 2000 hours per year has a constant failure rate

$$\lambda = 100 \times 10^{-6} \text{ failures/hour.}$$

The maintenance strategy is to replace the unit every 5000 operating hours.

In a five-year period:

a. How many spares will be necessary for preventive maintenance?

b. With a .90 probability, how many spares will be necessary for corrective maintenance?

Solution:

a. The operating time in a five-year period is $t = 5 \times 2000 = 10{,}000$ hours

The preventive maintenance cycle is 1/5000 hours

The number of spares required for preventive maintenance is

$$n = (1/5000) \times (10{,}000) = 2.$$

b. We want $P(r) \geq .90$

$\lambda t = (100 \times 10^{-6})(10{,}000) = 1.0$

From the cumulative Poisson tables in Appendix N, use the row with λ value of 1 in the left column and find the first element in this row that is greater than 0.90.

$$P(r \leq 2) = .92$$

$$\lambda t = 1.0$$

With a probability of .92, the number of spares needed for corrective maintenance is no more than two.

If preventive maintenance requires spare parts, the number of spares necessary is a function of the operating time and the maintenance cycle. The number of spares required for preventive maintenance is the product of the maintenance cycle and the total operating time. The total number of spares that must be kept in inventory is a function of the delivery time, the cost of maintaining the inventory, and the availability requirement.

Several methods are in use for forecasting spare parts requirements. Four methods are presented here. Experience will help determine which is best for a given part.

Moving Average

The forecasted value of the spare parts requirement is the average of the previous n requirements. The value of n is determined by experience. The formula is

$$R_t = \frac{X_{t-1} + X_{t-2} + \ldots + X_{t-n}}{n}$$

where

R_t = Forecasted spare parts requirement at time t

X_t = Actual spare parts requirement at time t

Example: If the best value of n has been found to be 5, and the last eight weeks' requirements were given as 23 25 22 26 21 20 23 24, then the predicted spare parts requirement for this part for the ninth week is

$$R_9 = \frac{26+21+20+23+24}{5} \approx 23$$

Weighted Moving Average

This method weights the most recent values higher than the previous values. The n-period weighted moving average uses the formula

$$R_t = \frac{nX_{t-1} + (n-1)X_{t-2} + \ldots + X_{t-n}}{n+n-1+\ldots+1}$$

Example: If the best value of n has been found to be 5, and the last eight weeks have seen spare part requirements for a particular part of 23 25 22 26 21 20 23 24, then the predicted spare parts requirement for this part for the ninth week is

$$R_9 = \frac{(5)24+(4)23+(3)20+(2)21+26}{5+4+3+2+1} \approx 23$$

Single Exponential Smoothing

The formula for predicting the spare parts requirement is

$$R_t = \alpha X_{t-1} + (1 - \alpha)R_{t-1}$$

where

R_t = Forecasted spare parts requirement at time t.

X_t = Actual spare parts requirement at time t.

α = Smoothing factor, usually between 0.1 and 0.4 with unstable demand situations using higher α-values. Experience dictates the best values for α.

(To get the time series started, it is common practice to use the forecast for time $t = 2$ to be the actual requirement for time $t = 1$.)

Example: Assume time periods of one week are used, and the spare parts requirement for the first week was 23 items. Assume further that an α-value of 0.2 is used, and that the actual requirement for the second week was 16 parts. The forecasted spare parts requirement for the third week would be

$$R_3 = 0.2(16) + 0.8(23) = 21.6 \text{ or } 22 \text{ items}$$

If the actual requirement for week three was 18 items, the forecasted requirement for week four would be

$$R_4 = 0.2(18) + 0.8(21.6) = 20.9 \text{ or } 21 \text{ items}$$

Croston Method

The Croston method applies exponential smoothing to both the magnitude of the spare parts requirement, Z, and the time period, P, between demands for the spare part. If no spare part is needed at the end of the review period, the values of Z and P are unchanged. If, however, X_t spare parts are needed at time t, the revised estimates for Z and P are given by the smoothing formulas

$$Z_{t+1} = \alpha X_{t+1} + (1 - \alpha)Z_t$$

$$P_{t+1} = \beta G_{t+1} + (1 - \beta)P_t$$

Where X_t = Actual value of demand at time t

Z_t = Forecasted value of demand at time t

G_t = Actual time between previous demand and this demand

P_t = Forecasted value of time between previous demand and this demand

α, β = smoothing factors between 0 and 1

Then, the forecast value for number of parts required per period for the next period would be

$$R_{t+2} = \frac{Z_{t+1}}{P_{t+1}}$$

Example: Assume time periods of one week are used, and the spare parts requirement for the first week was 23 items. Assume further that $\alpha = 0.2$ and $\beta = .03$, and that the actual requirement for the second week was 0 parts, and the actual requirement for the third week was 16 parts. The forecasted spare parts requirement for the fourth week would be calculated as follows:

$$Z_3 = 0.2(16) + 0.8(23) = 21.6$$

$$P_3 = 0.3(2) + 0.7(1) = 1.3$$

$$R_4 = \frac{21.6}{1.3} = 16.6$$

Part VII

Data Collection and Use

Chapter 15

A. Data Collection

1. TYPES OF DATA

> Identify and distinguish between various types of data (e.g., attributes vs. variable, discrete vs. continuous, censored vs. complete, univariate vs. multivariate). Select appropriate data types to meet various analysis objectives. (Evaluate)
>
> **Body of Knowledge VII.A.1**

Discrete (attributes) data are obtained when the characteristic being studied can have either a finite number or countably infinite number of possible values. For example, the results of a leak test might be designated as zero or one to indicate failed or passed. Another example would be the count of the number of scratches on an object. In this case the possible values are 0, 1, 2, . . . a so-called countably infinite set. Attribute control charts (for example, *p, np, u,* and *c*) are used to plot discrete data.

Continuous (variables) data are obtained when the characteristic being studied can have any value in a range of numbers. For example, the length of a part can be any value above zero. Between each two values on a continuous scale there are infinitely many other values. For example, between 2.350 and 2.351 inches the values 2.3502, 2.350786, and so on, occur. Variables control charts (for example, *X*-bar and *R*, ImR) are used to plot continuous data. These charts are used to plot a single variable and are referred to as *univariate* charts. Output variables such as a dimension, hardness, pH, and so on, and input variables such as pressure, temperature, and voltage are often plotted on these charts. If two or more input variables influence the process, it is usually best to use a *multivariate* chart rather than several univariate charts. The multivariate chart most like the *X*-bar and *R* chart is called the T^2 *generalized variance chart.* Multivariate charts are usually plotted with the help of software packages. A disadvantage of multivariate control charts is that the scale does not directly relate to the input variables individually.

When collecting failure data, there are several possible ways the tests may be administered. Consider the testing of several pumps. Timers might be placed on each pump so that *exact failure times* can be recorded. The test might be terminated while some pumps are still functioning. In this case it is known only that these pumps' failure times are longer than the test time. Such data are called *right censored.* Another testing protocol would require that the pumps be checked every 100 hours. If a pump was operating at hour 400 but not at hour 500, it would only be known that the failure occurred between these two times. These data are called *interval censored data.* If a pump is found to have failed at the first check at hour 100, it is known only that the failure occurred before hour 100. Such data are called *left censored* rather than interval censored because the failure may have occurred before the interval started, assuming a check of the pump was not conducted at time zero.

2. COLLECTION METHODS

> Identify appropriate methods and evaluate the results from surveys, automated tests, automated monitoring and reporting tools, etc., that are used to meet various data analysis objectives. (Evaluate)
>
> **Body of Knowledge VII.A.2**

One approach to data collection emphasizes the importance of the "cleanliness" of the information. This view holds that the best data are those that are uncontaminated by changes in outside factors. This can be accomplished by maintaining strict laboratory conditions, perhaps using an environmental chamber. The downside of this course of action is that very few products or processes operate in an environment in which all outside factors are controlled. This technique is most appropriate for early-stage investigation in which fundamental design cause-and-effect relationships are being determined. Some data, such as temperature, time, pressure, and so on, can be collected automatically. Data that attempt to measure qualities such as taste or pain are called *subjective* data, as opposed to *objective* data such as length or weight.

At the other end of the spectrum are *field service data.* This refers to information gathered from products in use by customers. This information is often impacted by differences in installation, environment, operator procedures, and similar factors that make analysis difficult. On the other hand, field service data represent realistic applications of the product and therefore must be taken into account as part of reliability analysis. In most cases this approach is used in the feedback process to aid in determining the accuracy of other reliability analysis methods and to provide input to future revision and design initiatives.

Various other methods of data collection have been devised. Perhaps the most powerful technique is to study the failure modes and mechanisms that can occur.

Details of various tools that have been designed for this purpose are outlined in Chapter 17, but the emphasis must be on providing adequate resources early enough in the design phases to impact the resultant product.

When considering data sources, special attention must be paid to the conditions under which the product will be transported, stored, and used. These conditions include differences in such factors as geographic location, operator habits, lubricants, possible chemical, radiological, or biologic exposure, vibration, electromagnetic environment, and so on.

Reliability data are often stated in binary terms, such as failed or non-failed. Some contemplation should also be given to any degradation of satisfaction from the user's viewpoint—short of failure—that may occur. This contemplation should occur in the light of rising user expectations.

Once the sources of data have been determined, it is necessary to form the data collection plan. This plan needs to provide answers to such questions as:

- *Who will collect the information?* Data collectors must be familiar with the measuring equipment and the product itself. If failure data are to be collected, a well-defined and understood definition of failure is essential. If possible, more than one person should be involved to provide some cross-checking and coverage when one must be away.
- *When will the data be collected?* The plan needs to specify at what stage of design or use the data will be collected. Dates and times help planning by other team members.
- *In what format will the data be collected?* The best step here is to design a data collection sheet and specify the format for summarizing the data.
- *What measurement equipment will be employed?* The plan should specify equipment and recording devices if appropriate.
- *What measures will be used to verify data accuracy and integrity?* The plan may specify units of measurement, calibration of equipment, or recording of ancillary information that might impact the data. If data are transmitted or stored digitally, the use of error correction systems may be specified.

3. DATA MANAGEMENT

> Describe key characteristics of a database (e.g., accuracy, completeness, update frequency). Specify the requirements for reliability-driven measurement systems and database plans, including consideration of the data collectors and users, and their functional responsibilities. (Evaluate)
>
> **Body of Knowledge VII.A.3**

Characteristics of a Good Database Design

1. *Adequate rules for data organization.* These are sometimes referred to as *normalization* rules. A typical list of such rules would include:

   - Each table contains a column that assigns a unique ID to each row
   - Principal entities in a table should be of the same type
   - Keep null entries to a minimum
   - Avoid redundancy

2. *Data integrity rules.* Rules that prevent invalid entries. For instance, a product number column would prohibit giving two products the same number.

3. *Data security rules.* Rules that restrict access to sensitive data to certain users.

   Rules that only permit certain users to perform certain activities or change certain data.

4. *Provision is made for updating of data in real time or via batch entry.* Error detection for automated entry, such as through sensors or measurement tools.

5. *Provision is made for maintenance.* Backup of the data as well as modifications to the data structures.

The product failure database should be one part of the organization's database management system, and as such must comply with that system's rules. The database needs a query system that will allow users to search, analyze, and update the database depending on their level of user access privileges. Rules for access and editing privileges should be controlled by a team with representation from all potential user groups as well as the reliability engineering function.

A guiding principle for the database should be that it becomes the appropriate repository of all data related to product failure. Therefore, effort should be expended to make the entry and retrieval of data as painless as possible. The database is typically accessible through a query language that allows the user to ask a question in a format that the database can respond to.

An essential step in the establishment of the product failure database is the determination of the attributes that will be used to access the data. These attributes should be determined by a team with representation from potential user groups. It is important that the attributes be defined early because adding attributes to an existing database can be labor-intensive. Table 15.1 lists possible user groups and some of their attribute requirements. The listing is by no means exhaustive and can perhaps best be used as a discussion starter.

Of course, each set of data must be clearly identified as to source, collection methods, date/time, product, responsible person, and so on.

Table 15.1 List of possible database user groups.

User group	Attributes	Typical queries
Production	Tooling and equipment used, process parameters, raw material identification	What is the failure rate when product ABC is run on machine 135 using stainless steel?
Research	Testing protocol, testing lab, material type, design type, environmental conditions, outside sources (for example, university studies)	What published literature covers the failure rates for die cast zinc in low-temperature applications?
Field service	Geographic location, installation types, shipping/handling parameters, user/operating conditions	What is the failure rate for item CDE when used on the East Coast?
Purchasing	Specification details, supplier identification, lot/batch identification	What is the failure rate for clips supplied by FGH?
Accounting	Cost breakdown of scrap/rework/ warranty due to failure	What were costs due to warranty for product JKL during 2007?

There is a natural tendency to want to restrict access to some failure data for proprietary or other reasons. This creates tension between users who desire access to all the failure data and those who hesitate to enter information out of concern that it will be misused. To counter this problem, it may be necessary to establish user privilege protocols and partitioned structures within the database.

Chapter 16

B. Data Use

1. DATA SUMMARY AND REPORTING

> Examine collected data for accuracy and usefulness. Analyze, interpret, and summarize data for presentation using techniques such as trend analysis, Weibull, graphic representation, etc., based on data types, sources, and required output. (Create)
>
> Body of Knowledge VII.B.1

Data analysis must be followed by communication of the results to decision makers. Good decisions are not made on the basis of poorly understood information. This makes the analysis → evaluate → summarize sequence the basis for any impact the data will have on products and processes.

The first steps in summarizing data are to construct a histogram and a run chart or a control chart. These help determine whether the process is stable as well as provide estimates of the mean and standard deviation. For example, if a sequence of batches of electrical components were tested in a chemical bath for one hour each, and the number of failures at one hour recorded, these numbers could be the basis for a p-chart as well as a histogram.

The control chart can be helpful in spotting a trend line. Linear regression may be used to fit a line to the data. Example 16.1 illustrates the use of a histogram, control chart, and linear regression.

Once failure data have been collected, it is often useful to display them in graphical format. Key reliability metrics include:

- *Probability density function* (PDF) can be plotted as a histogram showing the number of failures in each time block.

EXAMPLE 16.1

A reliability engineer has collected failure rates per thousand cycles as shown below:

Cycles	1000	2000	3000	4000	5000	6000	7000	8000	9000	10,000
Failure rate	.002	.003	.002	.002	.003	.004	.002	.003	.003	.004

Cycles	11,000	12,000	13,000	14,000	15,000	16,000	17,000	18,000	19,000	20,000
Failure rate	.003	.003	.004	.005	.004	.006	.005	.007	.007	.008

Cycles	21,000	22,000	23,000	24,000	25,000	26,000	27,000	28,000	29,000	30,000
Failure rate	.010	.009	.010	.011	.009	.008	.008	.011	.010	.011

Cycles	31,000	32,000	33,000	34,000
Failure rate	.009	.011	.010	.009

The engineer constructs a histogram:

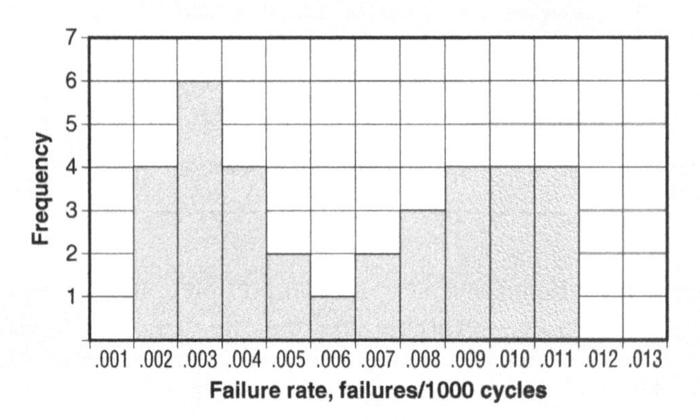

The histogram has an interesting bimodal pattern. The engineer decides to also construct a run chart:

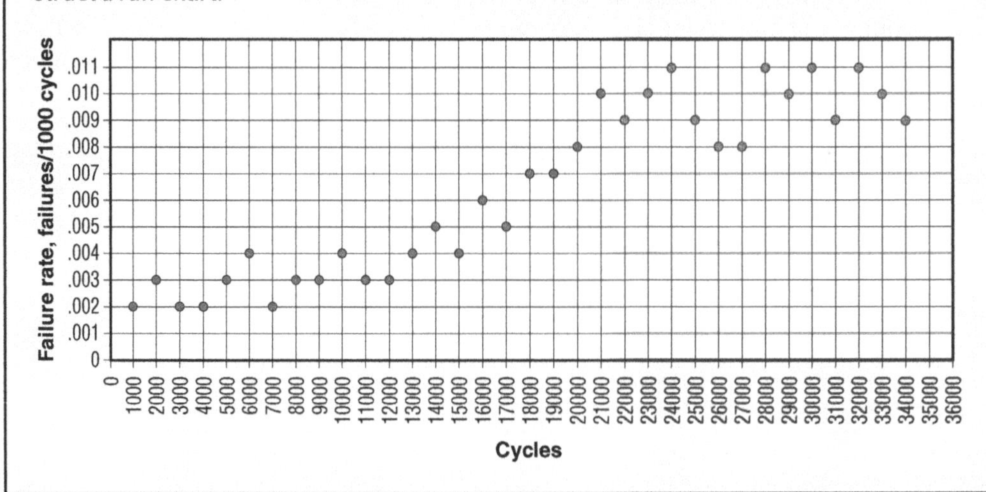

Continued

Continued

Collecting additional data seems to confirm a pattern of a low failure rate for about the first 12,000 cycles, increasing to a higher rate at about 24,000 cycles and continuing at that higher failure rate level. This would indicate that there is some additional failure mode that begins to occur at 12,000 cycles. The accountants working with warranty projections ask the engineer to provide a formula that can be used for predicting the failure rate during the transition period from 12,000 to 24,000 cycles. The engineer uses linear regression to produce the formula

$$\text{Failure rate} = 0.000\,000\,648 \times (\text{Cycles elapsed}) - 0.00482$$

(The linear regression formulas are given in Example 16.6.)

An important concern when using linear regression is that the procedure generates the best-fitting straight line even though the relationship may not be linear. The resultant line will not fit the data.

To determine how well the formula fits the given data, the engineer produces the following table:

	Failure rate		
Cycles	Predicted	Actual	Error
12000	0.0030	0.003	0.0000
13000	0.0036	0.004	0.0004
14000	0.0043	0.005	0.0007
15000	0.0049	0.004	0.0009
16000	0.0056	0.006	0.0004
17000	0.0062	0.005	0.0012
18000	0.0069	0.007	0.0001
19000	0.0075	0.007	0.0005
20000	0.0081	0.008	0.0001
21000	0.0088	0.010	0.0012
22000	0.0094	0.009	0.0004
23000	0.0100	0.010	0.0000
24000	0.0107	0.011	0.0003

This formula is provided to the warranty personnel with the caveat that it is only useful in the 12,000 to 24,000 cycle range and will not make valid predictions if it is extrapolated beyond these values.

Of course, the main concern at this point is to identify and correct the failure mode that apparently occurs at 12,000 cycles. We wish the engineer well in that endeavor.

- *The hazard function* shows the failure rate as a function of time. It is convenient to use the following formula for the hazard function:

$$\lambda(t) = \frac{\text{Fraction of failures during the time period}}{\text{Amount of time during the period}}$$

- *The reliability function* plots reliability as a function of time. The formula is

$$R(t) = \frac{\text{Number surviving at the end of the period}}{\text{Number of units tested}}$$

As noted in Chapter 4, the Weibull distribution takes on many shapes depending on the value of β, the shape parameter. This feature makes the distribution an extremely flexible tool for solving reliability problems and displaying results. Example 16.3 illustrates a method for plotting data on Weibull probability paper. Various versions of this graph paper can be downloaded from Weibull.com.

EXAMPLE 16.2

A total of 283 products are tested for 1100 hours. The number of failures in each 100-hour block of time is recorded.

Time	Number of failures	Number surviving	$\lambda(t)$	R(t)
0–99	0	283	0	1.00
100–199	2	281	.0001	.99
200–299	10	271	.0004	.96
300–399	30	241	.0011	.85
400–499	48	193	.0020	.68
500–599	60	133	.0031	.47
600–699	50	83	.0038	.29
700–799	42	41	.0051	.14
800–899	30	11	.0073	.04
900–999	8	3	.0073	.01
1000–1099	3	0	.0100	.00

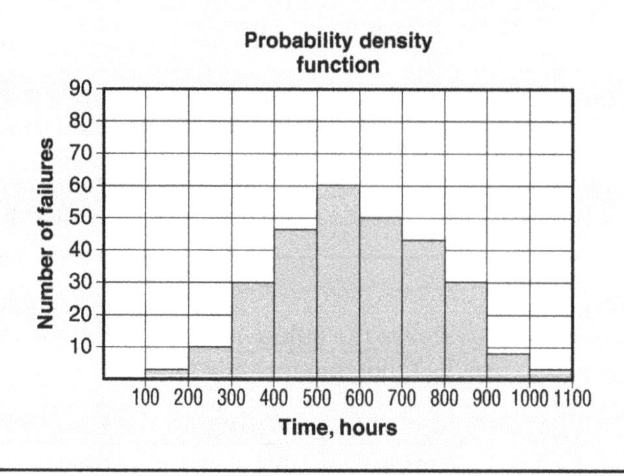

Continued

Continued

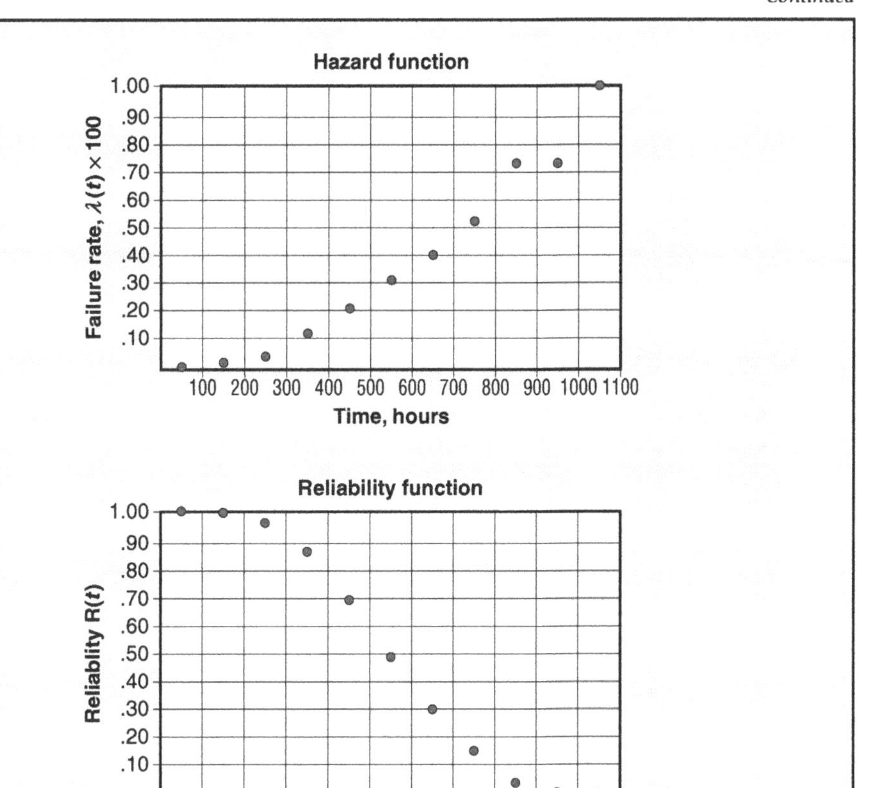

EXAMPLE 16.3

Twenty products are tested for 1000 hours. Fourteen products fail at the following times:

70, 128, 204, 291, 312, 377, 473, 549, 591, 663, 748, 827, 903, 955 hours respectively.

Estimate the shape parameter β and the characteristic life η.

Solution:

1. Form a table listing the times to failure in ascending order in the first row. Fill the second row from the median ranks table in Appendix Q using the first 14 entries in the column labeled "20."

Hrs.	70	128	204	291	312	377	473	549	591	663	748	827	903	955
MR	.034	.083	.131	.181	.230	.279	.328	.377	.426	.475	.525	.574	.623	.672

Continued

Continued

2. Plot the first column on the horizontal axis and the second column on the vertical axis of Weibull graph paper (see Figure 16.1).
3. Use a transparent straightedge to sketch a best-fit straight line for the points.
4. Draw a line parallel to the best-fit line that passes through the point labeled O along the left margin of the graph paper. This line is shown as a heavy dashed line in Figure 16.1. Note the value on the β scale where this line crosses the top line of the graph paper. This is the estimate for the shape parameter β. In this example $\beta \approx 1.3$.
5. Note that the vertical axis is labeled "unreliability." This means that the values on this axis are (1 – Reliability). Recall that we use .368 on the reliability scale to find η. This would translate to .623 on the unreliability scale. This value is indicated on the graph paper as a horizontal dashed line with η at its right end. The horizontal coordinate of the point where this line crosses the best-fit line is the estimated value for η.

In this example it appears that $\eta \approx 900$ hours.

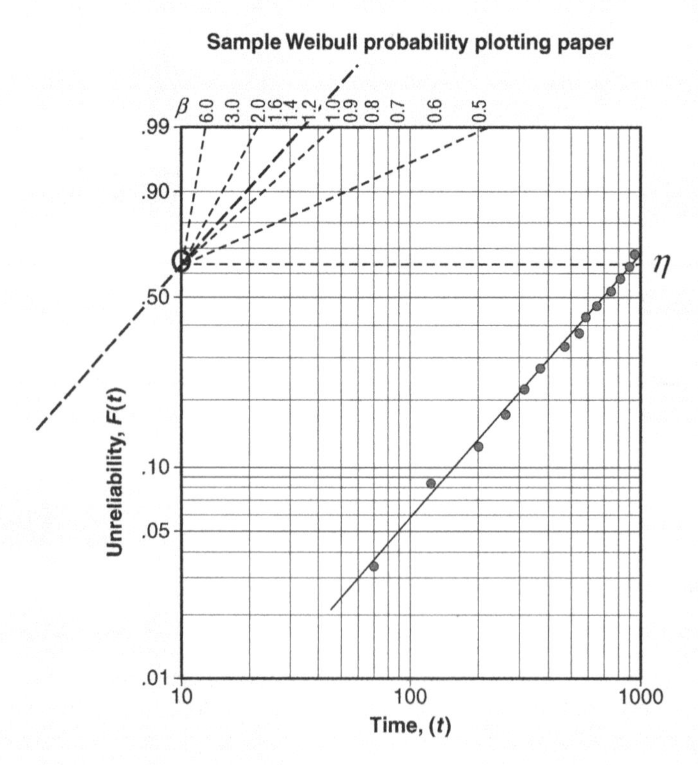

Figure 16.1 Weibull probability graph.

EXAMPLE 16.4

In the previous example, find:

- The reliability function
- The reliability at 200 hours, and compare with the graphical result in Figure 16.1
- The time at which 95 percent reliability will occur

Solution:

The reliability function with $\eta = 900$ and $\beta = 1.3$ is

$$R(t) = e^{-\left(\frac{t}{900}\right)^{1.3}}.$$

Substituting into this formula:

$$R(200) = e^{-(.222)^{1.3}} \approx .87$$

In order to solve for time (t), take the natural log of both sides of the reliability equation.

$$(t/\eta)^{\beta} = -\ln(.95)$$

$$t = \eta[-\ln(.95)]^{1.\beta}$$

$$t = 900[-\ln(.98)]^{1/1.3} = 91.6 \text{ hours}.$$

The Weibull reliability function is

$$R(t) = e^{-\left(\frac{t}{\eta}\right)^{\beta}}.$$

In Figure 16.1 the vertical line at 200 hours crosses the best-fit line at an unreliability of about 14%, making reliability ≈ 86%.

To find the time at which reliability of 95% occurs, locate 5% on the vertical axis. Move across to the best-fit line. The crossing point has a time value of about 90 hours.

2. PREVENTIVE AND CORRECTIVE ACTION

Select and use various root cause and failure analysis tools to determine the causes of degradation or failure, and identify appropriate preventive or corrective actions to take in specific situations. (Evaluate)

Body of Knowledge VII.B.2

Once defects or failures are identified, one of the most difficult and critical tasks in the entire enterprise begins, that of determining the root cause or causes. The fundamental tool for this purpose is the *cause-and-effect diagram,* also called the *fishbone* or *Ishikawa diagram.* This tool helps a team identify, explore, and communicate all the possible causes of the problem. It does this by dividing possible causes into broad categories that help stimulate inquiry as successive steps delve deeper. The general structure of the diagram, shown in Figure 16.2, illustrates why it is also called the fishbone diagram. The choice of categories or names of the main "bones" depends on the situation. Some alternatives might include *policies, technology, tradition, legislation,* and so on.

A team may use a cause-and-effect diagram to generate a number of potential causes in each category by going around the room and asking each person to suggest one cause and its associated category. As each cause is identified, it is shown as a subtopic of the main category by attaching a smaller line to the main "bone" for that category. The activity continues until the group is satisfied that all possible causes have been listed. Individual team members can then be assigned to collect data on various branches or sub-branches for presentation at a future meeting. Ideally, the data should be derived by changing the nature of the cause being investigated and observing the result. For example, if voltage variation is a suspected cause, put in a voltage regulator and see if the number of defects changes. One advantage of this approach is that it forces the team to work on the causes and not symptoms, personal feelings, history, and various other baggage. An alternative to the meeting format is to have an online fishbone diagram to which team members may post possible causes over a set period of time.

The fishbone diagram is an example of divergent thinking in that it seeks to go off in all directions, with the aim of finding all possible causes of failure. After completing the exercise, the next step is to perform a convergence exercise to prioritize the failure prevention activities. Various voting schemes can be used. Another option is to plug the results of the cause-and-effect exercise into a *cause-and-effect matrix,* which relates various causes to customer requirements. As shown in Figure 16.3, the causes are listed in a column along the left side, and customer requirements are listed along the top. An *importance* number is assigned to each customer requirement. The matrix relates product qualities to customer requirements similarly to the way it is done in the main body of a QFD matrix. A number is assigned to a cell in the matrix, which designates the relationship of the cause

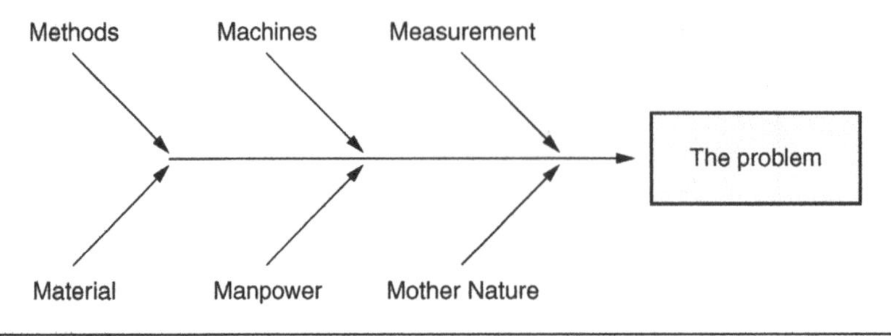

Figure 16.2 Traditional cause-and-effect diagram with the six M's.

Generating a corporate insurance proposal														
		Executive summary	Coverage alternatives	Easy to read	Pages remain in order	Figures accurate	Charts clear	Premium/coverage xref	Refd to OSHA requirements	Payment alternatives	Paper acid free	Tax considerations	Appearance of proposal	Row totals
	Importance:	5	8	9	4	9	7	3	5	2	3	8	5	
Machine	Printer speed varies			9			9							144
	Binder breaks comb				7									28
	Wrong comb used				7									28
Measurement	Cutter misses length												5	25
	Too little ink applied			8			7						6	151
Manpower	Summary missing	9		2										63
	Alternates unclear		9					3						81
	Government regs not covered								9			7		101
	Paper mis-loaded			7	3		3						5	121
Methods	Wrong weight specified			6									8	94
	Wrong paper specified			5			5						7	115
Material	Paper degrades			4			4						4	84
	Ink shade varies						5						3	50
Mother nature														

Figure 16.3 Cause-and-effect matrix.

in that line to the customer requirement in that column. The number in the cell is multiplied by the importance number for that column, and the row total is the sum of these products. For instance, in Figure 16.4, the total for the "Printer speed varies" row is $9 \times 9 + 9 \times 7$. The row totals can aid in prioritizing failure causes.

The *five whys* is another technique that helps dig deeper into a problem (see Example 16.5). This tool consists of repeating the question "Why does this happen?" as each answer surfaces.

EXAMPLE 16.5

Why is this part defective?

Because the hole is too large.

Why is the hole too large?

Because the bit "walked" during the drill operation.

Why did the bit walk?

Because the holding fixture had some play in it.

Why did the fixture have play?

Because the pneumatic clamps didn't apply enough pressure.

Why don't the clamps apply sufficient pressure?

Because of variation in the air pressure to the shop floor.

Needless to say, the number five is arbitrary, and in this case additional inquiry into air pressure variation would be appropriate.

An enhanced flowchart of the process will often aid in identifying root causes. Enhancements should include data on quality levels and possible causes at each step of the process. This tool can be especially helpful if the process is large or complex.

Scatter Diagrams

When several causes for a problem have been proposed, it may be necessary to collect some data to help determine which is a potential root cause. One way to analyze such data is with a scatter diagram. In this technique, measurements are taken for various levels of the variables suspected of being a cause. Then each variable is plotted against the measured value of the problem to get a rough idea of correlation or association.

EXAMPLE 16.6

An injection molding machine is producing parts with pitted surfaces, and four possible causes have been suggested: mold pressure, coolant temperature, mold cooldown time, and mold squeeze time. Values of each of these variables as well as the quality of the surface finish are collected on 10 batches. The data:

Continued

Continued

Batch number	Mold pressure	Coolant temperature	Cooldown time	Squeeze time	Surface finish
1	220	102.5	14.5	.72	37
2	200	100.8	16.0	.91	30
3	410	102.6	15.0	.90	40
4	350	101.5	16.2	.68	32
5	490	100.8	16.8	.85	27
6	360	101.4	14.8	.76	35
7	370	102.5	14.3	.94	43
8	330	99.8	16.5	.71	23
9	280	100.8	15.0	.65	32
10	400	101.2	16.6	.96	30

Four graphs have been plotted in Figure 16.4. In each graph, surface finish is on the vertical axis. The first graph plots mold pressure against surface finish. Batch #1 has a mold pressure of 220 and a surface finish of 37. Therefore one dot is plotted at 220 in the horizontal direction and 37 in the vertical direction. On each graph, one point is plotted for each batch. If the points tend to fall along a straight line, this indicates that there may

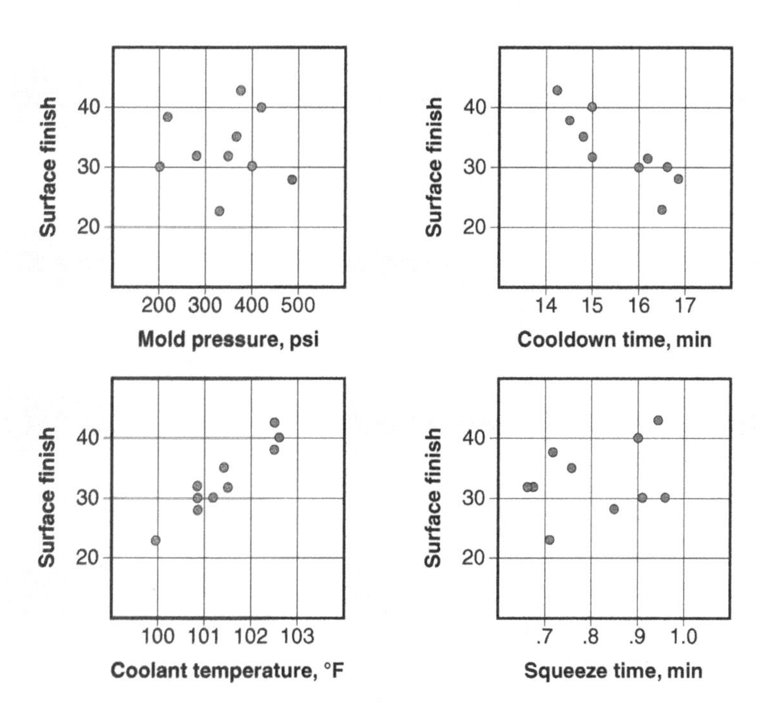

Figure 16.4 Scatter diagrams of variables in injection molding operation.

Continued

Continued

be a linear correlation or association between the two variables. If the points tend to closely follow a curve rather than a straight line, there may be a nonlinear relationship. Note that a high correlation does not imply a cause-and-effect relationship. A low correlation, however, does provide evidence that there is no such relationship, at least in the range of values considered. What variables can be eliminated as probable causes based on the above analysis?

The closer the points are to forming a straight line, the greater the linear correlation coefficient, denoted by the letter *r*. A positive correlation means that the line tips up on the right end. A negative correlation means the line tips down on its right end. If all the points fall exactly on a straight line that tips up on the right end, then $r = 1$. If all the points fall on a straight line that tips down on the right end, $r = -1$. In general $-1 \leq r \leq 1$.

The formula for the correlation coefficient is

$$r = \frac{S_{xy}}{\sqrt{S_{xx}S_{yy}}}$$

where *x* and *y* are the independent and dependent variables, respectively, and

$$S_{xx} = \sum x^2 - \frac{\left(\sum x\right)^2}{n}$$

$$S_{xy} = \sum xy - \frac{\sum x \sum y}{n}$$

$$S_{yy} = \sum y^2 - \frac{\left(\sum y\right)^2}{n}.$$

The formula for the best-fitting straight line is $y = mx + b$ where

$$m = \frac{S_{xy}}{S_{xx}} \text{ and } b = \bar{y} - b\bar{x}$$

3. MEASURES OF EFFECTIVENESS

Use various data analysis tools to evaluate the effectiveness of preventive and corrective actions in improving reliability. (Evaluate)

Body of Knowledge VII.B.3

The ultimate test of the effectiveness of a preventive or corrective action is the ability to turn the action on or off and observe the corresponding effect on the process.

Data should be collected before and after the installation of the preventive/corrective action. This section lists some tools for determining whether these data provide evidence that the action has been successful.

The *histogram* is probably the simplest tool to use. Suppose the proposed preventive/corrective action was designed to raise the mean effective lifetime of the item. Random samples from the process before and after the action should show a difference in their centers. If the two histograms look like those in Figure 16.5, one could conclude that the action was effective.

If, on the other hand, the two histograms looked like those shown in Figure 16.6, the results would be inconclusive. In this case it would be necessary to subject the data to the hypothesis test for means of two populations, as illustrated in Chapter 5, to determine whether there is a significant difference between the mean lifetimes of the two populations.

In a similar manner, if the preventive/corrective action was designed to reduce the variation in characteristic x, the histograms might look like those in Figure 16.7, in which case it would be safe to conclude that the action was effective.

Again, if the results from the histograms were ambiguous, it would be possible to reach a more definitive conclusion by using a hypothesis test, in this case the test for two population standard deviations, also illustrated in Chapter 5.

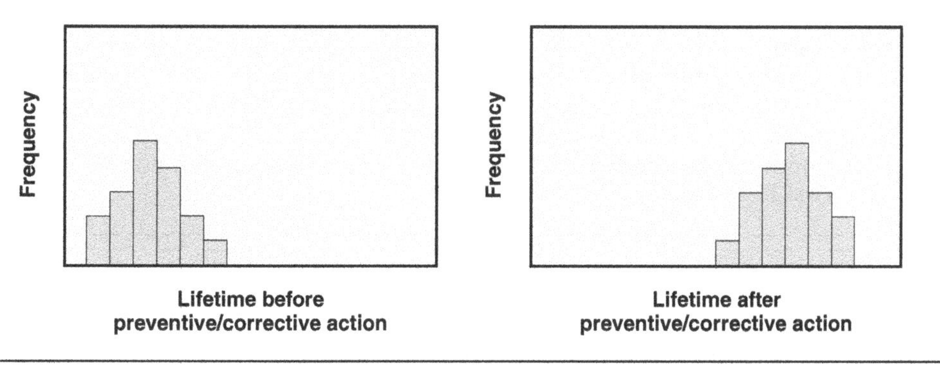

Figure 16.5 Histograms showing increase in mean lifetime.

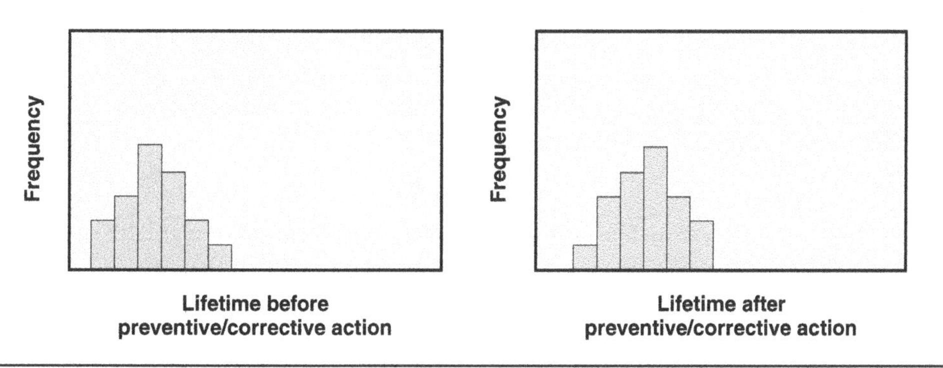

Figure 16.6 Inconclusive histograms.

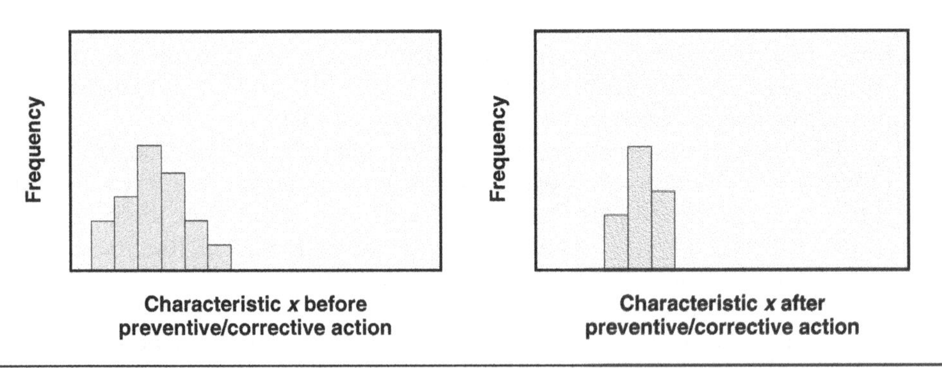

Figure 16.7 Histograms showing reduction of variation in characteristic *x*.

There is a caveat regarding the use of hypothesis tests for proportions: Consider avoiding the aggregation of defect types. It is sometimes best to delineate categories' defect types and study them individually.

EXAMPLE 16.7

There is excess variation in the moisture content of the starch leaving a dryer. In an attempt to reduce this variation, a speed control has been installed temporarily on the web drive motor. The problem-solving team collects moisture data with and without the speed control under various conditions. The team subjects each set of data to the hypothesis test for standard deviation of two populations to determine whether the variation has been reduced.

EXAMPLE 16.8

A quality improvement team is charged with reducing the percentage of units that are rejected due to surface scratches. The current reject rate is 15%. One possible cause of scratches is the design of a holding fixture. A sample of 1000 items is produced using a prototype of a new fixture design. The sample is inspected, and 138 units are rejected for scratches. The team uses a hypothesis test to determine whether there has been a reduction in the rejection rate at the 0.10 significance level. The hypothesis test for population proportion from Chapter 5 is used as follows:

$$n = 1000, p_0 = 0.15$$

1. The conditions are met because both 1000(.15) and 1000(.85) are greater than 5.
2. $H_0: p = 0.15$; $H_a: p < 0.15$.
3. $\alpha = 0.10$.

Continued

Continued

4. The critical value is $z_{0.10} = -1.28$ for the left-tailed test.
5. The test statistic is

$$z = \frac{0.138 - 0.15}{\sqrt{\frac{0.15(1-0.15)}{1000}}} \approx -1.06.$$

6. The null hypothesis can't be rejected at the 0.10 significance level.
7. The data do not support the hypothesis that the new fixture design reduces the rejection level at the 0.10 significance level, so the team does not spend the resources required to produce a new fixture.

The team decides to study the scratched units more carefully. They discover five categories of scratches. The categories and the approximate percentage of rejected units with those scratches are:

Tapered scratches	24%
Vertical scratches	27%
Diagonal scratches	23%
Parallel pair scratches	17%
J-shaped scratches	9%

(No units have more than one type of scratch.)

The team then reinspects the 1000 units produced with the prototype holding fixture and discovers that they had no J-shaped scratches. In other words it appears that the new fixture design completely eliminated the J-shaped scratches. The hypothesis test would, of course, have revealed this fact if the J-shaped scratches had been studied independently. By discovering defect categories, the team is able to eliminate a cause of about 9% of the defects and, more importantly, they have clues to other causes.

Thought question: What if cancer turns out to be many diseases? Have therapies that completely cured some of them been rejected because of the way the data were tested?

One of the frustrating aspects of preventive and corrective action problem solving is the tendency for the same problem to reoccur. This may be because the installed solution did not really solve the problem, or it may be because the solution, although correcting the problem, was discontinued for some reason. Therefore, a major part of preventive and corrective action activities is the installation of a system that monitors the process to ensure that the problem doesn't reoccur.

In situations where the problem occurs very rarely it is often best to establish a control chart to determine whether the desired change has occurred.

Documentation regarding problem-solving activity should be maintained and available to team members. Properly cross-referenced, this documentation can aid team members by providing a history of current or similar problems.

Chapter 17

C. Failure Analysis and Correction

1. FAILURE ANALYSIS METHODS

> Describe methods such as mechanical, materials, and physical analysis, scanning electron microscopy (SEM), etc., that are used to identify failure mechanisms. (Understand)
>
> **Body of Knowledge VII.C.1**

It is important to have as much information as possible about each failure. This section discusses some approaches that have proved useful.

The first step in failure analysis is to determine why the item has been classified as failed. This investigation should begin with the warranty report, but it should not end there. The item should be examined for failure in its construction, but further study should be made of the application and installation—things that should be included in a warranty report but often aren't. Traceability and other documentation should be examined for clues as to the root cause of the failure. If the item did not physically fail, thorough study of the environment in which it functioned is required.

If the part has become deformed or broken, two questions arise:

- Has the part failed to meet designed strength?
- Has the part been subjected to stress beyond its designed strength?

One approach to resolving this dichotomy is to subject a number of identical parts to various stresses. Of course the word "identical," while critical, is impossible to attain. This means a thorough examination of the failed part is necessary to determine whether it has some unique property that makes it more vulnerable to stress. This may involve techniques ranging from a visual study to chemical/metallurgical analysis. Nondestructive tools such as ultrasound, eddy current, X-ray, liquid penetrant, and magnetic particle testing may be helpful. In some cases grain structure analysis and electron microscope examination are appropriate.

If this analysis shows that the part has been subjected to stress outside its designed strength, new questions are raised, including:

- Is the design strength adequate?
- Was the part applied inappropriately?
- Did the environment (thermal, chemical, electromagnetic, vibration, and so on) contribute to the failure?

EXAMPLE 17.1

A residential window has failed and has been replaced by the local representative. The warranty report states "leaker." The following list shows questions that were raised, along with answers that were obtained with some effort.

Question	Answer
Where did the leak occur?	Two places along the vertical edges.
Was the glass cracked or chipped?	No.
Did the sash have flaws that could cause leaks?	No.
Was the caulking compound deteriorated?	Yes.
Was there evidence of abrasion on caulking?	No.
What type of deterioration was observed?	General pitting and some crumbling.
What direction did the window face?	North.
Where was the residence?	One block from the Oregon coastline.
Was the window rated for saline environment?	Yes.
Was the caulk rated for saline environment?	Yes.
Does the saline-rated caulk in stock resist saline deterioration?	No, upon testing, it degrades much as that on the failed window.
(To the caulk supplier) Why is your saline-rated caulk failing to resist salt?	We have mislabeled some regular caulk to indicate it is safe for saline environment.
(To the calk supplier) Do you want our continued business bad enough that you will provide a quick method for us to check each barrel of caulk for saline resistance since we apparently can't depend on your labeling?	Yes.
(To manufacturing) Can you locate and fix windows that were caulked with the mislabeled product?	We're on it.

The reliability program should support a traceability system that will identify materials, methods, processes, and supplier information for each item. This will help determine whether other items in the field are vulnerable.

After determining the root cause of a failure, a report should be generated that includes any recommendations for changes that will help reduce this type of failure in the future.

2. FAILURE REPORTING, ANALYSIS, AND CORRECTIVE ACTION SYSTEM (FRACAS)

> Identify the elements necessary for a FRACAS to be effective, and demonstrate the importance of a closed-loop process that includes root cause investigation and follow up. (Apply)
>
> Body of Knowledge VII.C.2

Despite the efforts that are put into identifying and preventing failure modes, some unanticipated failures occur. These failures are often of an insidious nature, having either eluded the team during FMEA activities or occurred even though corrective/preventive actions had been prescribed. For these reasons there may be a tendency toward lax reporting of these failures.

Failure reporting and corrective action systems (FRACAS) provide an organized, disciplined approach to this problem. The first step is to establish a *failure review board.* This group is responsible for implementing these guidelines:

- Require prompt and complete failure reporting.
- Begin the reporting with the earliest tests of products and processes, and continue through the life of the product.
- The analysis and corrective action phase of each report must determine root causes and appropriate corrective/preventive action. This phase should be completed within a prescribed period of time, usually 30 days.
- The corrective action must be closed-loop, as shown in Figure 17.1. Implement audits to determine whether corrective actions have been taken and whether they are effective. When corrective actions do not resolve the problem, the failure review board is responsible for a restudy of the failure mechanism and implementation of additional corrective actions.

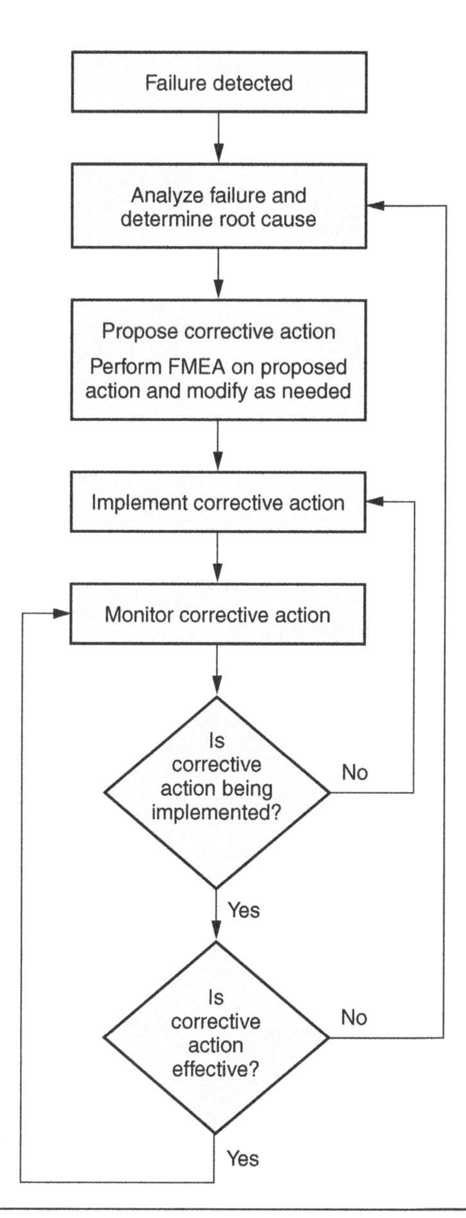

Figure 17.1 FRACAS flowchart.

Part VIII

Appendices

Appendix A

ASQ Certified Reliability Engineer Body of Knowledge

The topics in this Body of Knowledge include additional detail in the form of subtext explanations and the cognitive level at which the questions will be written. This information will provide useful guidance for both the Examination Development Committee and the candidates preparing to take the exam. The subtext is not intended to limit the subject matter or be all-inclusive of what might be covered in an exam. It is intended to clarify the type of content to be included in the exam. The descriptor in parentheses at the end of each entry refers to the highest cognitive level at which the topic will be tested. A more comprehensive description of cognitive levels is provided at the end of this document.

I. Reliability Management (18 Questions)

A. *Strategic management*

1. *Benefits of reliability engineering.* Describe how reliability engineering techniques and methods improve programs, processes, products, systems, and services. (Understand)
2. *Interrelationship of safety, quality, and reliability.* Define and describe the relationships among safety, reliability, and quality. (Understand)
3. *Role of the reliability function in the organization.* Describe how reliability techniques can be applied in other functional areas of the organization, such as marketing, engineering, customer/product support, safety and product liability, etc. (Apply)
4. *Reliability in product and process development.* Integrate reliability engineering techniques with other development activities, concurrent engineering, corporate improvement initiatives such as lean and six sigma methodologies, and emerging technologies. (Apply)
5. *Failure consequence and liability management.* Describe the importance of these concepts in determining reliability acceptance criteria. (Understand)
6. *Warranty management.* Define and describe warranty terms and conditions, including warranty period, conditions of use, failure criteria, etc., and identify the uses and limitations of warranty data. (Understand)

7. *Customer needs assessment.* Use various feedback methods (e.g., quality function deployment (QFD), prototyping, beta testing) to determine customer needs in relation to reliability requirements for products and services. (Apply)

8. *Supplier reliability.* Define and describe supplier reliability assessments that can be monitored in support of the overall reliability program. (Understand)

B. *Reliability program management*

1. *Terminology.* Explain basic reliability terms (e.g., MTTF, MTBF, MTTR, availability, failure rate, reliability, maintainability). (Understand)

2. *Elements of a reliability program.* Explain how planning, testing, tracking, and using customer needs and requirements are used to develop a reliability program, and identify various drivers of reliability requirements, including market expectations and standards, as well as safety, liability, and regulatory concerns. (Understand)

3. *Types of risk.* Describe the relationship between reliability and various types of risk, including technical, scheduling, safety, financial, etc. (Understand)

4. *Product lifecycle engineering.* Describe the impact various lifecycle stages (concept/design, introduction, growth, maturity, decline) have on reliability, and the cost issues (product maintenance, life expectation, software defect phase containment, etc.) associated with those stages. (Understand)

5. *Design evaluation.* Use validation, verification, and other review techniques to assess the reliability of a product's design at various lifecycle stages. (Analyze)

6. *Systems engineering and integration.* Describe how these processes are used to create requirements and prioritize design and development activities. (Understand)

C. *Ethics, safety, and liability*

1. *Ethical issues.* Identify appropriate ethical behaviors for a reliability engineer in various situations. (Evaluate)

2. *Roles and responsibilities.* Describe the roles and responsibilities of a reliability engineer in relation to product safety and liability. (Understand)

3. *System safety.* Identify safety-related issues by analyzing customer feedback, design data, field data, and other information. Use risk management tools (e.g., hazard analysis, FMEA, FTA, risk matrix) to identify and prioritize safety concerns, and identify steps that will minimize the misuse of products and processes. (Analyze)

II. Probability and Statistics for Reliability (27 Questions)

A. *Basic concepts*

1. *Statistical terms.* Define and use terms such as population, parameter, statistic, sample, the central limit theorem, etc., and compute their values. (Apply)

2. *Basic probability concepts.* Use basic probability concepts (e.g., independence, mutually exclusive, conditional probability) and compute expected values. (Apply)

3. *Discrete and continuous probability distributions.* Compare and contrast various distributions (binomial, Poisson, exponential, Weibull, normal, log-normal, etc.) and their functions (e.g., cumulative distribution functions (CDFs), probability density functions (PDFs), hazard functions), and relate them to the bathtub curve. (Analyze)

4. *Poisson process models.* Define and describe homogeneous and non-homogeneous Poisson process models (HPP and NHPP). (Understand)

5. *Non-parametric statistical methods.* Apply non-parametric statistical methods, including median, Kaplan-Meier, Mann-Whitney, etc., in various situations. (Apply)

6. *Sample size determination.* Use various theories, tables, and formulas to determine appropriate sample sizes for statistical and reliability testing. (Apply)

7. *Statistical process control (SPC) and process capability.* Define and describe SPC and process capability studies (C_p, C_{pk}, etc.), their control charts, and how they are all related to reliability. (Understand)

B. *Statistical inference*

1. *Point estimates of parameters.* Obtain point estimates of model parameters using probability plots, maximum likelihood methods, etc. Analyze the efficiency and bias of the estimators. (Evaluate)

2. *Statistical interval estimates.* Compute confidence intervals, tolerance intervals, etc., and draw conclusions from the results. (Evaluate)

3. *Hypothesis testing (parametric and non-parametric).* Apply hypothesis testing for parameters such as means, variance, proportions, and distribution parameters. Interpret significance levels and Type I and Type II errors for accepting/rejecting the null hypothesis. (Evaluate)

III. Reliability in Design and Development (26 Questions)

A. *Reliability design techniques*

1. *Environmental and use factors.* Identify environmental and use factors (e.g., temperature, humidity, vibration) and stresses (e.g., severity

of service, electrostatic discharge (ESD), throughput) to which a product may be subjected. (Apply)

2. *Stress-strength analysis.* Apply stress-strength analysis method of computing probability of failure, and interpret the results. (Evaluate)
3. *FMEA and FMECA.* Define and distinguish between failure mode and effects analysis and failure mode, effects, and criticality analysis and apply these techniques in products, processes, and designs. (Analyze)
4. *Common mode failure analysis.* Describe this type of failure (also known as common cause mode failure) and how it affects design for reliability. (Understand)
5. *Fault tree analysis (FTA) and success tree analysis (STA).* Apply these techniques to develop models that can be used to evaluate undesirable (FTA) and desirable (STA) events. (Analyze)
6. *Tolerance and worst-case analyses.* Describe how tolerance and worst-case analyses (e.g., root of sum of squares, extreme value) can be used to characterize variation that affects reliability. (Understand)
7. *Design of experiments.* Plan and conduct standard design of experiments (DOE) (e.g., full-factorial, fractional factorial, Latin square design). Implement robust-design approaches (e.g., Taguchi design, parametric design, DOE incorporating noise factors) to improve or optimize design. (Analyze)
8. *Fault tolerance.* Define and describe fault tolerance and the reliability methods used to maintain system functionality. (Understand)
9. *Reliability optimization.* Use various approaches, including redundancy, derating, trade studies, etc., to optimize reliability within the constraints of cost, schedule, weight, design requirements, etc. (Apply)
10. *Human factors.* Describe the relationship between human factors and reliability engineering. (Understand)
11. *Design for X (DFX).* Apply DFX techniques such as design for assembly, testability, maintainability environment (recycling and disposal), etc., to enhance a product's producibility and serviceability. (Apply)
12. *Reliability apportionment (allocation) techniques.* Use these techniques to specify subsystem and component reliability requirements. (Analyze)

B. *Parts and systems management*

1. *Selection, standardization, and reuse.* Apply techniques for materials selection, parts standardization and reduction, parallel modeling,

software reuse, including commercial off-the-shelf (COTS) software, etc. (Apply)

2. *Derating methods and principles.* Use methods such as S-N diagram, stress-life relationship, etc., to determine the relationship between applied stress and rated value, and to improve design. (Analyze)

3. *Parts obsolescence management.* Explain the implications of parts obsolescence and requirements for parts or system requalification. Develop risk mitigation plans such as lifetime buy, backwards compatibility, etc. (Apply)

4. *Establishing specifications.* Develop metrics for reliability, maintainability, and serviceability (e.g., MTBF, MTBR, MTBUMA, service interval) for product specifications. (Create)

IV. Reliability Modeling and Predictions (22 Questions)

A. *Reliability modeling*

1. *Sources and uses of reliability data.* Describe sources of reliability data (prototype, development, test, field, warranty, published, etc.), their advantages and limitations, and how the data can be used to measure and enhance product reliability. (Apply)

2. *Reliability block diagrams and models.* Generate and analyze various types of block diagrams and models, including series, parallel, partial redundancy, time-dependent, etc. (Create)

3. *Physics of failure models.* Identify various failure mechanisms (e.g., fracture, corrosion, memory corruption) and select appropriate theoretical models (e.g., Arrhenius, S-N curve) to assess their impact. (Apply)

4. *Simulation techniques.* Describe the advantages and limitations of the Monte Carlo and Markov models. (Apply)

5. *Dynamic reliability.* Describe dynamic reliability as it relates to failure criteria that change over time or under different conditions. (Understand)

B. *Reliability predictions*

1. *Part count predictions and part stress analysis.* Use parts failure rate data to estimate system- and subsystem-level reliability. (Apply)

2. *Reliability prediction methods.* Use various reliability prediction methods for both repairable and non-repairable components and systems, incorporating test and field reliability data when available. (Apply)

V. Reliability Testing (24 Questions)

A. *Reliability test planning*

1. *Reliability test strategies.* Create and apply the appropriate test strategies (e.g., truncation, test-to-failure, degradation) for various product development phases. (Create)

2. *Test environment.* Evaluate the environment in terms of system location and operational conditions to determine the most appropriate reliability test. (Evaluate)

B. *Testing during development.* Describe the purpose, advantages, and limitations of each of the following types of tests, and use common models to develop test plans, evaluate risks, and interpret test results. (Evaluate)

1. Accelerated life tests (e.g., single-stress, multiple-stress, sequential stress, step-stress)

2. Discovery testing (e.g., HALT, margin tests, sample size of 1)

3. Reliability growth testing (e.g., test, analyze, and fix (TAAF), Duane)

4. Software testing (e.g., white-box, black-box, operational profile, and fault-injection)

C. *Product testing.* Describe the purpose, advantages, and limitations of each of the following types of tests, and use common models to develop product test plans, evaluate risks, and interpret test results. (Evaluate)

1. Qualification/demonstration testing (e.g., sequential tests, fixed-length tests)

2. Product reliability acceptance testing (PRAT)

3. Ongoing reliability testing (e.g., sequential probability ratio test [SPRT])

4. Stress screening (e.g., ESS, HASS, burn-in tests)

5. Attribute testing (e.g., binomial, hypergeometric)

6. Degradation (wear-to-failure) testing

VI. Maintainability and Availability (15 Questions)

A. *Management strategies*

1. *Planning.* Develop plans for maintainability and availability that support reliability goals and objectives. (Create)

2. *Maintenance strategies.* Identify the advantages and limitations of various maintenance strategies (e.g., reliability-centered maintenance (RCM), predictive maintenance, repair or replace decision making), and determine which strategy to use in specific situations. (Apply).

3. *Availability tradeoffs.* Describe various types of availability (e.g., inherent, operational), and the tradeoffs in reliability and maintainability that might be required to achieve availability goals. (Apply)

B. *Maintenance and testing analysis*

1. *Preventive maintenance (PM) analysis.* Define and use PM tasks, optimum PM intervals, and other elements of this analysis, and identify situations in which PM analysis is not appropriate. (Apply)

2. *Corrective maintenance analysis.* Describe the elements of corrective maintenance analysis (e.g., fault-isolation time, repair/replace time, skill level, crew hours) and apply them in specific situations. (Apply)

3. *Non-destructive evaluation.* Describe the types and uses of these tools (e.g., fatigue, delamination, vibration signature analysis) to look for potential defects. (Understand)

4. *Testability.* Use various testability requirements and methods (e.g., built in tests (BITs), false-alarm rates, diagnostics, error codes, fault tolerance) to achieve reliability goals. (Apply)

5. *Spare parts analysis.* Describe the relationship between spare parts requirements and reliability, maintainability, and availability requirements. Forecast spare parts requirements using field data, production lead time data, inventory and other prediction tools, etc. (Analyze)

VII. Data Collection and Use (18 Questions)

A. *Data collection*

1. *Types of data.* Identify and distinguish between various types of data (e.g., attributes vs. variable, discrete vs. continuous, censored vs. complete, univariate vs. multivariate). Select appropriate data types to meet various analysis objectives. (Evaluate)

2. *Collection methods.* Identify appropriate methods and evaluate the results from surveys, automated tests, automated monitoring and reporting tools, etc., that are used to meet various data analysis objectives. (Evaluate)

3. *Data management.* Describe key characteristics of a database (e.g., accuracy, completeness, update frequency). Specify the requirements for reliability-driven measurement systems and database plans, including consideration of the data collectors and users, and their functional responsibilities. (Evaluate)

B. *Data use*

1. *Data summary and reporting.* Examine collected data for accuracy and usefulness. Analyze, interpret, and summarize data for presentation using techniques such as trend analysis, Weibull, graphic representation, etc., based on data types, sources, and required output. (Create)
2. *Preventive and corrective action.* Select and use various root cause and failure analysis tools to determine the causes of degradation or failure, and identify appropriate preventive or corrective actions to take in specific situations. (Evaluate)
3. *Measures of effectiveness.* Use various data analysis tools to evaluate the effectiveness of preventive and corrective actions in improving reliability. (Evaluate)

C. *Failure analysis and correction*

1. *Failure analysis methods.* Describe methods such as mechanical, materials, and physical analysis, scanning electron microscopy (SEM), etc., that are used to identify failure mechanisms. (Understand)
2. *Failure reporting, analysis, and corrective action system (FRACAS).* Identify the elements necessary for a FRACAS to be effective, and demonstrate the importance of a closed-loop process that includes root cause investigation and follow up. (Apply)

LEVELS OF COGNITION BASED ON BLOOM'S TAXONOMY—REVISED (2001)

In addition to content specifics, the subtext for each topic in this BOK also indicates the intended complexity level of the test questions for that topic. These levels are based on "Levels of Cognition" (from Bloom's Taxonomy—Revised, 2001) and are presented below in rank order, from least complex to most complex.

Remember

Recall or recognize terms, definitions, facts, ideas, materials, patterns, sequences, methods, principles, etc.

Understand

Read and understand descriptions, communications, reports, tables, diagrams, directions, regulations, etc.

Apply

Know when and how to use ideas, procedures, methods, formulas, principles, theories, etc.

Analyze

Break down information into its constituent parts and recognize their relationship to one another and how they are organized; identify sublevel factors or salient data from a complex scenario.

Evaluate

Make judgments about the value of proposed ideas, solutions, etc., by comparing the proposal to specific criteria or standards.

Create

Put parts or elements together in such a way as to reveal a pattern or structure not clearly there before; identify which data or information from a complex set is appropriate to examine further or from which supported conclusions can be drawn.

Appendix B
ASQ Code of Ethics

To uphold and advance the honor and dignity of the profession, and in keeping with high standards of ethical conduct I acknowledge that I:

FUNDAMENTAL PRINCIPLES

- Will be honest and impartial, and will serve with devotion my employer, my clients, and the public.
- Will strive to increase the competence and prestige of the profession.
- Will use my knowledge and skill for the advancement of human welfare, and in promoting the safety and reliability of products for public use.
- Will earnestly endeavor to aid the work of the Society.

RELATIONS WITH THE PUBLIC

1.1 Will do whatever I can to promote the reliability and safety of all products that come within my jurisdiction.

1.2 Will endeavor to extend public knowledge of the work of the Society and its members that relates to the public welfare.

1.3 Will be dignified and modest in explaining my work and merit.

1.4 Will preface any public statements that I may issue by clearly indicating on whose behalf they are made.

RELATIONS WITH EMPLOYERS AND CLIENTS

2.1 Will act in professional matters as a faithful agent or trustee for each employer or client.

2.2 Will inform each client or employer of any business connections, interests, or affiliations which might influence my judgment or impair the equitable character of my services.

2.3 Will indicate to my employer or client the adverse consequences to be expected if my professional judgment is overruled.

2.4 Will not disclose information concerning the business affairs or technical processes of any present or former employer or client without his consent.

2.5 Will not accept compensation from more than one party for the same service without the consent of all parties. If employed, I will engage in supplementary employment of consulting practice only with the consent of my employer.

RELATIONS WITH PEERS

3.1 Will take care that credit for the work of others is given to those whom it is due.

3.2 Will endeavor to aid the professional development and advancement of those in my employ or under my supervision.

3.3 Will not compete unfairly with others; will extend my friendship and confidence to all associates and those with whom I have business relations.

Appendix C

Control Limit Formulas

VARIABLES CHARTS

$\bar{x}$ and R chart:

$\textit{Averages chart}: \bar{\bar{x}} \pm A_2\bar{R}$ $\qquad$ $\textit{Range chart}: \text{LCL} = D_3\bar{R} \quad \text{UCL} = D_4\bar{R}$

$\bar{x}$ and s chart:

$\textit{Averages chart}: \bar{\bar{x}} \pm A_3\bar{s}$ $\qquad$ $\textit{Standard deviation chart}: \text{LCL} = B_3\bar{s} \quad \text{UCL} = B_4\bar{s}$

Individuals and moving range chart (two-value moving window):

$\textit{Individuals chart}: \bar{x} \pm 2.66\bar{R}$ $\quad$ $\textit{Moving range}: \text{UCL} = 3.267\bar{R}$

Moving average and moving range (two-value moving window):

$\textit{Moving average}: \bar{\bar{x}} \pm 1.88\bar{R}$ $\quad$ $\textit{Moving range}: \text{UCL} = 3.267\bar{R}$

Median chart:

$\textit{Median chart}: \bar{x}' \pm A_2'\bar{R}$ $\qquad$ $\textit{Range chart}: \text{LCL} = D_3\bar{R} \quad \text{UCL} = D_4\bar{R}$

ATTRIBUTE CHARTS

p-chart: $\bar{p} \pm 3\sqrt{\frac{\bar{p}(1-\bar{p})}{n}}$ $\qquad$ c-chart: $\bar{c} \pm 3\sqrt{\bar{c}}$

np-chart: $n\bar{p} \pm 3\sqrt{n\bar{p}(1-\bar{p})}$ $\qquad$ u-chart: $\bar{u} \pm 3\sqrt{\frac{\bar{u}}{n}}$

Appendix D

Constants for Control Charts

Subgroup size										A_2 for median charts			
N	A_2	d_2	D_3	D_4	A_3	c_4	B_3	B_4	E_2		A_4	D_5	D_6
2	1.880	1.128	–	3.267	2.659	0.798	–	3.267	2.660	1.880	2.224	–	3.865
3	1.023	1.693	–	2.574	1.954	0.886	–	2.568	1.772	1.187	1.091	–	2.745
4	0.729	2.059	–	2.282	1.628	0.921	–	2.266	1.457	0.796	0.758	–	2.375
5	0.577	2.326	–	2.114	1.427	0.940	–	2.089	1.290	0.691	0.594	–	2.179
6	0.483	2.534	–	2.004	1.287	0.952	0.030	1.970	1.184	0.548	0.495	–	2.055
7	0.419	2.704	0.076	1.924	1.182	0.959	0.118	1.882	1.109	0.508	0.429	0.078	1.967
8	0.373	2.847	0.136	1.864	1.099	0.965	0.185	1.815	1.054	0.433	0.380	0.139	1.901
9	0.337	2.970	0.184	1.816	1.032	0.969	0.239	1.761	1.010	0.412	0.343	0.187	1.850
10	0.308	3.078	0.223	1.777	0.975	0.973	0.284	1.716	0.975	0.362	0.314	0.227	1.809

Appendix E

Areas under Standard Normal Curve

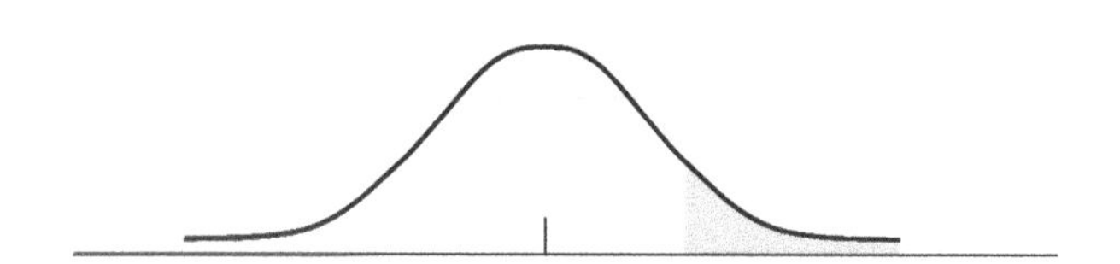

z	0.00	0.01	0.02	0.03	0.04	0.05	0.06	0.07	0.08	0.09
0.0	0.5000	0.4960	0.4920	0.4880	0.4840	0.4801	0.4761	0.4721	0.4681	0.4641
0.1	0.4602	0.4562	0.4522	0.448	0.4443	0.4404	0.4364	0.4325	0.4286	0.4247
0.2	0.4207	0.4168	0.4129	0.4090	0.4051	0.4013	0.3974	0.3936	0.3897	0.3859
0.3	0.3821	0.3783	0.3745	0.3707	0.3669	0.3632	0.3594	0.3557	0.3520	0.3483
0.4	0.3446	0.3409	0.3372	0.3336	0.3300	0.3264	0.3228	0.3192	0.3156	0.3121
0.5	0.3085	0.3050	0.3015	0.2981	0.2946	0.2912	0.2877	0.2843	0.2810	0.2776
0.6	0.2743	0.2709	0.2676	0.2643	0.2611	0.2578	0.2546	0.2514	0.2483	0.2451
0.7	0.2420	0.2389	0.2358	0.2327	0.2296	0.2266	0.2236	0.2206	0.2177	0.2148
0.8	0.2119	0.2090	0.2061	0.2033	0.2005	0.1977	0.1949	0.1922	0.1894	0.1867
0.9	0.1841	0.1814	0.1788	0.1762	0.1736	0.1711	0.1685	0.1660	0.1635	0.1611
1.0	0.1587	0.1562	0.1539	0.1515	0.1492	0.1469	0.1446	0.1423	0.1401	0.1379
1.1	0.1357	0.1335	0.1314	0.1292	0.1271	0.1251	0.1230	0.1210	0.1190	0.1170
1.2	0.1151	0.1131	0.1112	0.1093	0.1075	0.1056	0.1038	0.1020	0.1003	0.0985
1.3	0.0968	0.0951	0.0934	0.0918	0.0901	0.0885	0.0869	0.0853	0.0838	0.0823
1.4	0.0808	0.0793	0.0778	0.0764	0.0749	0.0735	0.0721	0.0708	0.0694	0.0681
1.5	0.0668	0.0655	0.0643	0.0630	0.0618	0.0606	0.0594	0.0582	0.0571	0.0559
1.6	0.0548	0.0537	0.0526	0.0516	0.0505	0.0495	0.0485	0.0475	0.0465	0.0455
1.7	0.0446	0.0436	0.0427	0.0418	0.0409	0.0401	0.0392	0.0384	0.0375	0.0367
1.8	0.0359	0.0351	0.0344	0.0336	0.0329	0.0322	0.0314	0.0307	0.0300	0.0294
1.9	0.0287	0.0281	0.0274	0.0268	0.0262	0.0256	0.0250	0.0244	0.0239	0.0233
2.0	0.0228	0.0222	0.0217	0.0212	0.0207	0.0202	0.0197	0.0192	0.0188	0.0183
2.1	0.0179	0.0174	0.0170	0.0166	0.0162	0.0158	0.0152	0.0150	0.0146	0.0143
2.2	0.0139	0.0136	0.0132	0.0129	0.0125	0.0122	0.0119	0.0116	0.0113	0.0110
2.3	0.0107	0.0104	0.0102	0.0099	0.0096	0.0094	0.0091	0.0089	0.0087	0.0084
2.4	0.0082	0.0080	0.0078	0.0075	0.0073	0.0071	0.0069	0.0068	0.0066	0.0064
2.5	0.0062	0.0060	0.0059	0.0057	0.0055	0.0054	0.0052	0.0051	0.0049	0.0048
2.6	0.0047	0.0045	0.0044	0.0043	0.0041	0.0040	0.0039	0.0038	0.0037	0.0036
2.7	0.0035	0.0034	0.0033	0.0032	0.0031	0.0030	0.0029	0.0028	0.0027	0.0026
2.8	0.0026	0.0025	0.0024	0.0023	0.0023	0.0022	0.0021	0.0021	0.0020	0.0019
2.9	0.0019	0.0018	0.0018	0.0017	0.0016	0.0016	0.0015	0.0015	0.0014	0.0014

Continued

Continued

z	0.00	0.01	0.02	0.03	0.04	0.05	0.06	0.07	0.08	0.09
3.0	0.0013	0.0013	0.0013	0.0012	0.0012	0.0011	0.0011	0.0011	0.0010	0.0010
3.1	0.0010	0.0009	0.0009	0.0009	0.0008	0.0008	0.0008	0.0008	0.0007	0.0007
3.2	0.0007	0.0007	0.0006	0.0006	0.0006	0.0006	0.0006	0.0005	0.0005	0.0005
3.3	0.0005	0.0005	0.0005	0.0004	0.0004	0.0004	0.0004	0.0004	0.0004	0.0003
3.4	0.0003	0.0003	0.0003	0.0003	0.0003	0.0003	0.0003	0.0003	0.0003	0.0002
3.5	0.0002	0.0002	0.0002	0.0002	0.0002	0.0002	0.0002	0.0002	0.0002	0.0002
3.6	0.0002	0.0002	0.0001	0.0001	0.0001	0.0001	0.0001	0.0001	0.0001	0.0001
3.7	0.0001	0.0001	0.0001	0.0001	0.0001	0.0001	0.0001	0.0001	0.0001	0.0001
3.8	0.0001	0.0001	0.0001	0.0001	0.0001	0.0001	0.0001	0.0001	0.0001	0.0001

For $z \geq 3.90$, areas are 0.0000 correct to four places.

Appendix F

F Distribution $F_{0.1}$

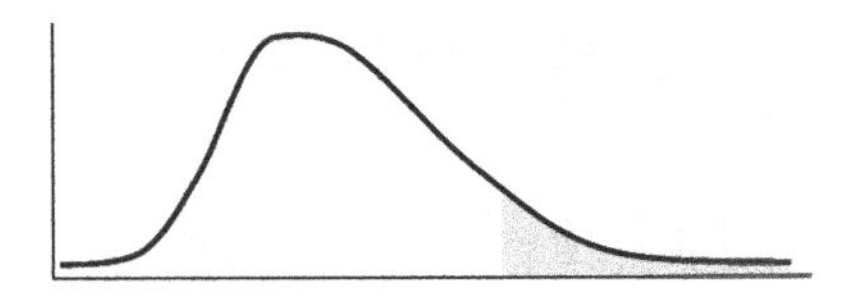

F distribution $F_{0.1}$

Numerator degrees of freedom (columns); Denominator degrees of freedom (rows)

	1	2	3	4	5	6	7	8	9	10	11
1	39.86	49.50	53.59	55.83	57.24	58.20	58.91	59.44	59.86	60.19	60.47
2	8.53	9.00	9.16	9.24	9.29	9.33	9.35	9.37	9.38	9.39	9.40
3	5.54	5.46	5.39	5.34	5.31	5.28	5.27	5.25	5.24	5.23	5.22
4	4.54	4.32	4.19	4.11	4.05	4.01	3.98	3.95	3.94	3.92	3.91
5	4.06	3.78	3.62	3.52	3.45	3.40	3.37	3.34	3.32	3.30	3.28
6	3.78	3.46	3.29	3.18	3.11	3.05	3.01	2.98	2.96	2.94	2.92
7	3.59	3.26	3.07	2.96	2.88	2.83	2.78	2.75	2.72	2.70	2.68
8	3.46	3.11	2.92	2.81	2.73	2.67	2.62	2.59	2.56	2.54	2.52
9	3.36	3.01	2.81	2.69	2.61	2.55	2.51	2.47	2.44	2.42	2.40
10	3.29	2.92	2.73	2.61	2.52	2.46	2.41	2.38	2.35	2.32	2.30
11	3.23	2.86	2.66	2.54	2.45	2.39	2.34	2.30	2.27	2.25	2.23
12	3.18	2.81	2.61	2.48	2.39	2.33	2.28	2.24	2.21	2.19	2.17
13	3.14	2.76	2.56	2.43	2.35	2.28	2.23	2.20	2.16	2.14	2.12
14	3.10	2.73	2.52	2.39	2.31	2.24	2.19	2.15	2.12	2.10	2.07
15	3.07	2.70	2.49	2.36	2.27	2.21	2.16	2.12	2.09	2.06	2.04
16	3.05	2.67	2.46	2.33	2.24	2.18	2.13	2.09	2.06	2.03	2.01
17	3.03	2.64	2.44	2.31	2.22	2.15	2.10	2.06	2.03	2.00	1.98
18	3.01	2.62	2.42	2.29	2.20	2.13	2.08	2.04	2.00	1.98	1.95
19	2.99	2.61	2.40	2.27	2.18	2.11	2.06	2.02	1.98	1.96	1.93
20	2.97	2.59	2.38	2.25	2.16	2.09	2.04	2.00	1.96	1.94	1.91
21	2.96	2.57	2.36	2.23	2.14	2.08	2.02	1.98	1.95	1.92	1.90
22	2.95	2.56	2.35	2.22	2.13	2.06	2.01	1.97	1.93	1.90	1.88
23	2.94	2.55	2.34	2.21	2.11	2.05	1.99	1.95	1.92	1.89	1.87
24	2.93	2.54	2.33	2.19	2.10	2.04	1.98	1.94	1.91	1.88	1.85
25	2.92	2.53	2.32	2.18	2.09	2.02	1.97	1.93	1.89	1.87	1.84
26	2.91	2.52	2.31	2.17	2.08	2.01	1.96	1.92	1.88	1.86	1.83
27	2.90	2.51	2.30	2.17	2.07	2.00	1.95	1.91	1.87	1.85	1.82
28	2.89	2.50	2.29	2.16	2.06	2.00	1.94	1.90	1.87	1.84	1.81
29	2.89	2.50	2.28	2.15	2.06	1.99	1.93	1.89	1.86	1.83	1.80
30	2.88	2.49	2.28	2.14	2.05	1.98	1.93	1.88	1.85	1.82	1.79
40	2.84	2.44	2.23	2.09	2.00	1.93	1.87	1.83	1.79	1.76	1.74
60	2.79	2.39	2.18	2.04	1.95	1.87	1.82	1.77	1.74	1.71	1.68
100	2.76	2.36	2.14	2.00	1.91	1.83	1.78	1.73	1.69	1.66	1.64
150	2.74	2.34	2.12	1.98	1.89	1.81	1.76	1.71	1.67	1.64	1.61
200	2.73	2.33	2.11	1.97	1.88	1.80	1.75	1.70	1.66	1.63	1.60

Continued

***F* distribution $F_{0.1}$** *(continued)*

Denominator degrees of freedom	Numerator degrees of freedom										
	12	**13**	**14**	**15**	**16**	**17**	**18**	**19**	**20**	**21**	**22**
1	60.71	60.90	61.07	61.22	61.35	61.46	61.57	61.66	61.74	61.81	61.88
2	9.41	9.41	9.42	9.42	9.43	9.43	9.44	9.44	9.44	9.44	9.45
3	5.22	5.21	5.20	5.20	5.20	5.19	5.19	5.19	5.18	5.18	5.18
4	3.90	3.89	3.88	3.87	3.86	3.86	3.85	3.85	3.84	3.84	3.84
5	3.27	3.26	3.25	3.24	3.23	3.22	3.22	3.21	3.21	3.20	3.20
6	2.90	2.89	2.88	2.87	2.86	2.85	2.85	2.84	2.84	2.83	2.83
7	2.67	2.65	2.64	2.63	2.62	2.61	2.61	2.60	2.59	2.59	2.58
8	2.50	2.49	2.48	2.46	2.45	2.45	2.44	2.43	2.42	2.42	2.41
9	2.38	2.36	2.35	2.34	2.33	2.32	2.31	2.30	2.30	2.29	2.29
10	2.28	2.27	2.26	2.24	2.23	2.22	2.22	2.21	2.20	2.19	2.19
11	2.21	2.19	2.18	2.17	2.16	2.15	2.14	2.13	2.12	2.12	2.11
12	2.15	2.13	2.12	2.10	2.09	2.08	2.08	2.07	2.06	2.05	2.05
13	2.10	2.08	2.07	2.05	2.04	2.03	2.02	2.01	2.01	2.00	1.99
14	2.05	2.04	2.02	2.01	2.00	1.99	1.98	1.97	1.96	1.96	1.95
15	2.02	2.00	1.99	1.97	1.96	1.95	1.94	1.93	1.92	1.92	1.91
16	1.99	1.97	1.95	1.94	1.93	1.92	1.91	1.90	1.89	1.88	1.88
17	1.96	1.94	1.93	1.91	1.90	1.89	1.88	1.87	1.86	1.86	1.85
18	1.93	1.92	1.90	1.89	1.87	1.86	1.85	1.84	1.84	1.83	1.82
19	1.91	1.89	1.88	1.86	1.85	1.84	1.83	1.82	1.81	1.81	1.80
20	1.89	1.87	1.86	1.84	1.83	1.82	1.81	1.80	1.79	1.79	1.78
21	1.87	1.86	1.84	1.83	1.81	1.80	1.79	1.78	1.78	1.77	1.76
22	1.86	1.84	1.83	1.81	1.80	1.79	1.78	1.77	1.76	1.75	1.74
23	1.84	1.83	1.81	1.80	1.78	1.77	1.76	1.75	1.74	1.74	1.73
24	1.83	1.81	1.80	1.78	1.77	1.76	1.75	1.74	1.73	1.72	1.71
25	1.82	1.80	1.79	1.77	1.76	1.75	1.74	1.73	1.72	1.71	1.70
26	1.81	1.79	1.77	1.76	1.75	1.73	1.72	1.71	1.71	1.70	1.69
27	1.80	1.78	1.76	1.75	1.74	1.72	1.71	1.70	1.70	1.69	1.68
28	1.79	1.77	1.75	1.74	1.73	1.71	1.70	1.69	1.69	1.68	1.67
29	1.78	1.76	1.75	1.73	1.72	1.71	1.69	1.68	1.68	1.67	1.66
30	1.77	1.75	1.74	1.72	1.71	1.70	1.69	1.68	1.67	1.66	1.65
40	1.71	1.70	1.68	1.66	1.65	1.64	1.62	1.61	1.61	1.60	1.59
60	1.66	1.64	1.62	1.60	1.59	1.58	1.56	1.55	1.54	1.53	1.53
100	1.61	1.59	1.57	1.56	1.54	1.53	1.52	1.50	1.49	1.48	1.48
150	1.59	1.57	1.55	1.53	1.52	1.50	1.49	1.48	1.47	1.46	1.45
200	1.58	1.56	1.54	1.52	1.51	1.49	1.48	1.47	1.46	1.45	1.44

Continued

F distribution $F_{0.1}$ *(continued)*

Denominator degrees of freedom	Numerator degrees of freedom										
	23	24	25	26	27	28	29	30	40	60	100
1	61.94	62.00	62.05	62.10	62.15	62.19	62.23	62.26	62.53	62.79	63.01
2	9.45	9.45	9.45	9.45	9.45	9.46	9.46	9.46	9.47	9.47	9.48
3	5.18	5.18	5.17	5.17	5.17	5.17	5.17	5.17	5.16	5.15	5.14
4	3.83	3.83	3.83	3.83	3.82	3.82	3.82	3.82	3.80	3.79	3.78
5	3.19	3.19	3.19	3.18	3.18	3.18	3.18	3.17	3.16	3.14	3.13
6	2.82	2.82	2.81	2.81	2.81	2.81	2.80	2.80	2.78	2.76	2.75
7	2.58	2.58	2.57	2.57	2.56	2.56	2.56	2.56	2.54	2.51	2.50
8	2.41	2.40	2.40	2.40	2.39	2.39	2.39	2.38	2.36	2.34	2.32
9	2.28	2.28	2.27	2.27	2.26	2.26	2.26	2.25	2.23	2.21	2.19
10	2.18	2.18	2.17	2.17	2.17	2.16	2.16	2.16	2.13	2.11	2.09
11	2.11	2.10	2.10	2.09	2.09	2.08	2.08	2.08	2.05	2.03	2.01
12	2.04	2.04	2.03	2.03	2.02	2.02	2.01	2.01	1.99	1.96	1.94
13	1.99	1.98	1.98	1.97	1.97	1.96	1.96	1.96	1.93	1.90	1.88
14	1.94	1.94	1.93	1.93	1.92	1.92	1.92	1.91	1.89	1.86	1.83
15	1.90	1.90	1.89	1.89	1.88	1.88	1.88	1.87	1.85	1.82	1.79
16	1.87	1.87	1.86	1.86	1.85	1.85	1.84	1.84	1.81	1.78	1.76
17	1.84	1.84	1.83	1.83	1.82	1.82	1.81	1.81	1.78	1.75	1.73
18	1.82	1.81	1.80	1.80	1.80	1.79	1.79	1.78	1.75	1.72	1.70
19	1.79	1.79	1.78	1.78	1.77	1.77	1.76	1.76	1.73	1.70	1.67
20	1.77	1.77	1.76	1.76	1.75	1.75	1.74	1.74	1.71	1.68	1.65
21	1.75	1.75	1.74	1.74	1.73	1.73	1.72	1.72	1.69	1.66	1.63
22	1.74	1.73	1.73	1.72	1.72	1.71	1.71	1.70	1.67	1.64	1.61
23	1.72	1.72	1.71	1.70	1.70	1.69	1.69	1.69	1.66	1.62	1.59
24	1.71	1.70	1.70	1.69	1.69	1.68	1.68	1.67	1.64	1.61	1.58
25	1.70	1.69	1.68	1.68	1.67	1.67	1.66	1.66	1.63	1.59	1.56
26	1.68	1.68	1.67	1.67	1.66	1.66	1.65	1.65	1.61	1.58	1.55
27	1.67	1.67	1.66	1.65	1.65	1.64	1.64	1.64	1.60	1.57	1.54
28	1.66	1.66	1.65	1.64	1.64	1.63	1.63	1.63	1.59	1.56	1.53
29	1.65	1.65	1.64	1.63	1.63	1.62	1.62	1.62	1.58	1.55	1.52
30	1.64	1.64	1.63	1.63	1.62	1.62	1.61	1.61	1.57	1.54	1.51
40	1.58	1.57	1.57	1.56	1.56	1.55	1.55	1.54	1.51	1.47	1.43
60	1.52	1.51	1.50	1.50	1.49	1.49	1.48	1.48	1.44	1.40	1.36
100	1.47	1.46	1.45	1.45	1.44	1.43	1.43	1.42	1.38	1.34	1.29
150	1.44	1.43	1.43	1.42	1.41	1.41	1.40	1.40	1.35	1.30	1.26
200	1.43	1.42	1.41	1.41	1.40	1.39	1.39	1.38	1.34	1.29	1.24

Appendix G

F Distribution $F_{0.05}$

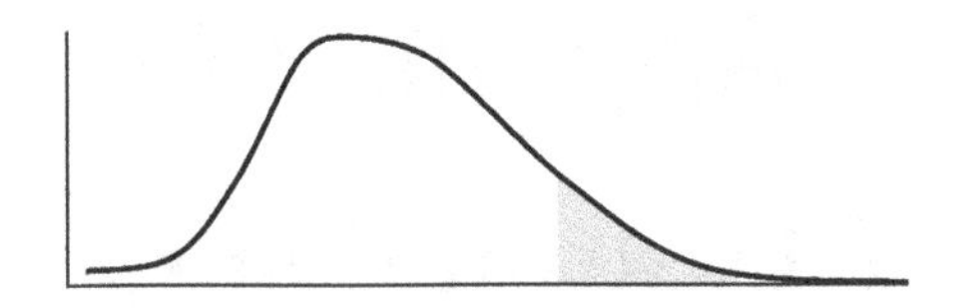

F distribution $F_{0.05}$

	Numerator degrees of freedom										
Denominator degrees of freedom	**1**	**2**	**3**	**4**	**5**	**6**	**7**	**8**	**9**	**10**	**11**
1	161.4	199.5	215.7	224.6	230.2	234.0	236.8	238.9	240.5	241.9	243.0
2	18.51	19.00	19.16	19.25	19.30	19.33	19.35	19.37	19.38	19.40	19.40
3	10.13	9.55	9.28	9.12	9.01	8.94	8.89	8.85	8.81	8.79	8.76
4	7.71	6.94	6.59	6.39	6.26	6.16	6.09	6.04	6.00	5.96	5.94
5	6.61	5.79	5.41	5.19	5.05	4.95	4.88	4.82	4.77	4.74	4.70
6	5.99	5.14	4.76	4.53	4.39	4.28	4.21	4.15	4.10	4.06	4.03
7	5.59	4.74	4.35	4.12	3.97	3.87	3.79	3.73	3.68	3.64	3.60
8	5.32	4.46	4.07	3.84	3.69	3.58	3.50	3.44	3.39	3.35	3.31
9	5.12	4.26	3.86	3.63	3.48	3.37	3.29	3.23	3.18	3.14	3.10
10	4.96	4.10	3.71	3.48	3.33	3.22	3.14	3.07	3.02	2.98	2.94
11	4.84	3.98	3.59	3.36	3.20	3.09	3.01	2.95	2.90	2.85	2.82
12	4.75	3.89	3.49	3.26	3.11	3.00	2.91	2.85	2.80	2.75	2.72
13	4.67	3.81	3.41	3.18	3.03	2.92	2.83	2.77	2.71	2.67	2.63
14	4.60	3.74	3.34	3.11	2.96	2.85	2.76	2.70	2.65	2.60	2.57
15	4.54	3.68	3.29	3.06	2.90	2.79	2.71	2.64	2.59	2.54	2.51
16	4.49	3.63	3.24	3.01	2.85	2.74	2.66	2.59	2.54	2.49	2.46
17	4.45	3.59	3.20	2.96	2.81	2.70	2.61	2.55	2.49	2.45	2.41
18	4.41	3.55	3.16	2.93	2.77	2.66	2.58	2.51	2.46	2.41	2.37
19	4.38	3.52	3.13	2.90	2.74	2.63	2.54	2.48	2.42	2.38	2.34
20	4.35	3.49	3.10	2.87	2.71	2.60	2.51	2.45	2.39	2.35	2.31
21	4.32	3.47	3.07	2.84	2.68	2.57	2.49	2.42	2.37	2.32	2.28
22	4.30	3.44	3.05	2.82	2.66	2.55	2.46	2.40	2.34	2.30	2.26
23	4.28	3.42	3.03	2.80	2.64	2.53	2.44	2.37	2.32	2.27	2.24
24	4.26	3.40	3.01	2.78	2.62	2.51	2.42	2.36	2.30	2.25	2.22
25	4.24	3.39	2.99	2.76	2.60	2.49	2.40	2.34	2.28	2.24	2.20
26	4.23	3.37	2.98	2.74	2.59	2.47	2.39	2.32	2.27	2.22	2.18
27	4.21	3.35	2.96	2.73	2.57	2.46	2.37	2.31	2.25	2.20	2.17
28	4.20	3.34	2.95	2.71	2.56	2.45	2.36	2.29	2.24	2.19	2.15
29	4.18	3.33	2.93	2.70	2.55	2.43	2.35	2.28	2.22	2.18	2.14
30	4.17	3.32	2.92	2.69	2.53	2.42	2.33	2.27	2.21	2.16	2.13
40	4.08	3.23	2.84	2.61	2.45	2.34	2.25	2.18	2.12	2.08	2.04
60	4.00	3.15	2.76	2.53	2.37	2.25	2.17	2.10	2.04	1.99	1.95
100	3.94	3.09	2.70	2.46	2.31	2.19	2.10	2.03	1.97	1.93	1.89

Continued

F* distribution $F_{0.05}$ *(continued)

	Numerator degrees of freedom										
Denominator degrees of freedom	12	13	14	15	16	17	18	19	20	21	22
1	243.9	244.7	245.4	245.9	246.5	246.9	247.3	247.7	248.0	248.3	248.6
2	19.41	19.42	19.42	19.43	19.43	19.44	19.44	19.44	19.45	19.45	19.45
3	8.74	8.73	8.71	8.70	8.69	8.68	8.67	8.67	8.66	8.65	8.65
4	5.91	5.89	5.87	5.86	5.84	5.83	5.82	5.81	5.80	5.79	5.79
5	4.68	4.66	4.64	4.62	4.60	4.59	4.58	4.57	4.56	4.55	4.54
6	4.00	3.98	3.96	3.94	3.92	3.91	3.90	3.88	3.87	3.86	3.86
7	3.57	3.55	3.53	3.51	3.49	3.48	3.47	3.46	3.44	3.43	3.43
8	3.28	3.26	3.24	3.22	3.20	3.19	3.17	3.16	3.15	3.14	3.13
9	3.07	3.05	3.03	3.01	2.99	2.97	2.96	2.95	2.94	2.93	2.92
10	2.91	2.89	2.86	2.85	2.83	2.81	2.80	2.79	2.77	2.76	2.75
11	2.79	2.76	2.74	2.72	2.70	2.69	2.67	2.66	2.65	2.64	2.63
12	2.69	2.66	2.64	2.62	2.60	2.58	2.57	2.56	2.54	2.53	2.52
13	2.60	2.58	2.55	2.53	2.51	2.50	2.48	2.47	2.46	2.45	2.44
14	2.53	2.51	2.48	2.46	2.44	2.43	2.41	2.40	2.39	2.38	2.37
15	2.48	2.45	2.42	2.40	2.38	2.37	2.35	2.34	2.33	2.32	2.31
16	2.42	2.40	2.37	2.35	2.33	2.32	2.30	2.29	2.28	2.26	2.25
17	2.38	2.35	2.33	2.31	2.29	2.27	2.26	2.24	2.23	2.22	2.21
18	2.34	2.31	2.29	2.27	2.25	2.23	2.22	2.20	2.19	2.18	2.17
19	2.31	2.28	2.26	2.23	2.21	2.20	2.18	2.17	2.16	2.14	2.13
20	2.28	2.25	2.22	2.20	2.18	2.17	2.15	2.14	2.12	2.11	2.10
21	2.25	2.22	2.20	2.18	2.16	2.14	2.12	2.11	2.10	2.08	2.07
22	2.23	2.20	2.17	2.15	2.13	2.11	2.10	2.08	2.07	2.06	2.05
23	2.20	2.18	2.15	2.13	2.11	2.09	2.08	2.06	2.05	2.04	2.02
24	2.18	2.15	2.13	2.11	2.09	2.07	2.05	2.04	2.03	2.01	2.00
25	2.16	2.14	2.11	2.09	2.07	2.05	2.04	2.02	2.01	2.00	1.98
26	2.15	2.12	2.09	2.07	2.05	2.03	2.02	2.00	1.99	1.98	1.97
27	2.13	2.10	2.08	2.06	2.04	2.02	2.00	1.99	1.97	1.96	1.95
28	2.12	2.09	2.06	2.04	2.02	2.00	1.99	1.97	1.96	1.95	1.93
29	2.10	2.08	2.05	2.03	2.01	1.99	1.97	1.96	1.94	1.93	1.92
30	2.09	2.06	2.04	2.01	1.99	1.98	1.96	1.95	1.93	1.92	1.91
40	2.00	1.97	1.95	1.92	1.90	1.89	1.87	1.85	1.84	1.83	1.81
60	1.92	1.89	1.86	1.84	1.82	1.80	1.78	1.76	1.75	1.73	1.72
100	1.85	1.82	1.79	1.77	1.75	1.73	1.71	1.69	1.68	1.66	1.65

Continued

F* distribution $F_{0.05}$ *(continued)

Denominator degrees of freedom	Numerator degrees of freedom										
	23	**24**	**25**	**26**	**27**	**28**	**29**	**30**	**40**	**60**	**100**
1	248.8	249.1	249.3	249.5	249.6	249.8	250.0	250.1	251.1	252.2	253.0
2	19.45	19.45	19.46	19.46	19.46	19.46	19.46	19.46	19.47	19.48	19.49
3	8.64	8.64	8.63	8.63	8.63	8.62	8.62	8.62	8.59	8.57	8.55
4	5.78	5.77	5.77	5.76	5.76	5.75	5.75	5.75	5.72	5.69	5.66
5	4.53	4.53	4.52	4.52	4.51	4.50	4.50	4.50	4.46	4.43	4.41
6	3.85	3.84	3.83	3.83	3.82	3.82	3.81	3.81	3.77	3.74	3.71
7	3.42	3.41	3.40	3.40	3.39	3.39	3.38	3.38	3.34	3.30	3.27
8	3.12	3.12	3.11	3.10	3.10	3.09	3.08	3.08	3.04	3.01	2.97
9	2.91	2.90	2.89	2.89	2.88	2.87	2.87	2.86	2.83	2.79	2.76
10	2.75	2.74	2.73	2.72	2.72	2.71	2.70	2.70	2.66	2.62	2.59
11	2.62	2.61	2.60	2.59	2.59	2.58	2.58	2.57	2.53	2.49	2.46
12	2.51	2.51	2.50	2.49	2.48	2.48	2.47	2.47	2.43	2.38	2.35
13	2.43	2.42	2.41	2.41	2.40	2.39	2.39	2.38	2.34	2.30	2.26
14	2.36	2.35	2.34	2.33	2.33	2.32	2.31	2.31	2.27	2.22	2.19
15	2.30	2.29	2.28	2.27	2.27	2.26	2.25	2.25	2.20	2.16	2.12
16	2.24	2.24	2.23	2.22	2.21	2.21	2.20	2.19	2.15	2.11	2.07
17	2.20	2.19	2.18	2.17	2.17	2.16	2.15	2.15	2.10	2.06	2.02
18	2.16	2.15	2.14	2.13	2.13	2.12	2.11	2.11	2.06	2.02	1.98
19	2.12	2.11	2.11	2.10	2.09	2.08	2.08	2.07	2.03	1.98	1.94
20	2.09	2.08	2.07	2.07	2.06	2.05	2.05	2.04	1.99	1.95	1.91
21	2.06	2.05	2.05	2.04	2.03	2.02	2.02	2.01	1.96	1.92	1.88
22	2.04	2.03	2.02	2.01	2.00	2.00	1.99	1.98	1.94	1.89	1.85
23	2.01	2.01	2.00	1.99	1.98	1.97	1.97	1.96	1.91	1.86	1.82
24	1.99	1.98	1.97	1.97	1.96	1.95	1.95	1.94	1.89	1.84	1.80
25	1.97	1.96	1.96	1.95	1.94	1.93	1.93	1.92	1.87	1.82	1.78
26	1.96	1.95	1.94	1.93	1.92	1.91	1.91	1.90	1.85	1.80	1.76
27	1.94	1.93	1.92	1.91	1.90	1.90	1.89	1.88	1.84	1.79	1.74
28	1.92	1.91	1.91	1.90	1.89	1.88	1.88	1.87	1.82	1.77	1.73
29	1.91	1.90	1.89	1.88	1.88	1.87	1.86	1.85	1.81	1.75	1.71
30	1.90	1.89	1.88	1.87	1.86	1.85	1.85	1.84	1.79	1.74	1.70
40	1.80	1.79	1.78	1.77	1.77	1.76	1.75	1.74	1.69	1.64	1.59
60	1.71	1.70	1.69	1.68	1.67	1.66	1.66	1.65	1.59	1.53	1.48
100	1.64	1.63	1.62	1.61	1.60	1.59	1.58	1.57	1.52	1.45	1.39

Appendix H

F Distribution $F_{0.01}$

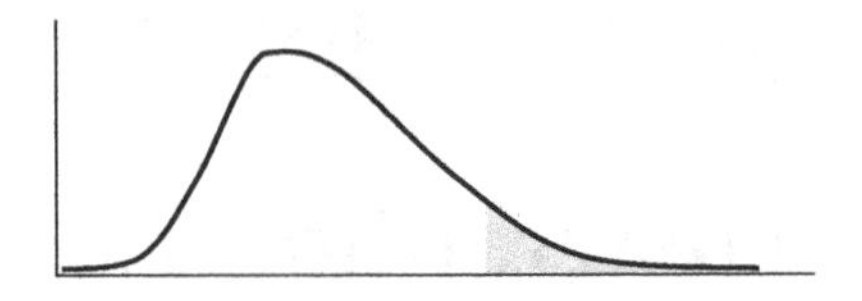

F distribution $F_{0.01}$

Numerator degrees of freedom (columns); Denominator degrees of freedom (rows)

	1	2	3	4	5	6	7	8	9	10	11
1	4052	4999	5404	5624	5764	5859	5928	5981	6022	6056	6083
2	98.5	99	99.16	99.25	99.3	99.33	99.36	99.38	99.39	99.4	99.41
3	34.12	30.82	29.46	28.71	28.24	27.91	27.67	27.49	27.34	27.23	27.13
4	21.2	18	16.69	15.98	15.52	15.21	14.98	14.8	14.66	14.55	14.45
5	16.26	13.27	12.06	11.39	10.97	10.67	10.46	10.29	10.16	10.05	9.963
6	13.75	10.92	9.78	9.148	8.746	8.466	8.26	8.102	7.976	7.874	7.79
7	12.25	9.547	8.451	7.847	7.46	7.191	6.993	6.84	6.719	6.62	6.538
8	11.26	8.649	7.591	7.006	6.632	6.371	6.178	6.029	5.911	5.814	5.734
9	10.56	8.022	6.992	6.422	6.057	5.802	5.613	5.467	5.351	5.257	5.178
10	10.04	7.559	6.552	5.994	5.636	5.386	5.2	5.057	4.942	4.849	4.772
11	9.646	7.206	6.217	5.668	5.316	5.069	4.886	4.744	4.632	4.539	4.462
12	9.33	6.927	5.953	5.412	5.064	4.821	4.64	4.499	4.388	4.296	4.22
13	9.074	6.701	5.739	5.205	4.862	4.62	4.441	4.302	4.191	4.1	4.025
14	8.862	6.515	5.564	5.035	4.695	4.456	4.278	4.14	4.03	3.939	3.864
15	8.683	6.359	5.417	4.893	4.556	4.318	4.142	4.004	3.895	3.805	3.73
16	8.531	6.226	5.292	4.773	4.437	4.202	4.026	3.89	3.78	3.691	3.616
17	8.4	6.112	5.185	4.669	4.336	4.101	3.927	3.791	3.682	3.593	3.518
18	8.285	6.013	5.092	4.579	4.248	4.015	3.841	3.705	3.597	3.508	3.434
19	8.185	5.926	5.01	4.5	4.171	3.939	3.765	3.631	3.523	3.434	3.36
20	8.096	5.849	4.938	4.431	4.103	3.871	3.699	3.564	3.457	3.368	3.294
21	8.017	5.78	4.874	4.369	4.042	3.812	3.64	3.506	3.398	3.31	3.236
22	7.945	5.719	4.817	4.313	3.988	3.758	3.587	3.453	3.346	3.258	3.184
23	7.881	5.664	4.765	4.264	3.939	3.71	3.539	3.406	3.299	3.211	3.137
24	7.823	5.614	4.718	4.218	3.895	3.667	3.496	3.363	3.256	3.168	3.094
25	7.77	5.568	4.675	4.177	3.855	3.627	3.457	3.324	3.217	3.129	3.056
26	7.721	5.526	4.637	4.14	3.818	3.591	3.421	3.288	3.182	3.094	3.021
27	7.677	5.488	4.601	4.106	3.785	3.558	3.388	3.256	3.149	3.062	2.988
28	7.636	5.453	4.568	4.074	3.754	3.528	3.358	3.226	3.12	3.032	2.959
29	7.598	5.42	4.538	4.045	3.725	3.499	3.33	3.198	3.092	3.005	2.931
30	7.562	5.39	4.51	4.018	3.699	3.473	3.305	3.173	3.067	2.979	2.906
40	7.314	5.178	4.313	3.828	3.514	3.291	3.124	2.993	2.888	2.801	2.727
60	7.077	4.977	4.126	3.649	3.339	3.119	2.953	2.823	2.718	2.632	2.559
100	6.895	4.824	3.984	3.513	3.206	2.988	2.823	2.694	2.59	2.503	2.43

Continued

F distribution $F_{0.01}$ *(continued)*

	Numerator degrees of freedom										
Denominator degrees of freedom	12	13	14	15	16	17	18	19	20	21	22
1	6107	6126	6143	6157	6170	6181	6191	6201	6208.7	6216.1	6223.1
2	99.42	99.42	99.43	99.43	99.44	99.44	99.44	99.45	99.448	99.451	99.455
3	27.05	26.98	26.92	26.87	26.83	26.79	26.75	26.72	26.69	26.664	26.639
4	14.37	14.31	14.25	14.2	14.15	14.11	14.08	14.05	14.019	13.994	13.97
5	9.888	9.825	9.77	9.722	9.68	9.643	9.609	9.58	9.5527	9.5281	9.5058
6	7.718	7.657	7.605	7.559	7.519	7.483	7.451	7.422	7.3958	7.3721	7.3506
7	6.469	6.41	6.359	6.314	6.275	6.24	6.209	6.181	6.1555	6.1324	6.1113
8	5.667	5.609	5.559	5.515	5.477	5.442	5.412	5.384	5.3591	5.3365	5.3157
9	5.111	5.055	5.005	4.962	4.924	4.89	4.86	4.833	4.808	4.7855	4.7651
10	4.706	4.65	4.601	4.558	4.52	4.487	4.457	4.43	4.4054	4.3831	4.3628
11	4.397	4.342	4.293	4.251	4.213	4.18	4.15	4.123	4.099	4.0769	4.0566
12	4.155	4.1	4.052	4.01	3.972	3.939	3.91	3.883	3.8584	3.8363	3.8161
13	3.96	3.905	3.857	3.815	3.778	3.745	3.716	3.689	3.6646	3.6425	3.6223
14	3.8	3.745	3.698	3.656	3.619	3.586	3.556	3.529	3.5052	3.4832	3.463
15	3.666	3.612	3.564	3.522	3.485	3.452	3.423	3.396	3.3719	3.3498	3.3297
16	3.553	3.498	3.451	3.409	3.372	3.339	3.31	3.283	3.2587	3.2367	3.2165
17	3.455	3.401	3.353	3.312	3.275	3.242	3.212	3.186	3.1615	3.1394	3.1192
18	3.371	3.316	3.269	3.227	3.19	3.158	3.128	3.101	3.0771	3.055	3.0348
19	3.297	3.242	3.195	3.153	3.116	3.084	3.054	3.027	3.0031	2.981	2.9607
20	3.231	3.177	3.13	3.088	3.051	3.018	2.989	2.962	2.9377	2.9156	2.8953
21	3.173	3.119	3.072	3.03	2.993	2.96	2.931	2.904	2.8795	2.8574	2.837
22	3.121	3.067	3.019	2.978	2.941	2.908	2.879	2.852	2.8274	2.8052	2.7849
23	3.074	3.02	2.973	2.931	2.894	2.861	2.832	2.805	2.7805	2.7582	2.7378
24	3.032	2.977	2.93	2.889	2.852	2.819	2.789	2.762	2.738	2.7157	2.6953
25	2.993	2.939	2.892	2.85	2.813	2.78	2.751	2.724	2.6993	2.677	2.6565
26	2.958	2.904	2.857	2.815	2.778	2.745	2.715	2.688	2.664	2.6416	2.6211
27	2.926	2.872	2.824	2.783	2.746	2.713	2.683	2.656	2.6316	2.609	2.5886
28	2.896	2.842	2.795	2.753	2.716	2.683	2.653	2.626	2.6018	2.5793	2.5587
29	2.868	2.814	2.767	2.726	2.689	2.656	2.626	2.599	2.5742	2.5517	2.5311
30	2.843	2.789	2.742	2.7	2.663	2.63	2.6	2.573	2.5487	2.5262	2.5055
40	2.665	2.611	2.563	2.522	2.484	2.451	2.421	2.394	2.3689	2.3461	2.3252
60	2.496	2.442	2.394	2.352	2.315	2.281	2.251	2.223	2.1978	2.1747	2.1533
10	2.368	2.313	2.265	2.223	2.185	2.151	2.12	2.092	2.0666	2.0431	2.0214

Continued

F distribution $F_{0.01}$ *(continued)*

Denominator degrees of freedom	Numerator degrees of freedom										
	23	24	25	26	27	28	29	30	40	60	100
1	6228.7	6234.3	6239.9	6244.5	6249.2	6252.9	6257.1	6260.4	6286.4	6313	6333.9
2	99.455	99.455	99.459	99.462	99.462	99.462	99.462	99.466	99.477	99.484	99.491
3	26.617	26.597	26.579	26.562	26.546	26.531	26.517	26.504	26.411	26.316	26.241
4	13.949	13.929	13.911	13.894	13.878	13.864	13.85	13.838	13.745	13.652	13.577
5	9.4853	9.4665	9.4492	9.4331	9.4183	9.4044	9.3914	9.3794	9.2912	9.202	9.13
6	7.3309	7.3128	7.296	7.2805	7.2661	7.2528	7.2403	7.2286	7.1432	7.0568	6.9867
7	6.092	6.0743	6.0579	6.0428	6.0287	6.0156	6.0035	5.992	5.9084	5.8236	5.7546
8	5.2967	5.2793	5.2631	5.2482	5.2344	5.2214	5.2094	5.1981	5.1156	5.0316	4.9633
9	4.7463	4.729	4.713	4.6982	4.6845	4.6717	4.6598	4.6486	4.5667	4.4831	4.415
10	4.3441	4.3269	4.3111	4.2963	4.2827	4.27	4.2582	4.2469	4.1653	4.0819	4.0137
11	4.038	4.0209	4.0051	3.9904	3.9768	3.9641	3.9522	3.9411	3.8596	3.7761	3.7077
12	3.7976	3.7805	3.7647	3.7501	3.7364	3.7238	3.7119	3.7008	3.6192	3.5355	3.4668
13	3.6038	3.5868	3.571	3.5563	3.5427	3.53	3.5182	3.507	3.4253	3.3413	3.2723
14	3.4445	3.4274	3.4116	3.3969	3.3833	3.3706	3.3587	3.3476	3.2657	3.1813	3.1118
15	3.3111	3.294	3.2782	3.2636	3.2499	3.2372	3.2253	3.2141	3.1319	3.0471	2.9772
16	3.1979	3.1808	3.165	3.1503	3.1366	3.1238	3.1119	3.1007	3.0182	2.933	2.8627
17	3.1006	3.0835	3.0676	3.0529	3.0392	3.0264	3.0145	3.0032	2.9204	2.8348	2.7639
18	3.0161	2.999	2.9831	2.9683	2.9546	2.9418	2.9298	2.9185	2.8354	2.7493	2.6779
19	2.9421	2.9249	2.9089	2.8942	2.8804	2.8675	2.8555	2.8442	2.7608	2.6742	2.6023
20	2.8766	2.8594	2.8434	2.8286	2.8148	2.8019	2.7898	2.7785	2.6947	2.6077	2.5353
21	2.8183	2.801	2.785	2.7702	2.7563	2.7434	2.7313	2.72	2.6359	2.5484	2.4755
22	2.7661	2.7488	2.7328	2.7179	2.704	2.691	2.6789	2.6675	2.5831	2.4951	2.4218
23	2.7191	2.7017	2.6857	2.6707	2.6568	2.6438	2.6316	2.6202	2.5355	2.4471	2.3732
24	2.6764	2.6591	2.643	2.628	2.614	2.601	2.5888	2.5773	2.4923	2.4035	2.3291
25	2.6377	2.6203	2.6041	2.5891	2.5751	2.562	2.5498	2.5383	2.453	2.3637	2.2888
26	2.6022	2.5848	2.5686	2.5535	2.5395	2.5264	2.5142	2.5026	2.417	2.3273	2.2519
27	2.5697	2.5522	2.536	2.5209	2.5069	2.4937	2.4814	2.4699	2.384	2.2938	2.218
28	2.5398	2.5223	2.506	2.4909	2.4768	2.4636	2.4513	2.4397	2.3535	2.2629	2.1867
29	2.5121	2.4946	2.4783	2.4631	2.449	2.4358	2.4234	2.4118	2.3253	2.2344	2.1577
30	2.4865	2.4689	2.4526	2.4374	2.4233	2.41	2.3976	2.386	2.2992	2.2079	2.1307
40	2.3059	2.288	2.2714	2.2559	2.2415	2.228	2.2153	2.2034	2.1142	2.0194	1.9383
60	2.1336	2.1154	2.0984	2.0825	2.0677	2.0538	2.0408	2.0285	1.936	1.8363	1.7493
100	2.0012	1.9826	1.9651	1.9489	1.9337	1.9194	1.9059	1.8933	1.7972	1.6918	1.5977

Appendix I

Chi Square Distribution

Chi square distribution

df	$\chi^2_{0.995}$	$\chi^2_{0.99}$	$\chi^2_{0.975}$	$\chi^2_{0.95}$	$\chi^2_{0.90}$	$\chi^2_{0.10}$	$\chi^2_{0.05}$	$\chi^2_{0.025}$	$\chi^2_{0.01}$	$\chi^2_{0.005}$
1	0.000	0.000	0.001	0.004	0.016	2.706	3.841	5.024	6.635	7.879
2	0.010	0.020	0.051	0.103	0.211	4.605	5.991	7.378	9.210	10.597
3	0.072	0.115	0.216	0.352	0.584	6.251	7.815	9.348	11.345	12.838
4	0.207	0.297	0.484	0.711	1.064	7.779	9.488	11.143	13.277	14.860
5	0.412	0.554	0.831	1.145	1.610	9.236	11.070	12.832	15.086	16.750
6	0.676	0.872	1.237	1.635	2.204	10.645	12.592	14.449	16.812	18.548
7	0.989	1.239	1.690	2.167	2.833	12.017	14.067	16.013	18.475	20.278
8	1.344	1.647	2.180	2.733	3.490	13.362	15.507	17.535	20.090	21.955
9	1.735	2.088	2.700	3.325	4.168	14.684	16.919	19.023	21.666	23.589
10	2.156	2.558	3.247	3.940	4.865	15.987	18.307	20.483	23.209	25.188
11	2.603	3.053	3.816	4.575	5.578	17.275	19.675	21.920	24.725	26.757
12	3.074	3.571	4.404	5.226	6.304	18.549	21.026	23.337	26.217	28.300
13	3.565	4.107	5.009	5.892	7.041	19.812	22.362	24.736	27.688	29.819
14	4.075	4.660	5.629	6.571	7.790	21.064	23.685	26.119	29.141	31.319
15	4.601	5.229	6.262	7.261	8.547	22.307	24.996	27.488	30.578	32.801
16	5.142	5.812	6.908	7.962	9.312	23.542	26.296	28.845	32.000	34.267
17	5.697	6.408	7.564	8.672	10.085	24.769	27.587	30.191	33.409	35.718
18	6.265	7.015	8.231	9.390	10.865	25.989	28.869	31.526	34.805	37.156
19	6.844	7.633	8.907	10.117	11.651	27.204	30.144	32.852	36.191	38.582
20	7.434	8.260	9.591	10.851	12.443	28.412	31.410	34.170	37.566	39.997
21	8.034	8.897	10.283	11.591	13.240	29.615	32.671	35.479	38.932	41.401
22	8.643	9.542	10.982	12.338	14.041	30.813	33.924	36.781	40.289	42.796
23	9.260	10.196	11.689	13.091	14.848	32.007	35.172	38.076	41.638	44.181
24	9.886	10.856	12.401	13.848	15.659	33.196	36.415	39.364	42.980	45.558
25	10.520	11.524	13.120	14.611	16.473	34.382	37.652	40.646	44.314	46.928
26	11.160	12.198	13.844	15.379	17.292	35.563	38.885	41.923	45.642	48.290
27	11.808	12.878	14.573	16.151	18.114	36.741	40.113	43.195	46.963	49.645
28	12.461	13.565	15.308	16.928	18.939	37.916	41.337	44.461	48.278	50.994

Continued

Chi square distribution *(continued)*

df	$\chi^2_{0.995}$	$\chi^2_{0.99}$	$\chi^2_{0.975}$	$\chi^2_{0.95}$	$\chi^2_{0.90}$	$\chi^2_{0.10}$	$\chi^2_{0.05}$	$\chi^2_{0.025}$	$\chi^2_{0.01}$	$\chi^2_{0.005}$
29	13.121	14.256	16.047	17.708	19.768	39.087	42.557	45.722	49.588	52.335
30	13.787	14.953	16.791	18.493	20.599	40.256	43.773	46.979	50.892	53.672
31	14.458	15.655	17.539	19.281	21.434	41.422	44.985	48.232	52.191	55.002
32	15.134	16.362	18.291	20.072	22.271	42.585	46.194	49.480	53.486	56.328
33	15.815	17.073	19.047	20.867	23.110	43.745	47.400	50.725	54.775	57.648
34	16.501	17.789	19.806	21.664	23.952	44.903	48.602	51.966	56.061	58.964
35	17.192	18.509	20.569	22.465	24.797	46.059	49.802	53.203	57.342	60.275
40	20.707	22.164	24.433	26.509	29.051	51.805	55.758	59.342	63.691	66.766
45	24.311	25.901	28.366	30.612	33.350	57.505	61.656	65.410	69.957	73.166
50	27.991	29.707	32.357	34.764	37.689	63.167	67.505	71.420	76.154	79.490
55	31.735	33.571	36.398	38.958	42.060	68.796	73.311	77.380	82.292	85.749
60	35.534	37.485	40.482	43.188	46.459	74.397	79.082	83.298	88.379	91.952
65	39.383	41.444	44.603	47.450	50.883	79.973	84.821	89.177	94.422	98.105
70	43.275	45.442	48.758	51.739	55.329	85.527	90.531	95.023	100.425	104.215
75	47.206	49.475	52.942	56.054	59.795	91.061	96.217	100.839	106.393	110.285
80	51.172	53.540	57.153	60.391	64.278	96.578	101.879	106.629	112.329	116.321
85	55.170	57.634	61.389	64.749	68.777	102.079	107.522	112.393	118.236	122.324
90	59.196	61.754	65.647	69.126	73.291	107.565	113.145	118.136	124.116	128.299
95	63.250	65.898	69.925	73.520	77.818	113.038	118.752	123.858	129.973	134.247
100	67.328	70.065	74.222	77.929	82.358	118.498	124.342	129.561	135.807	140.170

Appendix J

Values of the t-Distribution

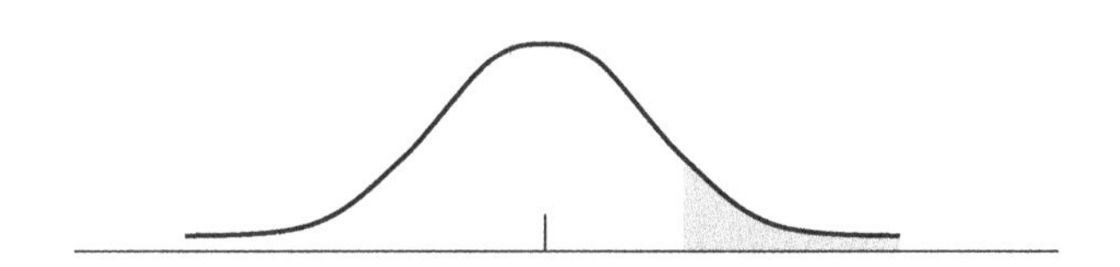

Values of t-distribution

ν	$t_{0.100}$	$t_{0.050}$	$t_{0.025}$	$t_{0.010}$	$t_{0.005}$	ν
1	3.078	6.314	12.706	31.821	63.656	1
2	1.886	2.920	4.303	6.965	9.925	2
3	1.638	2.353	3.182	4.541	5.841	3
4	1.533	2.132	2.776	3.747	4.604	4
5	1.476	2.015	2.571	3.365	4.032	5
6	1.440	1.943	2.447	3.143	3.707	6
7	1.415	1.895	2.365	2.998	3.499	7
8	1.397	1.860	2.306	2.896	3.355	8
9	1.383	1.833	2.262	2.821	3.250	9
10	1.372	1.812	2.228	2.764	3.169	10
11	1.363	1.796	2.201	2.718	3.106	11
12	1.356	1.782	2.179	2.681	3.055	12
13	1.350	1.771	2.160	2.650	3.012	13
14	1.345	1.761	2.145	2.624	2.977	14
15	1.341	1.753	2.131	2.602	2.947	15
16	1.337	1.746	2.120	2.583	2.921	16
17	1.333	1.740	2.110	2.567	2.898	17
18	1.330	1.734	2.101	2.552	2.878	18
19	1.328	1.729	2.093	2.539	2.861	19
20	1.325	1.725	2.086	2.528	2.845	20
21	1.323	1.721	2.080	2.518	2.831	21
22	1.321	1.717	2.074	2.508	2.819	22
23	1.319	1.714	2.069	2.500	2.807	23
24	1.318	1.711	2.064	2.492	2.797	24
25	1.316	1.708	2.060	2.485	2.787	25
26	1.315	1.706	2.056	2.479	2.779	26
27	1.314	1.703	2.052	2.473	2.771	27
28	1.313	1.701	2.048	2.467	2.763	28

Continued

Values of *t*-distribution *(continued)*

ν	$t_{0.10}$	$t_{0.05}$	$t_{0.025}$	$t_{0.01}$	$t_{0.005}$	ν
29	1.311	1.699	2.045	2.462	2.756	29
30	1.310	1.697	2.042	2.457	2.750	30
31	1.309	1.696	2.040	2.453	2.744	31
32	1.309	1.694	2.037	2.449	2.738	32
33	1.308	1.692	2.035	2.445	2.733	33
34	1.307	1.691	2.032	2.441	2.728	34
35	1.306	1.690	2.030	2.438	2.724	35
40	1.303	1.684	2.021	2.423	2.704	40
45	1.301	1.679	2.014	2.412	2.690	45
50	1.299	1.676	2.009	2.403	2.678	50
55	1.297	1.673	2.004	2.396	2.668	55
60	1.296	1.671	2.000	2.390	2.660	60
70	1.294	1.667	1.994	2.381	2.648	70
80	1.292	1.664	1.990	2.374	2.639	80
90	1.291	1.662	1.987	2.368	2.632	90
100	1.290	1.660	1.984	2.364	2.626	100
200	1.286	1.653	1.972	2.345	2.601	200
400	1.284	1.649	1.966	2.336	2.588	400
600	1.283	1.647	1.964	2.333	2.584	600
800	1.283	1.647	1.963	2.331	2.582	800
999	1.282	1.646	1.962	2.330	2.581	999

Appendix K

Statistical Tolerance Factors for at Least 99 Percent of the Population

("k-Values")

	One-sided tolerance Confidence level		
n	0.90	0.95	0.99
10	3.532	3.981	5.075
11	3.444	3.852	4.828
12	3.371	3.747	4.633
13	3.310	3.659	4.472
14	3.257	3.585	4.336
15	3.212	3.520	4.224
16	3.172	3.463	4.124
17	3.136	3.415	4.038
18	3.106	3.370	3.961
19	3.078	3.331	3.893
20	3.052	3.295	3.832
21	3.028	3.262	3.776
22	3.007	3.233	3.727
23	2.987	3.206	3.680
24	2.969	3.181	3.638
25	2.952	3.158	3.601
30	2.884	3.064	3.446
40	2.793	2.941	3.250
50	2.735	2.863	3.124

	Two-sided tolerance Confidence level		
n	0.90	0.95	0.99
10	3.959	4.433	5.594
11	3.849	4.277	5.308
12	3.758	4.150	5.079
13	3.682	4.044	4.893
14	3.618	3.955	4.737
15	3.562	3.878	4.605
16	3.514	3.812	4.492
17	3.471	3.754	4.393
18	3.433	3.702	4.307
19	3.399	3.656	4.230
20	3.368	3.615	4.161
21	3.340	3.577	4.100
22	3.315	3.543	4.044
23	3.292	3.512	3.993
24	3.270	3.483	3.947
25	3.251	3.457	3.904
30	3.170	3.350	3.733
40	3.066	3.213	3.518
50	3.001	3.126	3.385

Appendix L

Critical Values for the Mann-Whitney Test

Critical values for a one-tailed Mann-Whitney test with $\alpha = 0.05$ or a two-tailed test with $\alpha = 0.10$.

n_2 \ n_1	3		4		5		6		7		8		9		10	
	M_L	M_R	M_L	M_R	M_L	M_R	M_L	M_R	M_L	M_R	M_L	M_R	M_L	M_R	M_L	M_R
3	6	15														
4	7	17	12	24												
5	7	20	13	27	19	36										
6	8	22	14	30	20	40	28	50								
7	9	24	15	33	22	43	30	54	39	66						
8	9	27	16	36	24	46	32	58	41	71	52	84				
9	10	29	17	39	25	50	33	63	43	76	54	90	66	105		
10	11	31	18	42	26	54	35	67	46	80	57	95	69	111	83	127

Critical values for a one-tailed Mann-Whitney test with $\alpha = 0.025$ or a two-tailed test with $\alpha = 0.05$.

n_2 \ n_1	3		4		5		6		7		8		9		10	
	M_L	M_R	M_L	M_R	M_L	M_R	M_L	M_R	M_L	M_R	M_L	M_R	M_L	M_R	M_L	M_R
3	–	–														
4	6	18	11	25												
5	6	21	12	28	18	37										
6	7	23	12	32	19	41	26	52								
7	7	26	13	35	20	45	28	56	37	68						
8	8	28	14	38	21	49	29	61	39	73	49	87				
9	8	31	15	41	22	53	31	65	41	78	51	93	63	108		
10	9	33	16	44	24	56	32	70	43	83	54	98	66	114	79	131

Appendix M

Critical Values for Wilcoxon Signed Rank Test

	Approx. α value		Critical value	
n	1 tail	2 tail	W_l	W_r
7	0.01	0.02	0	28
	0.025	0.05	2	26
	0.05	0.1	4	24
	0.1	0.2	6	32
8	0.005	0.01	0	36
	0.01	0.02	2	34
	0.025	0.05	4	32
	0.05	0.1	6	30
	0.1	0.2	8	28
9	0.005	0.01	2	43
	0.01	0.02	3	42
	0.025	0.05	6	39
	0.05	0.1	8	37
	0.1	0.2	11	34
10	0.005	0.01	3	52
	0.01	0.02	5	50
	0.025	0.05	8	47
	0.05	0.1	11	44
	0.1	0.2	14	41
11	0.005	0.01	5	61
	0.01	0.02	7	59
	0.025	0.05	11	55
	0.05	0.1	14	52
	0.1	0.2	18	48
12	0.005	0.01	7	71
	0.01	0.02	10	68
	0.025	0.05	14	64
	0.05	0.1	17	61
	0.1	0.2	22	56
13	0.005	0.01	10	81
	0.01	0.02	13	78
	0.025	0.05	17	74
	0.05	0.1	21	70
	0.1	0.2	26	65

	Approx. α value		Critical value	
n	1 tail	2 tail	W_l	W_r
14	0.005	0.01	13	92
	0.01	0.02	16	89
	0.025	0.05	21	84
	0.05	0.1	26	79
	0.1	0.2	31	74
15	0.005	0.01	16	104
	0.01	0.02	20	100
	0.025	0.05	25	95
	0.05	0.1	30	90
	0.1	0.2	37	83
16	0.005	0.01	19	117
	0.01	0.02	24	112
	0.025	0.05	30	106
	0.05	0.1	36	100
	0.1	0.2	42	94
17	0.005	0.01	23	130
	0.01	0.02	28	125
	0.025	0.05	35	118
	0.05	0.1	41	112
	0.1	0.2	49	104
18	0.005	0.01	28	143
	0.01	0.02	33	138
	0.025	0.05	40	131
	0.05	0.1	47	124
	0.1	0.2	55	116
19	0.005	0.01	32	158
	0.01	0.02	38	152
	0.025	0.05	46	144
	0.05	0.1	54	136
	0.1	0.2	62	128
20	0.005	0.01	37	173
	0.01	0.02	43	167
	0.025	0.05	52	158
	0.05	0.1	60	150
	0.1	0.2	70	140

Appendix N

Poisson Distribution

Probability of x or fewer occurrences of an event

Poisson distribution

$\lambda\downarrow x\rightarrow$	0	1	2	3	4	5	6	7	8	9	10	11	12	13	14	15	16	17
0.005	0.995	1.000	1.000	1.000	1.000	1.000	1.000	1.000	1.000	1.000	1.000	1.000	1.000	1.000	1.000	1.000	1.000	1.000
0.01	0.990	1.000	1.000	1.000	1.000	1.000	1.000	1.000	1.000	1.000	1.000	1.000	1.000	1.000	1.000	1.000	1.000	1.000
0.02	0.980	1.000	1.000	1.000	1.000	1.000	1.000	1.000	1.000	1.000	1.000	1.000	1.000	1.000	1.000	1.000	1.000	1.000
0.03	0.970	1.000	1.000	1.000	1.000	1.000	1.000	1.000	1.000	1.000	1.000	1.000	1.000	1.000	1.000	1.000	1.000	1.000
0.04	0.961	0.999	1.000	1.000	1.000	1.000	1.000	1.000	1.000	1.000	1.000	1.000	1.000	1.000	1.000	1.000	1.000	1.000
0.05	0.951	0.999	1.000	1.000	1.000	1.000	1.000	1.000	1.000	1.000	1.000	1.000	1.000	1.000	1.000	1.000	1.000	1.000
0.06	0.942	0.998	1.000	1.000	1.000	1.000	1.000	1.000	1.000	1.000	1.000	1.000	1.000	1.000	1.000	1.000	1.000	1.000
0.07	0.932	0.998	1.000	1.000	1.000	1.000	1.000	1.000	1.000	1.000	1.000	1.000	1.000	1.000	1.000	1.000	1.000	1.000
0.08	0.923	0.997	1.000	1.000	1.000	1.000	1.000	1.000	1.000	1.000	1.000	1.000	1.000	1.000	1.000	1.000	1.000	1.000
0.09	0.914	0.996	1.000	1.000	1.000	1.000	1.000	1.000	1.000	1.000	1.000	1.000	1.000	1.000	1.000	1.000	1.000	1.000
0.1	0.905	0.995	1.000	1.000	1.000	1.000	1.000	1.000	1.000	1.000	1.000	1.000	1.000	1.000	1.000	1.000	1.000	1.000
0.15	0.861	0.990	0.999	1.000	1.000	1.000	1.000	1.000	1.000	1.000	1.000	1.000	1.000	1.000	1.000	1.000	1.000	1.000
0.2	0.819	0.982	0.999	1.000	1.000	1.000	1.000	1.000	1.000	1.000	1.000	1.000	1.000	1.000	1.000	1.000	1.000	1.000
0.25	0.779	0.974	0.998	1.000	1.000	1.000	1.000	1.000	1.000	1.000	1.000	1.000	1.000	1.000	1.000	1.000	1.000	1.000
0.3	0.741	0.963	0.996	1.000	1.000	1.000	1.000	1.000	1.000	1.000	1.000	1.000	1.000	1.000	1.000	1.000	1.000	1.000
0.35	0.705	0.951	0.994	1.000	1.000	1.000	1.000	1.000	1.000	1.000	1.000	1.000	1.000	1.000	1.000	1.000	1.000	1.000
0.4	0.670	0.938	0.992	0.999	1.000	1.000	1.000	1.000	1.000	1.000	1.000	1.000	1.000	1.000	1.000	1.000	1.000	1.000
0.5	0.607	0.910	0.986	0.998	1.000	1.000	1.000	1.000	1.000	1.000	1.000	1.000	1.000	1.000	1.000	1.000	1.000	1.000
0.6	0.549	0.878	0.977	0.997	1.000	1.000	1.000	1.000	1.000	1.000	1.000	1.000	1.000	1.000	1.000	1.000	1.000	1.000
0.7	0.497	0.844	0.966	0.994	0.999	1.000	1.000	1.000	1.000	1.000	1.000	1.000	1.000	1.000	1.000	1.000	1.000	1.000
0.8	0.449	0.809	0.953	0.991	0.999	1.000	1.000	1.000	1.000	1.000	1.000	1.000	1.000	1.000	1.000	1.000	1.000	1.000
0.9	0.407	0.772	0.937	0.987	0.998	1.000	1.000	1.000	1.000	1.000	1.000	1.000	1.000	1.000	1.000	1.000	1.000	1.000
1	0.368	0.736	0.920	0.981	0.996	0.999	1.000	1.000	1.000	1.000	1.000	1.000	1.000	1.000	1.000	1.000	1.000	1.000
1.2	0.301	0.663	0.879	0.966	0.992	0.998	1.000	1.000	1.000	1.000	1.000	1.000	1.000	1.000	1.000	1.000	1.000	1.000
1.4	0.247	0.592	0.833	0.946	0.986	0.997	0.999	1.000	1.000	1.000	1.000	1.000	1.000	1.000	1.000	1.000	1.000	1.000
1.6	0.202	0.525	0.783	0.921	0.976	0.994	0.999	1.000	1.000	1.000	1.000	1.000	1.000	1.000	1.000	1.000	1.000	1.000
1.8	0.165	0.463	0.731	0.891	0.964	0.990	0.997	0.999	1.000	1.000	1.000	1.000	1.000	1.000	1.000	1.000	1.000	1.000
2	0.135	0.406	0.677	0.857	0.947	0.983	0.995	0.999	1.000	1.000	1.000	1.000	1.000	1.000	1.000	1.000	1.000	1.000

Continued

Poisson distribution *(continued)*

$\lambda\downarrow x\rightarrow$	0	1	2	3	4	5	6	7	8	9	10	11	12	13	14	15	16	17
2.2	0.111	0.355	0.623	0.819	0.928	0.975	0.993	0.998	1.000	1.000	1.000	1.000	1.000	1.000	1.000	1.000	1.000	1.000
2.4	0.091	0.308	0.570	0.779	0.904	0.964	0.988	0.997	0.999	1.000	1.000	1.000	1.000	1.000	1.000	1.000	1.000	1.000
2.6	0.074	0.267	0.518	0.736	0.877	0.951	0.983	0.995	0.999	1.000	1.000	1.000	1.000	1.000	1.000	1.000	1.000	1.000
2.8	0.061	0.231	0.469	0.692	0.848	0.935	0.976	0.992	0.998	0.999	1.000	1.000	1.000	1.000	1.000	1.000	1.000	1.000
3	0.050	0.199	0.423	0.647	0.815	0.916	0.966	0.988	0.996	0.999	1.000	1.000	1.000	1.000	1.000	1.000	1.000	1.000
3.2	0.041	0.171	0.380	0.603	0.781	0.895	0.955	0.983	0.994	0.998	1.000	1.000	1.000	1.000	1.000	1.000	1.000	1.000
3.4	0.033	0.147	0.340	0.558	0.744	0.871	0.942	0.977	0.992	0.997	0.999	1.000	1.000	1.000	1.000	1.000	1.000	1.000
3.6	0.027	0.126	0.303	0.515	0.706	0.844	0.927	0.969	0.988	0.996	0.999	1.000	1.000	1.000	1.000	1.000	1.000	1.000
3.8	0.022	0.107	0.269	0.473	0.668	0.816	0.909	0.960	0.984	0.994	0.998	0.999	1.000	1.000	1.000	1.000	1.000	1.000
4	0.018	0.092	0.238	0.433	0.629	0.785	0.889	0.949	0.979	0.992	0.997	0.999	1.000	1.000	1.000	1.000	1.000	1.000
4.5	0.011	0.061	0.174	0.342	0.532	0.703	0.831	0.913	0.960	0.983	0.993	0.998	0.999	1.000	1.000	1.000	1.000	1.000
5	0.007	0.040	0.125	0.265	0.440	0.616	0.762	0.867	0.932	0.968	0.986	0.995	0.998	0.999	1.000	1.000	1.000	1.000
5.5	0.004	0.027	0.088	0.202	0.358	0.529	0.686	0.809	0.894	0.946	0.975	0.989	0.996	0.998	0.999	1.000	1.000	1.000
6	0.002	0.017	0.062	0.151	0.285	0.446	0.606	0.744	0.847	0.916	0.957	0.980	0.991	0.996	0.999	0.999	1.000	1.000
6.5	0.002	0.011	0.043	0.112	0.224	0.369	0.527	0.673	0.792	0.877	0.933	0.966	0.984	0.993	0.997	0.999	1.000	1.000
7	0.001	0.007	0.030	0.082	0.173	0.301	0.450	0.599	0.729	0.830	0.901	0.947	0.973	0.987	0.994	0.998	0.999	1.000
7.5	0.001	0.005	0.020	0.059	0.132	0.241	0.378	0.525	0.662	0.776	0.862	0.921	0.957	0.978	0.990	0.995	0.998	0.999
8	0.000	0.003	0.014	0.042	0.100	0.191	0.313	0.453	0.593	0.717	0.816	0.888	0.936	0.966	0.983	0.992	0.996	0.998
8.5	0.000	0.002	0.009	0.030	0.074	0.150	0.256	0.386	0.523	0.653	0.763	0.849	0.909	0.949	0.973	0.986	0.993	0.997
9	0.000	0.001	0.006	0.021	0.055	0.116	0.207	0.324	0.456	0.587	0.706	0.803	0.876	0.926	0.959	0.978	0.989	0.995
9.5	0.000	0.001	0.004	0.015	0.040	0.089	0.165	0.269	0.392	0.522	0.645	0.752	0.836	0.898	0.940	0.967	0.982	0.991
10	0.000	0.000	0.003	0.010	0.029	0.067	0.130	0.220	0.333	0.458	0.583	0.697	0.792	0.864	0.917	0.951	0.973	0.986
10.5	0.000	0.000	0.002	0.007	0.021	0.050	0.102	0.179	0.279	0.397	0.521	0.639	0.742	0.825	0.888	0.932	0.960	0.978

Appendix O

Binomial Distribution

Probability of x or fewer occurrences in a sample of size n

Binomial distribution

		← p →																	
n	*x*	0.01	0.02	0.03	0.04	0.05	0.06	0.07	0.08	0.09	0.10	0.15	0.20	0.25	0.30	0.35	0.40	0.45	0.50
2	0	0.980	0.960	0.941	0.922	0.903	0.884	0.865	0.846	0.828	0.810	0.723	0.640	0.563	0.490	0.423	0.360	0.303	0.250
2	1	1.000	1.000	0.999	0.998	0.998	0.996	0.995	0.994	0.992	0.990	0.978	0.960	0.938	0.910	0.878	0.840	0.798	0.750
3	0	0.970	0.941	0.913	0.885	0.857	0.831	0.804	0.779	0.754	0.729	0.614	0.512	0.422	0.343	0.275	0.216	0.166	0.125
3	1	1.000	0.999	0.997	0.995	0.993	0.990	0.986	0.982	0.977	0.972	0.939	0.896	0.844	0.784	0.718	0.648	0.575	0.500
3	2	1.000	1.000	1.000	1.000	1.000	1.000	1.000	0.999	0.999	0.999	0.997	0.992	0.984	0.973	0.957	0.936	0.909	0.875
4	0	0.961	0.922	0.885	0.849	0.815	0.781	0.748	0.716	0.686	0.656	0.522	0.410	0.316	0.240	0.179	0.130	0.092	0.063
4	1	0.999	0.998	0.995	0.991	0.986	0.980	0.973	0.966	0.957	0.948	0.890	0.819	0.738	0.652	0.563	0.475	0.391	0.313
4	2	1.000	1.000	1.000	1.000	1.000	0.999	0.999	0.998	0.997	0.996	0.988	0.973	0.949	0.916	0.874	0.821	0.759	0.688
4	3	1.000	1.000	1.000	1.000	1.000	1.000	1.000	1.000	1.000	1.000	0.999	0.998	0.996	0.992	0.985	0.974	0.959	0.938
5	0	0.951	0.904	0.859	0.815	0.774	0.734	0.696	0.659	0.624	0.590	0.444	0.328	0.237	0.168	0.116	0.078	0.050	0.031
5	1	0.999	0.996	0.992	0.985	0.977	0.968	0.958	0.946	0.933	0.919	0.835	0.737	0.633	0.528	0.428	0.337	0.256	0.188
5	2	1.000	1.000	1.000	0.999	0.999	0.998	0.997	0.995	0.994	0.991	0.973	0.942	0.896	0.837	0.765	0.683	0.593	0.500
5	3	1.000	1.000	1.000	1.000	1.000	1.000	1.000	1.000	1.000	1.000	0.998	0.993	0.984	0.969	0.946	0.913	0.869	0.813
5	4	1.000	1.000	1.000	1.000	1.000	1.000	1.000	1.000	1.000	1.000	1.000	1.000	0.999	0.998	0.995	0.990	0.982	0.969
6	0	0.941	0.886	0.833	0.783	0.735	0.690	0.647	0.606	0.568	0.531	0.377	0.262	0.178	0.118	0.075	0.047	0.028	0.016
6	1	0.999	0.994	0.988	0.978	0.967	0.954	0.939	0.923	0.905	0.886	0.776	0.655	0.534	0.420	0.319	0.233	0.164	0.109
6	2	1.000	1.000	0.999	0.999	0.998	0.996	0.994	0.991	0.988	0.984	0.953	0.901	0.831	0.744	0.647	0.544	0.442	0.344
6	3	1.000	1.000	1.000	1.000	1.000	1.000	1.000	0.999	0.999	0.999	0.994	0.983	0.962	0.930	0.883	0.821	0.745	0.656
6	4	1.000	1.000	1.000	1.000	1.000	1.000	1.000	1.000	1.000	1.000	1.000	0.998	0.995	0.989	0.978	0.959	0.931	0.891
6	5	1.000	1.000	1.000	1.000	1.000	1.000	1.000	1.000	1.000	1.000	1.000	1.000	1.000	0.999	0.998	0.996	0.992	0.984
7	0	0.932	0.868	0.808	0.751	0.698	0.648	0.602	0.558	0.517	0.478	0.321	0.210	0.133	0.082	0.049	0.028	0.015	0.008
7	1	0.998	0.992	0.983	0.971	0.956	0.938	0.919	0.897	0.875	0.850	0.717	0.577	0.445	0.329	0.234	0.159	0.102	0.063
7	2	1.000	1.000	0.999	0.998	0.996	0.994	0.990	0.986	0.981	0.974	0.926	0.852	0.756	0.647	0.532	0.420	0.316	0.227
7	3	1.000	1.000	1.000	1.000	1.000	1.000	0.999	0.999	0.998	0.997	0.988	0.967	0.929	0.874	0.800	0.710	0.608	0.500
7	4	1.000	1.000	1.000	1.000	1.000	1.000	1.000	1.000	1.000	1.000	0.999	0.995	0.987	0.971	0.944	0.904	0.847	0.773
7	5	1.000	1.000	1.000	1.000	1.000	1.000	1.000	1.000	1.000	1.000	1.000	1.000	0.999	0.996	0.991	0.981	0.964	0.938
7	6	1.000	1.000	1.000	1.000	1.000	1.000	1.000	1.000	1.000	1.000	1.000	1.000	1.000	1.000	0.999	0.998	0.996	0.992

Continued

Binomial distribution *(continued)*

n	x	0.01	0.02	0.03	0.04	0.05	0.06	0.07	0.08	0.09	0.10	0.15	0.20	0.25	0.30	0.35	0.40	0.45	0.50
8	0	0.923	0.851	0.784	0.721	0.663	0.610	0.560	0.513	0.470	0.430	0.272	0.168	0.100	0.058	0.032	0.017	0.008	0.004
8	1	0.997	0.990	0.978	0.962	0.943	0.921	0.897	0.870	0.842	0.813	0.657	0.503	0.367	0.255	0.169	0.106	0.063	0.035
8	2	1.000	1.000	0.999	0.997	0.994	0.990	0.985	0.979	0.971	0.962	0.895	0.797	0.679	0.552	0.428	0.315	0.220	0.145
8	3	1.000	1.000	1.000	1.000	1.000	0.999	0.999	0.998	0.997	0.995	0.979	0.944	0.886	0.806	0.706	0.594	0.477	0.363
8	4	1.000	1.000	1.000	1.000	1.000	1.000	1.000	1.000	1.000	1.000	0.997	0.990	0.973	0.942	0.894	0.826	0.740	0.637
8	5	1.000	1.000	1.000	1.000	1.000	1.000	1.000	1.000	1.000	1.000	1.000	0.999	0.996	0.989	0.975	0.950	0.912	0.855
8	6	1.000	1.000	1.000	1.000	1.000	1.000	1.000	1.000	1.000	1.000	1.000	1.000	1.000	0.999	0.996	0.991	0.982	0.965
8	7	1.000	1.000	1.000	1.000	1.000	1.000	1.000	1.000	1.000	1.000	1.000	1.000	1.000	1.000	1.000	0.999	0.998	0.996
9	0	0.914	0.834	0.760	0.693	0.630	0.573	0.520	0.472	0.428	0.387	0.232	0.134	0.075	0.040	0.021	0.010	0.005	0.002
9	1	0.997	0.987	0.972	0.952	0.929	0.902	0.873	0.842	0.809	0.775	0.599	0.436	0.300	0.196	0.121	0.071	0.039	0.020
9	2	1.000	0.999	0.998	0.996	0.992	0.986	0.979	0.970	0.960	0.947	0.859	0.738	0.601	0.463	0.337	0.232	0.150	0.090
9	3	1.000	1.000	1.000	1.000	0.999	0.999	0.998	0.996	0.994	0.992	0.966	0.914	0.834	0.730	0.609	0.483	0.361	0.254
9	4	1.000	1.000	1.000	1.000	1.000	1.000	1.000	1.000	0.999	0.999	0.994	0.980	0.951	0.901	0.828	0.733	0.621	0.500
9	5	1.000	1.000	1.000	1.000	1.000	1.000	1.000	1.000	1.000	1.000	0.999	0.997	0.990	0.975	0.946	0.901	0.834	0.746
9	6	1.000	1.000	1.000	1.000	1.000	1.000	1.000	1.000	1.000	1.000	1.000	1.000	0.999	0.996	0.989	0.975	0.950	0.910
9	7	1.000	1.000	1.000	1.000	1.000	1.000	1.000	1.000	1.000	1.000	1.000	1.000	1.000	1.000	0.999	0.996	0.991	0.980
9	8	1.000	1.000	1.000	1.000	1.000	1.000	1.000	1.000	1.000	1.000	1.000	1.000	1.000	1.000	1.000	1.000	0.999	0.998
10	0	0.904	0.817	0.737	0.665	0.599	0.539	0.484	0.434	0.389	0.349	0.197	0.107	0.056	0.028	0.013	0.006	0.003	0.001
10	1	0.996	0.984	0.965	0.942	0.914	0.882	0.848	0.812	0.775	0.736	0.544	0.376	0.244	0.149	0.086	0.046	0.023	0.011
10	2	1.000	0.999	0.997	0.994	0.988	0.981	0.972	0.960	0.946	0.930	0.820	0.678	0.526	0.383	0.262	0.167	0.100	0.055
10	3	1.000	1.000	1.000	1.000	0.999	0.998	0.996	0.994	0.991	0.987	0.950	0.879	0.776	0.650	0.514	0.382	0.266	0.172
10	4	1.000	1.000	1.000	1.000	1.000	1.000	1.000	0.999	0.999	0.998	0.990	0.967	0.922	0.850	0.751	0.633	0.504	0.377
10	5	1.000	1.000	1.000	1.000	1.000	1.000	1.000	1.000	1.000	1.000	0.999	0.994	0.980	0.953	0.905	0.834	0.738	0.623

Appendix P

Exponential Distribution

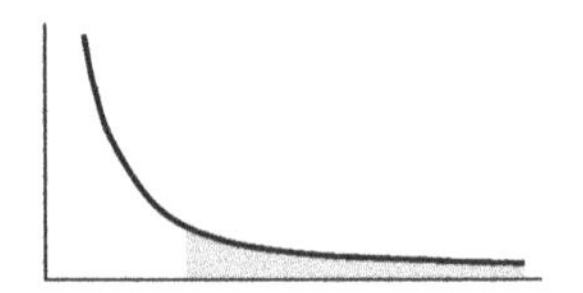

Exponential distribution

X	Area to left of *X*	Area to right of *X*
0	0.00000	1.00000
0.1	0.09516	0.90484
0.2	0.18127	0.81873
0.3	0.25918	0.74082
0.4	0.32968	0.67032
0.5	0.39347	0.60653
0.6	0.45119	0.54881
0.7	0.50341	0.49659
0.8	0.55067	0.44933
0.9	0.59343	0.40657
1	0.63212	0.36788
1.1	0.66713	0.33287
1.2	0.69881	0.30119
1.3	0.72747	0.27253
1.4	0.75340	0.24660
1.5	0.77687	0.22313
1.6	0.79810	0.20190
1.7	0.81732	0.18268
1.8	0.83470	0.16530
1.9	0.85043	0.14957
2	0.86466	0.13534
2.1	0.87754	0.12246
2.2	0.88920	0.11080
2.3	0.89974	0.10026
2.4	0.90928	0.09072
2.5	0.91792	0.08208
2.6	0.92573	0.07427

Continued

Exponential distribution *(continued)*

X	Area to left of *X*	Area to right of *X*
2.7	0.93279	0.06721
2.8	0.93919	0.06081
2.9	0.94498	0.05502
3	0.95021	0.04979
3.1	0.95495	0.04505
3.2	0.95924	0.04076
3.3	0.96312	0.03688
3.4	0.96663	0.03337
3.5	0.96980	0.03020
3.6	0.97268	0.02732
3.7	0.97528	0.02472
3.8	0.97763	0.02237
3.9	0.97976	0.02024
4	0.98168	0.01832
4.1	0.98343	0.01657
4.2	0.98500	0.01500
4.3	0.98643	0.01357
4.4	0.98772	0.01228
4.5	0.98889	0.01111
4.6	0.98995	0.01005
4.7	0.99090	0.00910
4.8	0.99177	0.00823
4.9	0.99255	0.00745
5	0.99326	0.00674
5.1	0.99390	0.00610
5.2	0.99448	0.00552
5.3	0.99501	0.00499
5.4	0.99548	0.00452
5.5	0.99591	0.00409
5.6	0.99630	0.00370
5.7	0.99665	0.00335
5.8	0.99697	0.00303
5.9	0.99726	0.00274
6	0.99752	0.00248

Appendix Q

Median Ranks

Median ranks

n	1	2	3	4	5	6	7	8	9	10	11	12
1	0.500	0.292	0.206	0.159	0.130	0.109	0.095	0.083	0.074	0.067	0.061	0.056
2		0.708	0.500	0.386	0.315	0.266	0.230	0.202	0.181	0.163	0.149	0.137
3			0.794	0.614	0.500	0.422	0.365	0.321	0.287	0.260	0.237	0.218
4				0.841	0.685	0.578	0.500	0.440	0.394	0.356	0.325	0.298
5					0.870	0.734	0.635	0.560	0.500	0.452	0.412	0.379
6						0.891	0.770	0.679	0.606	0.548	0.500	0.460
7							0.905	0.798	0.713	0.644	0.588	0.540
8								0.917	0.819	0.740	0.675	0.621
9									0.926	0.837	0.763	0.702
10										0.933	0.851	0.782
11											0.939	0.863
12												0.944

n	13	14	15	16	17	18	19	20	21	22	23	24
1	0.052	0.049	0.045	0.043	0.040	0.038	0.036	0.034	0.033	0.031	0.030	0.029
2	0.127	0.118	0.110	0.104	0.098	0.092	0.088	0.083	0.079	0.076	0.073	0.070
3	0.201	0.188	0.175	0.165	0.155	0.147	0.139	0.132	0.126	0.121	0.115	0.111
4	0.276	0.257	0.240	0.226	0.213	0.201	0.191	0.181	0.173	0.165	0.158	0.152
5	0.351	0.326	0.305	0.287	0.270	0.255	0.242	0.230	0.220	0.210	0.201	0.193
6	0.425	0.396	0.370	0.348	0.328	0.310	0.294	0.279	0.266	0.254	0.244	0.234
7	0.500	0.465	0.435	0.409	0.385	0.364	0.345	0.328	0.313	0.299	0.286	0.275
8	0.575	0.535	0.500	0.470	0.443	0.418	0.397	0.377	0.360	0.344	0.329	0.316
9	0.649	0.604	0.565	0.530	0.500	0.473	0.448	0.426	0.407	0.388	0.372	0.357
10	0.724	0.674	0.630	0.591	0.557	0.527	0.500	0.475	0.453	0.433	0.415	0.398
11	0.799	0.743	0.695	0.652	0.615	0.582	0.552	0.525	0.500	0.478	0.457	0.439
12	0.873	0.813	0.760	0.713	0.672	0.636	0.603	0.574	0.547	0.522	0.500	0.480

Continued

Median ranks *(continued)*

n	13	14	15	16	17	18	19	20	21	22	23	24
13	0.948	0.882	0.825	0.774	0.730	0.690	0.655	0.623	0.593	0.567	0.543	0.520
14		0.951	0.890	0.835	0.787	0.745	0.706	0.672	0.640	0.612	0.585	0.561
15			0.955	0.896	0.845	0.799	0.758	0.721	0.687	0.656	0.628	0.602
16				0.957	0.902	0.853	0.809	0.770	0.734	0.701	0.671	0.643
17					0.960	0.908	0.861	0.819	0.780	0.746	0.714	0.684
18						0.962	0.912	0.868	0.827	0.790	0.756	0.725
19							0.964	0.917	0.874	0.835	0.799	0.766
20								0.966	0.921	0.879	0.842	0.807
21									0.967	0.924	0.885	0.848
22										0.969	0.927	0.889
23											0.970	0.930
24												0.971

Glossary

A

accelerated life testing—A technique in which units are tested at stress levels higher than they were designed for in an effort to cause failures sooner.

accuracy—The closeness of agreement between a test result or measurement result and the true value.

alias—Effect that is completely confounded with another effect due to the nature of the designed experiment. Aliases are the result of confounding, which may or may not be deliberate.

α **(alpha)—**1: The maximum probability, or risk, of making a type I error when dealing with the significance level of a test. 2: The probability or risk of incorrectly deciding that a shift in the process mean has occurred when the process is unchanged (when referring to α in general or as the *p*-value obtained in the test). 3: α is usually designated as producer's risk.

alternative hypothesis, H_a—A hypothesis that is accepted if the null hypothesis (H_0) is rejected. Example 1: Consider the null hypothesis that the statistical model for a population is a normal distribution. The alternative hypothesis to this null hypothesis is that the statistical model of the population is *not* a normal distribution. Note 1: The alternative hypothesis is a statement that contradicts the null hypothesis. The corresponding test statistic is used to decide between the null and alternative hypotheses. Note 2: The alternative hypothesis can also be denoted H_1, H_A, or H^A, with no clear preference as long as the symbolism parallels the null hypothesis notation.

analysis of covariance (ANCOVA)—A technique for estimating and testing the effects of treatments when one or more concomitant variables influence the response variable. Note: Analysis of covariance can be viewed as a combination of regression analysis and analysis of variance.

analysis of variance (ANOVA)—A technique to determine if there are statistically significant differences between group means by analyzing group variances.

Arrhenius model—A technique in accelerated life testing using relationships between temperature and failure rates.

attribute—A countable or categorized quality characteristic that is qualitative rather than quantitative in nature.

availability—The probability that a system or equipment is operating satisfactorily at any point in time when used under stated conditions. The total time considered includes operating time, active repair time, administrative time and logistic time.

B

balanced design—A design where all treatment combinations have the same number of observations. If replication in a design exists, it would be balanced only if the replication was consistent across all the treatment combinations. In other words, the number of replicates of each treatment combination is the same.

balanced incomplete block (BIB) design—Incomplete block design in which each block contains the same number (k) of different levels from the (l) levels of the principal factor arranged so that every pair of levels occurs in the same number (l) of blocks from the b blocks. Note: This design implies that every level of the principal factor appears the same number of times in the experiment.

batch—A definite quantity of some product accumulated under conditions considered uniform, or accumulated from a common source. This term is sometimes synonymous with *lot.*

β (beta)—The maximum probability, or risk, of making a type II error (see comment on α (alpha). The probability or risk of incorrectly deciding that a shift in the process mean has not occurred when the process has changed. β is usually designated as consumer's risk. See *power curve.*

bias—A systematic difference between the mean of a test result or measurement result and a true value.

BIB—See *balanced incomplete block design.*

bimodal—Having two distinct statistical modes.

binomial distribution—A two-parameter discrete distribution involving the mean μ and the variance σ^2, of the variable x with probability p, where p is a constant $0 \leq p \leq 1$, and sample size n. Mean = np and variance = $np(1 - p)$.

blemish—An imperfection that causes awareness but does not impair function or usage.

block—A collection of experimental units more homogeneous than the full set of experimental units. Blocks are usually selected to allow for special causes, in addition to those introduced as factors to be studied. These special causes may be avoidable within blocks, thus providing a more homogeneous experimental subspace.

block diagram (or **reliability block diagram [RBD]**)—A diagram that displays the relationships of components of a system whether in serial, parallel, or some other configuration.

block effect—An effect resulting from a block in an experimental design. Existence of a block effect generally means that the method of blocking was appropriate and that an assignable cause has been found.

blocking—Method of including blocks in an experiment in order to broaden the applicability of the conclusions or to minimize the impact of selected assignable causes. The randomization of the experiment is restricted and occurs within blocks.

B*X* life—The time at which *X* percent of the units in a population will have failed. For example, if an item has a B10 life of 100 hours, that means that 10 percent of the population will have failed by 100 hours of operation.

C

***c* (count)**—The number of events (often nonconformities) of a given classification occurring in a sample of fixed size.

capability—Performance of a process demonstrated to be in a state of statistical control. See *process capability* and *process performance.*

capability index—See *process capability index.*

cause—A cause is an identified reason for the presence of a symptom, defect, or problem. See *effect.*

chi square distribution (χ^2 distribution)—A positively skewed distribution that varies with the degrees of freedom with a minimum value of zero. See Appendix I.

chi square statistic (χ^2 statistic)—A value obtained from the χ^2 distribution at a given percentage point and specified degrees of freedom.

chi square test (χ^2 test)—A statistic used in testing a hypothesis concerning the discrepancy between observed and expected results.

coefficient of determination (R^2)—A measure of the part of the variance for one variable that can be explained by its linear relationship with another variable (or variables). The coefficient of determination is the square of the correlation between the observed y values and the fitted y values, and is also the fraction of the variation in y that is explained by the fitted equation.

coefficient of variation (CV)—Measures relative dispersion. It is the standard deviation divided by the mean, and is commonly reported as a percentage.

complete block—Block that accommodates a complete set of treatment combinations.

completely randomized design—A design in which the treatments are assigned at random to the full set of experimental units. No blocks are involved in a completely randomized design.

completely randomized factorial design—A factorial design in which all the treatments are assigned at random to the full set of experimental units. *See completely randomized design.*

concomitant variable—A variable or factor that cannot be accounted for in the data analysis or design of the experiment but whose effect on the results should be accounted for.

confidence coefficient (1 – α)—*See confidence level.*

confidence interval—A confidence interval is an estimate of the interval between two statistics that includes the true value of the parameter with some probability.

confidence level (confidence coefficient) (1 – α)—The probability that the confidence interval described by a set of confidence limits actually includes the population parameter.

confidence limits—The endpoints of the interval about the sample statistic that is believed, with a specified confidence level, to include the population parameter. See *confidence interval.*

confounding—Indistinguishably combining an effect with other effects or blocks.

consumer's risk (β)—Probability of acceptance when the quality level has a value stated by the acceptance sampling plan as unsatisfactory. Note 1: Such acceptance is a type II error. Note 2: Consumer's risk is usually designated as β (beta).

continuous distribution—Distribution where data is from a continuous scale. Examples of continuous scales are the normal, *t*, and *F* distributions.

continuous scale—A scale with a continuum of possible values. Note: A continuous scale can be transformed into a discrete scale by grouping values, but this leads to some loss of information.

control plan—A document describing the system elements to be applied to control variation of processes, products, and services in order to minimize deviation from their preferred values.

correlation—Correlation measures the linear association between two variables. It is commonly measured by the correlation coefficient, *r*. See also *regression analysis.*

correlation coefficient (*r*)—A number between –1 and 1 that indicates the degree of linear relationship between two sets of numbers.

covariance—Measures the relationship between pairs of observations from two variables.

C_p (process capability index)—Index describing process capability in relation to specified tolerance of a characteristic divided by a measure of the length of the reference interval for a process in a state of statistical control.

C_{pk} (minimum process capability index)—Smaller of C_{pk_U} (upper process capability index) and C_{pk_L} (lower process capability index).

C_{pk_L} (lower process capability index; C_{p_L})—Index describing process capability in relation to the lower specification limit.

C_{pk_U} (upper process capability index; C_{p_U})—Index describing process capability in relation to the upper specification limit.

critical to quality (CTQ)—Characteristic of a product or service that is essential to ensure customer satisfaction.

critical value—The numerical values of the test statistic that determine the rejection region.

CTQ—See *critical to quality.*

cube point—In a design, experimental runs that are in the corner points of the design space. In general, a factorial design consists of cube points only. They also exist in fractional factorial designs and some central composite designs.

cumulative frequency distribution—The sum of the frequencies accumulated up to the upper boundary of a class in the distribution.

cumulative sum chart (CUSUM chart)—The CUSUM control chart calculates the cumulative sum of deviations from target to detect shifts in the level of the measurement.

CV—See *coefficient of variation.*

D

defect—Nonfulfillment of a requirement related to an intended or specified use.

defective (defective unit)—Unit with one or more defects.

defects per million opportunities (DPMO)—Measure of capability for discrete (attribute) data found by dividing the number of defects by the opportunities for defects times one million. It allows for comparison of different types of product.

defects per unit (DPU)—Measure of capability for discrete (attribute) data found by dividing the number of defects by the number of units.

degrees of freedom (ν, *df*)—In general, the number of independent comparisons available to estimate a specific parameter, which serves as a means of entering certain statistical tables.

demerit—A weighting assigned to a classification of an event or events to provide a means of obtaining a weighted quality score.

dependability—Measure of the degree to which an item is operable and capable of performing its required function at any (random) time during a specified mission profile, given item availability at the start of the mission

dependent variable—See *response variable.*

design of experiments (DOE; DOX)—The arrangement in which an experimental program is to be conducted, including the selection of factor combinations and their levels.

design resolution—See *resolution.*

design space—The multi-dimensional region of possible treatment combinations formed by the selected factors and their levels.

designated imperfection (Δ)—A category of imperfection that, because of the type and/or magnitude or severity, is to be treated as an event for control purposes.

deviation (measurement usage)—The difference between a measurement and its stated value or intended level.

discrete distribution—Probability distribution where data is from a discrete scale. Examples of discrete distributions are the binomial and Poisson distributions. Attribute data involve discrete distributions.

discrete scale—A scale with only a set or sequence of distinct values. Examples: Defects per unit, events in a given time period, types of defects, number of orders on a truck.

discrimination—See *resolution.*

dispersion—A term synonymous with variation.

dispersion effect—Influence of a single factor on the variance of the response variable.

DOE—See *design of experiments.*

dot plot—A plot of a frequency distribution where the values are plotted on the x-axis. The y-axis is a count. Each time a value occurs, the point is plotted according to the count for the value.

DOX—See *design of experiments.*

durability—Measure of useful life.

E

EDA—See *exploratory data analysis.*

effect—The result of taking an action; the expected or predicted impact when an action is to be taken or is proposed. An effect is the symptom, defect, or problem. See *cause.*

effect (design of experiments usage)—A relationship between factor(s) and a response variable(s). Specific types include main effect, dispersion effect, or interaction effect.

element—See *unit.*

event—An occurrence of some attribute or outcome. In the quality field, events are often nonconformities.

evolutionary operation (EVOP)—A sequential form of experimentation conducted in production facilities during regular production. The range of variation of the factors is usually quite small in order to avoid extreme changes in settings, so it often requires considerable replication and time.

EVOP—See *evolutionary operation.*

experiment space—See *design space.*

experimental error—Variation in the response variable beyond that accounted for by the factors, blocks, or other assignable sources in the conduct of the experiment.

exploratory data analysis (EDA)—Exploratory data analysis isolates patterns and features of the data and reveals these forcefully to the analyst.

Eyring model—A model used in accelerated life testing using temperature as the accelerant.

F

$F_{1,\nu 2}$—*F* test statistic. See *F test.*

$F_{1,\nu 2,\alpha}$—Critical value for *F* test. See *F test.*

***F* distribution**—A continuous distribution that is a useful reference for assessing the ratio of independent variances. See Appendices F, G, and H for the actual values of the distribution.

***F* test**—A statistical test that uses the *F* distribution. It is most often used when dealing with a hypothesis related to the ratio of independent variances.

factor—Predictor variable that is varied with the intent of assessing its effect on the response variable.

factor level—See *level.*

factorial design—Experimental design consisting of all possible treatments formed from two or more factors, each being studied at two or more levels. When all combinations are run, the interaction effects as well as main effects can be estimated.

failure—The inability, because of defect(s), of an item, product, or service to perform its required functions as needed.

failure mechanism—The physical, chemical, or mechanical process that caused the defect or failure.

failure mode—The type of defect contributing to a failure.

failure rate—The number of failures per unit time (for equal time intervals).

first quartile (Q_1 or lower quartile)—One quarter of the data lies below. See *quartiles.*

fixed factor—A factor that only has a limited number of levels that are of interest.

fixed model—A model that contains only fixed factors.

flowchart—A basic quality tool that uses graphical representation for the steps in a process. Effective flowcharts include decisions, inputs, outputs, as well as process steps.

fractional factorial design—Experimental design consisting of a subset (fraction) of the factorial design.

frequency—Number of occurrences or observed values in a specified class, sample, or population.

frequency distribution—A set of all the various values that individual observations may have and the frequency of their occurrence in the sample or population.

G

gage R&R study—A type of measurement system analysis done to evaluate the performance of a test method or measurement system. Such a study quantifies the capabilities and limitations of a measurement instrument, often estimating its repeatability and reproducibility.

Gaussian distribution—See *normal distribution.*

generator—In experimental design, a generator is used to determine the level of confounding and the pattern of aliases in a fractional factorial design.

geometric distribution—Case of the negative binomial distribution where $c = 1$ (c is integer parameter). The geometric distribution is a discrete distribution.

Gopertz model—Use in calculating reliability growth.

H

H_0—See *null hypothesis.*

H_1—See *alternative hypothesis.*

H_A—See *alternative hypothesis.*

Hawthorne effect—The effect in an experiment that occurs when humans perform better than they normally would because they are being measured or observed.

hazard rate—Instantaneous failure rate at some specified time t.

histogram—A plot of a frequency distribution in the form of rectangles (cells) whose bases are equal to the class interval and whose areas are proportional to the frequencies.

hypothesis—Statement about a population to be tested. See *null hypothesis, alternative hypothesis,* and *hypothesis testing.*

hypothesis testing—A statistical hypothesis is a conjecture about a population parameter. There are two statistical hypotheses for each situation—the null hypothesis (H_0) and the alternative hypothesis (H_a). The null hypothesis proposes that there is no difference between the population of the sample and the specified population; the alternative hypothesis proposes that there is a difference between the sample and the specified population.

I

i—See *moving average.*

independent variable—See *predictor variable.*

inherent process variation—Variation in a process when the process is operating in a state of statistical control.

input variable—Variable that can contribute to the variation in a process.

interaction effect—Effect for which the apparent influence of one factor on the response variable depends upon one or more other factors. Existence of an interaction effect means that the factors can not be changed independently of each other.

interaction plot—Plot providing the average responses at the combinations of levels of two distinct factors.

intercept—See *regression analysis.*

interquartile range (IQR)—The middle 50 percent of the data obtained by Q_3–Q_1.

IQR—See *interquartile range.*

isolated lot—A unique lot or one separated from the sequence of lots in which it was produced or collected.

isolated sequence of lots—Group of lots in succession but not forming part of a larger sequence or produced by a continuing process.

K

Kaplan-Meier estimator—Used to approximate unreliability for probability plotting.

kurtosis—A measure of peakedness or flattening of a distribution near its center in comparison to the normal.

L

λ (lambda)—Failure rate.

Latin square design—A design involving three factors in which the combination of the levels of any one of them with the levels of the other two appears once and only once.

least squares, method of—A technique of estimating a parameter that minimizes the sum of the difference squared, where the difference is between the observed value and the predicted value (residual) derived from the model.

level—Potential setting, value, or assignment of a factor or the value of the predictor variable.

level of significance—See *significance level.*

linear regression coefficients—The numbers associated with each predictor variable in a linear regression equation that tells how the response variable changes with each unit increase in the predictor variable. See *regression analysis.*

linear regression equation—A function that indicates the linear relationship between a set of predictor variables and a response variable. See *regression analysis.*

linearity (general sense)—The degree to which a pair of variables follows a straight-line relationship. Linearity can be measured by the correlation coefficient.

linearity (measurement system sense)—The difference in bias through the range of measurement. A measurement system that has good linearity will have a constant bias no matter the magnitude of measurement. If one views the relation between the observed measurement result on the y-axis and the true value on the x-axis, an ideal measurement system would have a line of slope = 1.

Lloyd-Lipow model—Used in reliability growth when fatigue is the major cause of failure.

lognormal distribution—If log x is normally distributed, it is a lognormal distribution. See *normal distribution.*

M

main effect—Influence of a single factor on the mean of the response variable.

main effects plot—Plot giving the average responses at the various levels of individual factors.

maintainability—Measure of the ability of an item to be retained or restored to specified condition when maintenance is performed by personnel having specified skill levels, using prescribed procedures and resources at each prescribed level of maintenance and repair.

mean life—The arithmetic average of the lifetimes of all items considered. A lifetime may consist of time between malfunctions, time between repairs, time to removal or replacement of parts, or any other desired interval of observation.

mean time between failures (MTBF)—Average time between failure events. The mean number of life units during which all parts of the item perform within their specified limits, during a particular measurement interval under stated conditions.

mean time to failure (MTTF)—Measure of system reliability for nonrepairable items: The total number of life units of an item divided by the total number of failures within that population, during a particular measurement interval under stated conditions. Expectation of the time to failure.

mean time to repair (MTTR)—Measure of maintainability: The sum of corrective maintenance times at any specific level of repair, divided by the total number of failures within an item repaired at that level, during a particular interval under stated conditions.

means, tests for—Testing for means includes computing a confidence interval and hypothesis testing by comparing means to a population mean (known or unknown) or to other sample means.

median—The value for which half the data is larger and half is smaller. The median provides an estimator that is insensitive to very extreme values in a data set, whereas the average is affected by extreme values. Note: For an odd number of units, the median is the middle measurement; for an even number of units, the median is the average of the two middle units.

meta-analysis—Use of statistical methods to combine the results of multiple studies into a single conclusion.

midrange—(Highest value + Lowest value)/2.

mistake-proofing—The use of process or design features to prevent manufacture of nonconforming product.

mixture design—A design constructed to handle the situation in which the predictor variables are constrained to sum to a fixed quantity, such as proportions of ingredients that make up a formulation or blend.

model—Description relating the response variable to predictor variable(s) and including attendant assumptions.

moving average—Let $x_1, x_2, \ldots$ denote individual observations. The moving average of span w at time i:

$$M_i = \frac{x_i + x_{i-1} + \ldots + x_{i-w+1}}{w}$$

MTBF—See *mean time between failures*.

MTTF—See *mean time to failure.*

MTTR—See *mean time to repair.*

μ (mu)—See *population mean.*

multimodal—More than one mode.

multiple linear regression—See *regression analysis.*

multivariate control chart—A variables control chart that allows plotting of more than one variable. These charts make use of the T^2 statistic to combine information from the dispersion and mean of several variables.

N

negative binomial distribution—A two-parameter, discrete distribution.

noise factor—In robust parameter design, a noise factor is a predictor variable that is hard to control or is not desired to control as part of the standard experimental conditions.

nominal scale—Scale with unordered, labeled categories, or a scale ordered by convention.

normal distribution (Gaussian distribution)—A continuous, symmetrical, bell-shaped frequency distribution for variables that is the basis for the control charts for variables.

null hypothesis, H_0—The hypothesis that there is no difference (null) between the population of the sample and the specified population (or between the populations associated with each sample). The null hypothesis can never be proved true, but it can be shown (with specified risks of error) to be untrue; that is, that a difference exists between the populations. Example: in a random sample of independent random variables with the same normal distribution with unknown mean and unknown standard deviation, a typical null hypothesis for the mean μ is that the mean is less than or equal to a given value μ_0. The hypothesis is written as: $H_0 = \mu \leq \mu_0$.

O

OC curve—See operating characteristic curve.

ogive—A type of graph that represents the cumulative frequencies for the classes in a frequency distribution.

$1 - \alpha$—See *confidence level.*

$1 - \beta$—The power of testing a hypothesis is $1 - \beta$. It is the probability of correctly rejecting the null hypothesis, H_0.

one-tailed test—A hypothesis test that involves only one of the tails of a distribution. Example: We wish to reject the null hypothesis H_0 only if the true mean is *larger* than μ_0.

$$H_0: \mu = \mu_0$$

$$H_a: \mu < \mu_0$$

A one-tailed test is either right-tailed or left-tailed, depending on the direction of the inequality of the alternative hypothesis.

operating characteristic curve (OC curve)—A curve showing the relationship between the probability of acceptance of product and the incoming quality level for a given acceptance sampling plan.

ordinal scale—Scale with ordered labeled categories.

orthogonal design—A design in which all pairs of factors at particular levels appear together an equal number of times.

outlier—An extremely high or an extremely low data value compared to the rest of the data values. Great caution must be used when trying to identify an outlier.

output variable—Variable representing the outcome of the process.

P

parameter—A constant or coefficient describing some characteristic of a population (examples: standard deviation, mean).

Pareto chart—A graphical tool based on the Pareto principle for ranking causes from most significant to least significant.

Pareto principle—The principle, named after 19th century economist Vilfredo Pareto, suggests that most effects come from relatively few causes; that is, about 80 percent of the effects come from about 20 percent of the possible causes.

parts per million (PPM or ppm)—One part per million (or one part per 10^6).

Pearson's correlation coefficient—See *correlation coefficient.*

percentile—Division of the data set into 100 equal groups

Poisson distribution—The Poisson distribution describes occurrences of isolated events in a continuum of time or space. It is a one-parameter, discrete distribution depending only on the mean.

pooled standard deviation—A standard deviation value resulting from some combination of individual standard deviation values.

population—Entire set (totality) of units, quantity of material, or observations under consideration. A population may be real and finite, real and infinite, or completely hypothetical. See *sample.*

population mean (μ)—The true mean of the population, represented by μ (mu). The *sample mean*, $\bar{x}$, is a common estimator of the population mean.

population standard deviation—See *standard deviation.*

population variance—See *variance.*

power—Equivalent to one minus the probability of a type II error ($1 - \beta$). A higher power is associated with a higher probability of finding a statistically significant difference. Lack of power usually occurs with smaller sample sizes.

power curve—The curve showing the relationship between the probability ($1 - \beta$) of rejecting the hypothesis that a sample belongs to a given population with a given characteristic(s) and the actual population value of that characteristic(s).

P_p (process performance index)—Index describing process performance in relation to specified tolerance:

$$P_p = \frac{U - L}{6s}$$

s is used for standard deviation instead of σ since both random and special causes may be present. Note: A state of statistical control is not required.

P_{pk} (minimum process performance index)—Smaller of upper process performance index and lower process performance index.

P_{pk_L} (lower process performance index or P_{P_L})—Index describing process performance in relation to the lower specification limit. For a symmetrical normal distribution:

$$P_{pk_L} = \frac{\bar{x} - L}{3s}$$

where s is defined under P_p.

PPM (or ppm)—See *parts per million.*

predicted value—The prediction of future observations based on the formulated model.

prediction interval—Similar to a confidence interval. It is an interval based on the predicted value that is likely to contain the values of future observations. It will be wider than the confidence interval because it contains bounds on individual observations rather than a bound on the mean of a group of observations.

predictor variable—Variable that can contribute to the explanation of the outcome of an experiment.

probability distribution—A function that completely describes the probabilities with which specific values occur. The values may be from a discrete scale or a continuous scale.

probability plot—Plot of ranked data versus the sample cumulative frequency on a special vertical scale. The special scale is chosen (that is, normal, lognormal, and so on) so that the cumulative distribution is a straight line.

process—A series of steps that work together to a common end. It consists of interrelated resources and activities to transform inputs into outputs. A process can be graphically represented using a flowchart.

process capability—Calculated inherent variability of a characteristic of a product. It represents the best performance of the process over a period of stable operations.

process capability index—A single-number assessment of ability to meet specification limits on the quality characteristic(s) of interest. The indices compare the variability of the characteristic to the specification limits. Three basic process capability indices are C_p, C_{pk}, and C_{pm}.

process control—Process management focused on fulfilling process requirements. Process control is also the methodology for keeping a process within boundaries and minimizing the variation of a process.

process performance—Statistical measure of the outcome of a characteristic from a process that may *not* have been demonstrated to be in a state of statistical control.

process performance index—A single-number assessment of ability to meet specification limits on the quality characteristic(s) of interest. The indices compare the variability of the characteristic to the specification limits. Three basic process capability indices are P_p, P_{pk}, and P_{pm}.

process quality—A statistical measure of the quality of product from a given process. The measure may be an attribute (qualitative) or a variable (quantitative). A common measure of process quality is the fraction or proportion of nonconforming units in the process.

producer's risk (α)—The probability of non-acceptance when the quality level has a value stated by the acceptance sampling plan as acceptable.

proportions, tests for—Tests for proportions include the binomial distribution. The standard deviation for proportions is given by

$$s = \sqrt{\frac{p(1-p)}{n}}$$

where p is the population proportion and n is the sample size.

***p*-value**—Probability of observing the test statistic value or any other value at least as unfavorable to the null hypothesis.

Q

Q_1—See *first quartile.*

qualitative data—See *attribute data.*

quality—Degree to which a set of inherent characteristics fulfils requirements.

quality management—Coordinated activities to direct and control an organization with regard to quality. Such activities generally include establishment of the quality policy, quality objectives, quality planning, quality control, quality assurance, and quality improvement.

quartiles—Division of the distribution into four groups, denoted by Q_1 (first quartile), Q_2(second quartile), and Q_3 (third quartile). Note that Q_1 is the same as the 25th percentile, Q_2 is the same as the 50th percentile and the median, and Q_3 corresponds to the 75th percentile.

R

r—See *correction coefficient.*

R—See *range.*

$\bar{R}$ (pronounced r-bar)—The average range calculated from the set of subgroup ranges under consideration. See *range.*

R^2—See *coefficient of determination.*

***R* chart**—See *range chart.*

random cause—Source of process variation that is inherent in a process over time. Also called *common cause* or *chance cause.*

random sampling—Sampling where a sample of *n* sampling units is taken from a population in such a way that each of the possible combinations of *n* sampling units has a particular probability of being taken.

random variation—Variation from random causes.

randomization—Process used to assign treatments to experimental units so that each experimental unit has an equal chance of being assigned a particular treatment.

randomized block design—Experimental design consisting of *b* blocks with *t* treatments assigned via randomization to the experimental units within each block.

range (*R*)—A measure of dispersion that is the absolute difference between the highest and lowest value in a given subgroup: R = highest observed value – lowest observed value.

range chart (*R* chart)—A variables control chart that plots the range of a subgroup to detect shifts in the subgroup range. See *range* (*R*).

rational subgroup—Subgroup wherein the variation is presumed to be only from random causes.

redundancy—The existence of more than one means for accomplishing a given function. Each means of accomplishing the function need not necessarily be identical.

regression—See *regression analysis.*

regression analysis—A technique that uses predictor variable(s) to predict the variation in a response variable. Regression analysis uses the method of least squares to determine the values of the linear regression coefficients and the corresponding model.

rejection region—The numerical values of the test statistic for which the null hypothesis will be rejected.

relative frequency—Number of occurrences or observed values in a specified class divided by the total number of occurrences or observed values.

reliability—The probability that an item can perform its intended function for a specified interval under stated conditions.

repairability—The probability that a failed system will be restored to operable condition in a specified active repair time.

replicate—A single repetition of the experiment. See *also replication.*

replication—Performance of an experiment more than once for a given set of predictor variables. Each of the repetitions of the experiment is called a replicate. Replication differs from repeated measures in that it is a repeat of the entire experiment for a given set of predictor variables, not just a repeat of measurements on the same experiment.

representative sample—Sample that by itself or as part of a sampling system or protocol exhibits characteristics and properties of the population sampled.

reproducibility—Precision under conditions where independent measurement results are obtained with the same method on identical measurement items with different operators using different equipment.

residual analysis—Method of using residuals to determine appropriateness of assumptions made by a statistical method.

residual plot—A plot used in residual analysis to determine appropriateness of assumptions made by a statistical method.

residuals—The difference between the observed result and the predicted value (estimated treatment response) for that based on an empirically determined model.

resolution—1. The smallest measurement increment that can be detected by the measurement system. 2. In the context of experimental design, resolution refers to the level of confounding in a fractional factorial design. For example, in a resolution III design, the main effects are confounded with the two-way interaction effects.

response surface design—A design intended to investigate the functional relationship between the response variable and a set of predictor variables. It is generally most useful when the predictor variables are continuous.

response surface methodology—A methodology that uses design of experiments, regression analysis, and optimization techniques to determine the best relationship between the response variable and a set of predictor variables.

response variable—Variable representing the outcome of an experiment.

resubmitted lot—A lot that previously has been designated as not acceptable and that is submitted again for acceptance inspection after having been further tested, sorted, reprocessed, and so on.

risk, consumer's (β)—See consumer's risk, β.

risk, producer's (α)—See producer's risk, α.

robust—A characteristic of a statistic or statistical method. A robust statistical method still gives reasonable results even though the standard assumptions are not met. A robust statistic is unchanged by the presence of unusual data points or outliers.

robust parameter design—A design that aims at reducing the performance variation of a product or process by choosing the setting of its control factors to make it less sensitive to the variability from noise factors.

root cause analysis—The process of identifying causes. Many systems are available for analyzing data to ultimately determine the root cause.

RSM—See *response surface methodology.*

S

s—See *standard deviation.*

s^2—See *variance.*

sample—A group of units, portions or material, or observations taken from a larger collection of units, quantity of material, or observations, that serves to provide information that may be used for making a decision concerning the larger quantity (the population).

sample mean—The sample mean (or average) is the sum of random variables in a random sample divided by the number in the sum.

sample size (n)—Number of sampling units in a sample.

sample standard deviation—See *standard deviation.*

sample variance—See *variance.*

sampling interval—In systematic sampling, the fixed interval of time, output, running hours, and so on, between samples.

sampling plan (acceptance sampling usage)—A specific plan that states the sample size(s) to be used and the associated criteria for accepting the lot. Note: the sampling plan does not contain the rules on how to take the sample.

scatter plot or **diagram**—A plot of two variables, one on the *y*-axis and the other on the *x*-axis. The resulting graph allows visual examination for patterns to determine if the variables show any relationship or if there is just random "scatter." This pattern or lack thereof aids in choosing the appropriate type of model for estimation.

serviceability—The ease or difficulty with which equipment can be repaired.

σ (sigma)—See *standard deviation.*

σ^2 (sigma square)—See *variance.*

$\sigma_{\bar{x}}$ (sigma x-bar)—The standard deviation (or standard error) of $\bar{x}$.

$\hat{\sigma}$ (sigma-hat)—In general, any estimate of the population standard deviation. There are various ways to get this estimate depending on the particular application.

signal—An indication on a control chart that a process is not stable or that a shift has occurred. Typical indicators are points outside control limits, runs, trends, cycles, patterns, and so on.

significance level—Maximum probability of rejecting the null hypothesis when in fact it is true. Note: the significance level is usually designated by α and should be set before beginning the test.

Six Sigma—A methodology that provides businesses with the tools to improve the capability of their business processes.

skewness—A measure of symmetry about the mean. For the normal distribution, skewness is zero since it is symmetric.

slope—See *regression analysis.*

special cause—Source of process variation other than inherent process variation.

specification limit(s)—Limiting value(s) stated for a characteristic. See *tolerance.*

spread—A term sometimes synonymous with variation or dispersion.

stable process—A process that is predictable within limits; a process that is subject only to random causes. (This is also known as a state of statistical control.)

standard deviation—A measure of the spread of the process output or the spread of a sampling statistic from the process. When working with the population, the standard deviation is usually denoted by σ(sigma). When working with a sample, the standard deviation is usually denoted by *s*.

standard error—The standard deviation of a sample statistic or estimator. When dealing with sample statistics, we either refer to the standard deviation of the sample statistic or to its standard error.

standard error of predicted values—A measure of the variation of individual predicted values of the dependent variable about the population value for a given value of the predictor variable. This includes the variability of individuals about the sample line about the population line. It measures the variability of individual observations and can be used to calculate a prediction interval.

statistic—A value calculated from or based on sample data (for example, a subgroup average or range), used to make inferences about the process that produced the output from which the sample came. A quantity calculated from a sample of observations, most often to form an estimate of some population parameter.

statistical measure—A statistic or mathematical function of a statistic.

statistical thinking—A philosophy of learning and action based on the fundamental principles:

- All work occurs in a system of interconnected processes.
- Variation exists in all processes.
- Understanding and reducing variation are keys to success.

statistical tolerance interval—Interval estimator determined from a random sample so as to provide a specified level of confidence that the interval covers at least a specified proportion of the sampled population.

T

***t* distribution**—A theoretical distribution widely used in practice to evaluate the sample mean when the population standard deviation is estimated from the data. Also known as *Student's* t *distribution.*

Taguchi design—See *robust parameter design.*

target value—Preferred reference value of a characteristic stated in a specification.

temperature–humidity model—Used in accelerated life testing when temperature and humidity are the major accelerants.

temperature–nonthermal models—Used in accelerated life testing when temperature and another factor are the major accelerants.

test statistic—A statistic calculated using data from a sample. It is used to determine whether the null hypothesis will be rejected.

testing—A means of determining the ability of an item to meet specified requirements by subjecting the item to a set of physical, chemical, environmental, or operating actions and conditions.

time series—Sequence of successive time intervals.

tolerance—Difference between upper and lower specification limits.

tolerance limits—See *specification limit(s).*

transformation—A reexpression of the data aimed toward achieving normality.

treatment—The specific setting of factor levels for an experimental unit.

true value—A value for a quantitative characteristic that does not contain any sampling or measurement variability. (The true value is never exactly known; it is a hypothetical concept.)

***t*-test**—A test of significance that uses the *t* distribution to compare a sample statistic to a hypothesized population mean or to compare two means.

2^n factorial design—Factorial design in which n factors are studied, each of them at exactly two levels.

two-tailed test—A hypothesis test that involves two tails of a distribution. Example: we wish to reject the null hypothesis H_0 if the true mean is within minimum and maximum (two tails) limits.

$$H_0: \mu = \mu_0$$

$$H_a: \mu \neq \mu_0$$

type I error—The probability or risk of rejecting a hypothesis that is true. This probability is represented by α (alpha). See *operating characteristic curve* and *producer's risk.*

type II error—The probability or risk or accepting a hypothesis that is false. This probability is represented by β (beta). See *power curve* and *consumer's risk.*

U

uncertainty—A parameter that characterizes the dispersion of the values that could reasonably be attributed to the particular quantity subject to measurement or characteristic. Uncertainty indicates the variability of the measured value or characteristic that considers two major components of error: 1) bias and 2) the random error from the imprecision of the measurement process.

unique lot—Lot formed under conditions peculiar to that lot and not part of a routine sequence.

unit—A quantity of product, material, or service forming a cohesive entity on which a measurement or observation can be made.

universe—A group of populations, often reflecting different characteristics of the items or material under consideration.

V

variance—A measure of the variation in the data. When working with the entire population, the population variance is used; when working with a sample, the sample variance is used.

variances, tests for—A formal statistical test based on the null hypothesis that the variances of different groups are equal. Many times in regression analysis a formal test of variances is not done. Instead, residual analysis checks the assumption of equal variance across the values of the response variable in the model.

variation—Difference between values of a characteristic. Variation can be measured and calculated in different ways—such as range, standard deviation, or variance. Also known as *dispersion* or *spread.*

W

warning limits—There is a high probability that the statistic under consideration is in a state of statistical control when it is within the warning limits (generally 2σ) of a control chart.

References

Ireson, W. G., C. F. Combs, Jr., and R. Y. Moss. 1995. *The Handbook of Reliability Engineering and Management,* 2nd ed. New York: McGraw Hill.

Kececioglu, D. 1993. *Reliability and Life Testing Handbook.* Volume I. NJ: Prentice Hall.

———. 2001. *Reliability and Life Testing Handbook.* Volume II. Tucson, AZ: University of Arizona.

McLean, H. W. 2002. *HALT, HASS, and HASA Explained: Accelerated Reliability Techniques.* Milwaukee: ASQ Quality Press.

Meeker, W. G., and G. J. Hahn. 1985. *How to Plan an Accelerated Life Test—Some Practical Guidelines.* Volume 10. Milwaukee: ASQ Quality Press.

O'Connor, P. 2002. *Practical Reliability Engineering,* 4th ed. England: John Wiley.

Pecht, M., and J. Gu. 2009. "Physics-of-Failure-Based Prognostics for Electronic Product." *Transactions of the Institute for Measurement and Control* 31 (3/4): 309–22.

Tseng, S-T., and Z-C. Wen. 2000. "Step-Stress Accelerated Degradation Analysis for Highly Reliable Products." *Journal of Quality Technology* 32 (3): 209–16.

Index

K

L

M

N

O

P

Q

R

S